Understanding Behavior

UNDERSTANDING BEHAVIOR

What Primate Studies Tell Us About Human Behavior

EDITED BY
James D. Loy
Calvin B. Peters

University of Rhode Island
Kingston, R.I.

New York   Oxford
OXFORD UNIVERSITY PRESS
1991

Oxford University Press

Oxford   New York   Toronto
Delhi   Bombay   Calcutta   Madras   Karachi
Petaling Jaya   Singapore   Hong Kong   Tokyo
Nairobi   Dar es Salaam   Cape Town
Melbourne   Auckland

and associated companies in
Berlin   Ibadan

Copyright © 1991 by Oxford University Press, Inc.

Published by Oxford University Press, Inc.,
200 Madison Avenue, New York, New York 10016

Oxford is a registered trademark of Oxford University Press

Library of Congress Cataloging-in-Publication Data
Understanding behavior: what primate studies tell us about human behavior
/ edited by James D. Loy and Calvin B. Peters.
p. cm. Includes bibliographical references and index.
ISBN 0-19-506020-2
1. Psychology, Comparative.   2. Behavior evolution.
3. Primates—Behavior.
I. Loy, James D.   II. Peters, Calvin B.
BF671.T54 1991 156—dc20 90-39539

9 8 7 6 5 4 3 2 1

Printed in the United States of America
on acid-free paper

To
Kent Matthews Loy
Bette LaSere Erickson

Preface

... a man, however well behaved,
At best is only a monkey shaved!
W. S. GILBERT

The possibility that humans and nonhuman primates are connected in some intimate way has the capacity to captivate professional primatologists and the public alike. Consider for a moment the titles of some recent non-technical books on primate behavior. Jane Goodall's first popular work on the chimpanzees of the Gombe Stream Reserve was called *In The Shadow of Man,* and a few years later John MacKinnon described the behavior of living apes and the evolutionary emergence of humanity in a little volume entitled *The Ape Within Us.* More recently, Shirley Strum described the behavior of wild baboons in a book called *Almost Human.* Such titles promise much more than interesting descriptions of charming animals. They hold out the hope that by studying our nonhuman relations we can learn something of value about ourselves.

Primatologists have held fast to that hope. Over the past half century, hundreds of specialists have studied the behaviors of apes, monkeys, and prosimians in laboratories and in the field. Most of these investigators were from the traditional human sciences, and many fully expected to obtain information about humans through their animal studies. The question that is addressed by the essays in this volume is straightforward: Has the approach worked?

As might be expected, the answer turns out to be: "Yes and No." As several contributors to this book note in their chapters, it is impossible to learn about humans directly by studying their collateral relatives, the nonhuman primates. But, as those same authors then proceed to demonstrate, comparative primatology has produced many indirect insights into human behavior and evolution. Although the comparative method has taken its lumps from specialists and nonspecialists alike (satirist Will Cuppy once noted: "The psychology of the Orang-utan has been thoroughly described by scientists from their observations of the Sea-urchin"), the chapters that follow describe how primate studies have led to greater understanding of several aspects of human behavior, from attachment theory to aggression, hormones to homosexuality, concealed ovulation to culture, ontogeny to orgasms.

The insights have not come easily. It has taken primatologists some years (and several failed attempts) to learn to apply properly the comparative method to humans. Presumably, few would now claim that anything useful was learned by Desmond Morris's comparison (in *The Naked Ape*) of women's lipsticked mouths and the brightly colored labia on some primate posteriors. Despite their dangers and limitations, comparisons of humans and nonhuman primates will, of course, continue to be made. And the process does have its rewards. The interested (and quantitatively inclined) reader can glean dozens of insights reported in the following chapters.

Simplified behavioral models can and do illuminate the workings of complex human behaviors. Further, interspecific studies allow the construction of heuristic models of hominid evolution and the identification of the "emergent" characteristics of humans (those traits that define humanness and separate us from our nonhuman brethren). For some people, the primate data reveal comforting similarities between humans and the other primates; for others, the revealed differences are what matter.

The contributors to this volume were faced with a formidable task as they constructed their comparative analyses. As noted in the introductory chapter, primatologists have long been aware that such analyses can "make us forget what we already know" about humans and may also lead to oversimplification of human data. We believe our contributors have, in the main, successfully avoided these dangers. Other than helping our authors avoid these pitfalls, however, it has been our editorial policy to let them construct their own comparative arguments and emphasize those points they felt were most important. Readers will no doubt find that all of the following chapters contain some conclusions with which they agree and others about which they would argue. In our view, the ability of this book to generate heated discussions about comparative primatology is one of its principal strengths.

Finally, we believe the book reflects the fact that members of the traditional human disciplines need not fear an imperial primatology. The gap noted long ago by Solly Zuckerman between humans and the nonhuman primates is real and, to a great extent, the refrain that "if you want to learn about humans, you must study humans" is true. Nonetheless, the human disciplines neglect information derived from primate studies at their peril. A complete understanding of the human condition seems to demand that we make sense of our relationship to the apes and monkeys, a relationship that continues to both fascinate and mortify.

ACKNOWLEDGMENTS

This book began as an idea in an idle hallway conversation. Had it not been for the participation and help of many people, it would have gone the way of most hallway ideas. That you hold in your hands a printed version of our musings is largely due to their efforts.

Don Symons, Meredith Small, Sarah Hrdy, Marquisa LaVelle, Helen Mederer, Sally Gouzoules, and Frans de Waal reviewed portions of the manuscript and provided useful and insightful comments on a number of issues. The book is better because of their help.

Fred Smith of the University of Tennessee and Joe Hendricks of Oregon State University rescued us from a tight spot and in so doing gave the notion of interlibrary loan new meaning. Joanne Lawrence, Carla Rattenni, and Marion Houston combed snarls out of a manuscript that often promised to get the better of us.

This book really belongs to the authors of the essays that appear here. They have our most sincere thanks. Their timely submissions, cheerful revisions, and enlightened visions made our job much easier.

Kingston, R.I.   J.D.L.
June 1990   C.B.P.

Contents

Contributors

Bernard Chapais
Département d'Anthropologie
Université de Montréal
Montréal, Québec

Janice Chism
Department of Biology
Winthrop College
Rock Hill, S.C.

James D. Loy
Department of Sociology and
    Anthropology
University of Rhode Island
Kingston, R.I.

Patrick Mehlman
Laboratory Animal Breeders and
    Services
Yemassee Primate Center
Yemassee, S.C.
    and
Département d'Anthropologie
Université de Montréal
Montréal, Québec

Ronald D. Nadler
Yerkes Regional Primate Research
    Center
Emory University
Atlanta, Ga.

Nancy A. Nicolson
Neuropsychology and Psychobiology
University of Limburg
The Netherlands

Calvin B. Peters
Department of Sociology and
    Anthropology
University of Rhode Island
Kingston, R.I.

Charles H. Phoenix
Oregon Regional Primate Research
    Center
505 N.W. 185th Avenue
Beaverton, Oreg.

Donald Stone Sade
Department of Anthropology
Northwestern University
Evanston, Ill.

David Taub
Laboratory Animal Breeders and
    Services
Yemassee Primate Center
Yemassee, S.C.
    and
Department of Psychiatry and
    Behavioral Sciences
Medical University of South Carolina
Charleston, S.C.

Linda D. Wolfe
Department of Anthropology
University of Florida
Gainesville, Fla.

Understanding Behavior

Mortifying Reflections: Primatology and the Human Disciplines

JAMES D. LOY
AND CALVIN B. PETERS

> I confess to you, I could never look long upon a monkey, without very mortifying reflections.
>
> WILLIAM CONGREVE

> The big baboon is found upon the plains of Cariboo.
> He goes about with nothing on, a shocking thing to do.
> But if he dressed respectably and let his whiskers grow,
> How like this big baboon would be to Mister So-and-so!
>
> HILAIRE BELLOC

How like indeed! We flock as visitors to the monkeys and apes in zoological parks because in them we see the reflections of our own private Mister So-and-sos. Some of their antics fill us with a mixture of amusement and satisfying superiority; others mortify us almost as much as if we had been caught doing them ourselves. Primatologists, of course, share this general fascination and identification with our near relatives. (At a recent primatology meeting, the conventioneers wore buttons announcing "I Monkey Around!" The double or triple entendre was, no doubt, the source of much amusement.) Primatologists, however, are not satisfied with amusement or fascination. They seek to move beyond such affective responses to try to utilize the similarities and differences between humans and nonhumans in order to gain insight into our evolutionary past or to derive solutions to puzzles of human behavior and society. The questions posed in this volume are straightforward: Has primatology succeeded in producing useful and significant information about humans? Have primatologists been able to move beyond the insights gained by the average zoo visitor? Is there a response based in primatological data to the semirhetorical question, "If we want to understand humans, shouldn't we study humans?"

These questions appear simple to the uninitiated; a list of insights produced by primatologists would seem sufficient to settle the issue. But to primatologists and social scientists these problems are so perplexing that they are rarely addressed. The gaps between humans and nonhuman primates are, quite frankly, immense. Humans last shared a common ancestor with prosimians about 55 million years ago, with New World monkeys about 40 million years ago, and with Old World monkeys about 25 million years ago (Gingerich, 1984). It has been at least 6–8 million years since humans shared a common ancestor with the living apes (Napier and Napier, 1985). Even assuming that the nonhuman species were behaviorally "frozen" at the time of divergence (which of course they were not) and that all subsequent differences accumulated in the line leading to humans, how can we hope to bridge the gaps and learn about Us by watching Them? Anthropologist Donald Symons has put the primatological dilemma this way:

> It is not clear how meaningful comparisons are to be made between humans and other animals, and seemingly straightforward comparisons may mislead more often than they enlighten; many times biology not only fails to increase our understanding of human beings but seems to have the magical power to make us forget what we already know (Symons, 1979:128).

There are similar caveats. In the effort to make comparisons with humans, ethologist Robert Hinde has warned that primatologists run the risk of grossly oversimplifying human phenomena in order to bridge the vast gaps between monkeys and people (Hinde, 1987).

A BRIEF HISTORY OF PRIMATE STUDIES

Despite its dangers, the idea that we can learn about ourselves by studying other animals is an old one. Aristotle justified his inquiries into the natural history of other animal species on the ground that humanity was, to some extent, a continuation of the nonhuman world (Mayer, 1982; Midgley, 1980). In more modern times, Darwin, too, argued for the value of human–nonhuman behavioral comparisons. In *The Expression of the Emotions in Man and Animals* (1872), he paid special attention to the behavior patterns of animals as evidence concerning the evolutionary origins of human expressions:

> I have attended, as closely as I could, to the expression of the several passions in some of the commoner animals; and this I believe to be of paramount importance, not of course for deciding how far in man certain expressions are characteristic of certain states of mind, but as affording the safest basis for generalization on the causes, or origin, of the various movements of Expression (Darwin, 1872:17).

Further, Darwin was fascinated by the expressions and vocalizations of monkeys and apes and noted that the expressive actions of these animals are "closely analogous to those of man" (1872:131). In Darwin's view similarities in behavior patterns among humans and the other primates could be used to speculate about the expressions of human forebears.

> It is a curious, though perhaps an idle speculation, how early in the long line of our progenitors the various expressive movements, now exhibited by man, were succes-

sively acquired. . . . We may confidently believe that laughter, as a sign of pleasure or enjoyment, was practiced by our progenitors long before they deserved to be called human; for very many kinds of monkeys, when pleased, utter a reiterated sound, clearly analogous to our laughter, often accompanied by vibratory movements of their jaws or lips, with the corners of the mouth drawn backwards and upwards, by the wrinkling of the cheeks, and even by the brightening of the eyes (1872:360).

The historic attachment to primate–human connections laid the foundation for the modern age of primate behavioral studies that was ushered in by the work of Robert Yerkes, Solly Zuckerman, and others in the early twentieth century. Yerkes, working in the United States, concerned himself primarily with studies of the great apes. For Yerkes, the apes, especially chimpanzees, provided simplified human behavioral models, and his research was justified with the promise of results applicable to human problems.

It is the basic idea in the organization of the [Yale] laboratories to use anthropoid subjects as extensively as practicable, but also intensively, for the fruitful increase of knowledge of life and as a means of approaching the solutions of various human problems for which we may not freely and effectively use ourselves as materials of observation and experiment (Yerkes, 1932:9).

Sir Solly Zuckerman, best known for his book *The Social Life of Monkeys and Apes* (1932) took a distinctly different view. Zuckerman argued that, although behavioral studies of monkeys and apes may shed light on humans' evolutionary past, such investigations would provide little if any information about the behavior of living people. Zuckerman's position was clear:

In the life of the monkey one may see a crude picture of the social life from which our earliest human ancestors emerged. But only that. The behaviour of the sub-human primate represents a pre-human social level, a level which, though without culture itself, seems to have contained the seeds that grew into the culture of primitive man (1932:26).

However, with regard to the behavior of living people, Zuckerman believed that "the ethnological approach is at present the only practicable way to knowledge of human social behaviour" (1932:17). Further, he maintained that

It is not legitimate to infer . . . that comparison of animal and human social life will enable us "to discover some of the basic instincts and impulses upon which the whole edifice of human society is reared." It is an unsafe assumption, born of anthropomorphic generalizations about animal behaviour and vague thought about evolution, that if human behaviour itself will not reveal those "basic instincts and impulses," comparison with the lives of other organisms will (Zuckerman, 1932:23).

Zuckerman's conviction that the gap between nonhuman primate and human behavior was unbridgeable because "animal behaviour is almost entirely on a physiological plane, while almost all human behaviour is condi-

Fig. 1–1 Robert Yerkes used the great apes as simplified models for studying human behavior. (Photo: Yerkes Regional Primate Research Center of Emory University.)

tioned by culture" (1932:22) never waned. As recently as 1981, he reaffirmed his belief that interspecific behavioral studies involving humans had been and would continue to be of little use: "I totally fail to see how any analogical comparisons with the ways of monkeys and apes can help in the understanding of what some see as the major problems of human behaviour today" (Zuckerman, 1981:394).

Zuckerman's insistent warning about the error of using data from nonhuman primates in an effort to learn about human behavior fell on deaf ears. Among the deafest were those of U.S. psychologist C. R. Carpenter. Carpenter was convinced that systematic behavioral comparisons between humans and nonhumans could be productive because "human behavioral characteristics have anlagen, analogies and homologies in non-human invertebrate and vertebrate froms" (Carpenter, 1942b, in Carpenter, 1964:342). The nonhuman species with the greatest morphological similarities to humans (that is, monkeys and apes) were expected to yield the most valuable data. In Carpenter's view, behavioral studies of monkeys and apes would reveal archaic (evolutionarily old) "behavior mechanisms" that still work among living humans:

> The study of the phylogenesis of behavior and social relations on the non-human primate level will facilitate the understanding and control of human activities. This approach is likely to reveal archaic behavior mechanisms which are organically functional on the human level. The kinship of non-human primates to man makes it reasonable to assume that the study of these types may yield data which can be readily employed in understanding basic kinds of human motivation (Carpenter, 1942a, in Carpenter, 1964:358).

Carpenter's belief that humans might still be influenced by "archaic behavior mechanisms" that could be seen clearly among monkeys and apes stood in sharp contrast to Zuckerman's contention that the behaviors of humans and nonhumans were separated by an "immense gap." Although both men thought research on the nonhuman primates would allow insights into human evolution, only Carpenter believed that modern human societal problems might be approached as well. Carpenter's more ambitious goals were spelled out in a clarion call for the development of a fully comparative primatology:

> The task of scientists who would investigate comparatively the social interactions of a wide range of types of organisms, including the primates, becomes that of collecting information on the basis of which tentative generalizations or principles may be formulated. Then, when these principles have been formulated, and the limits to which they can be generalized determined, the application of them can be made for purposes of aiding in investigating, understanding and controlling human social interactions and human societies (Carpenter, 1952, in Carpenter, 1964:368).

Carpenter's vision of a primatology that would aid in reconstructing the behavioral and social evolution of humans, and in so doing point the way toward the solution of contemporary problems, found a formidable champion in Harvard's Earnest Hooton. In his introduction to a volume that linked primate research with anthropology, Hooton ran the risk of offending his ethnological colleagues with an explicit endorsement of Carpenter's hopes:

> I shall no doubt evoke the indignant disagreement of social anthropologists when I suggest that more is to be learned about the genesis of the human family and the beginning of social organization and community life in early man by the study of

Fig. 1–2 C. R. Carpenter (right) believed that studies of nonhuman primates would provide insights into the solution of human problems. (Photo: *Puerto Rico Health Sciences Journal.)*

contemporary infra-human primates living under natural conditions than by the studies of retarded human groups living today under conditions variously described as "primitive," "uncivilized" or "savage" (Hooton, 1955:7).

Hooton's appreciation for the anthropological value of primate studies was of great significance. In what amounts to the beginning of the genealogy of modern primatology, Hooton begat Sherwood Washburn. Washburn begat students who published 15 of the first 19 modern primatology dissertations written in the United States. In turn, these students most probably begat over half of today's American primate behaviorists (Gilmore, 1981).

Washburn's link to Hooton and through him to the belief in the human significance of primate studies was underscored in 1961:

The behavior of monkeys and apes has always held great fascination for men. In recent years plain curiosity about their behavior has been reinforced by the desire to understand human behavior. Anthropologists have come to understand that the evolution of man's behavior, particularly his social behavior, has played an integral role in his biological evolution. In the attempt to reconstruct the life of man as it was

Fig. 1–3 Earnest Hooton's academic progeny strongly influenced the course of American primatology in the middle and late twentieth century. (Photo: Peabody Museum, Harvard, © President and Fellows of Harvard College, 1950. All rights reserved.)

shaped through the ages, many studies of primate behavior are now under way in the laboratory and in the field. As the contrasts and similarities between the behavior of primates and man—especially preagricultural, primitive man—become clearer, they should give useful insights into the kind of social behavior that characterized the ancestors of man a million years ago (Washburn and DeVore, 1961, reprinted in Southwick, 1963:98).

Just after the middle of the twentieth century, a worldwide surge of interest in primate behavior occurred. Provisioning and intensive study of Japanese macaques began about this time. Several new U.S. workers (many of them students of Washburn) began field studies, and, in his Wisconsin laboratories, Harry Harlow marked his third decade of primate research by beginning his well known studies of rhesus infants reared with man-made surrogate mothers. English workers who took to the field at this time included Jane Goodall, John Crook, and K.R.L. Hall. Hall stated his views on the objectives of primate studies, reiterating those of the Washburn lineage: ". . . one may hope that the [primatological] investigator may derive significant hypotheses for experimental and observational testing on the human material" (Hall, 1963:297).

The foundation of modern primatology, therefore, rests firmly on the belief that we can learn about humans by comparing them with prosimians, monkeys, and apes. Arguments about *how* one can validly compare humans and nonhumans, as well as about *what* one can learn from such comparisons, have continued over the years, but primatologists have never flagged in their belief that their work will provide insights into the human condition.

PRIMATOLOGISTS AMONG THE HUMAN DISCIPLINES

Despite this belief, primatologists have not been warmly received by the human disciplines—sociology, anthropology, psychology, and political science. To be sure, a goodly number of primatologists hold appointments in departments of anthropology and psychology, and even more have earned advanced degrees under the rubric of the human disciplines. But primatologists live a peculiar, resident-alien-like existence among their human-oriented colleagues. They speak a different language—macaques not Marx, vervets not values—and generally publish in journals seldom read and often unknown outside the primatological community. Of course, in some ways the status of primatologists is not different from that of all other academic specialists. We all inhabit semicloistered communities, speak a language unfamiliar to outsiders, and publish in places known only to the cognoscenti in whose ranks we claim membership.

In the case of primatology, however, this "normal" development has come at a high cost. As their specialty has crystallized, primatologists have increasingly tended to speak to and write for one another. To a great extent, many of them have abandoned the effort to extrapolate from their findings to human society and behavior. This movement away from the human disciplines and toward a kind of "pure" primatology is reflected in the contents of journals devoted to primatological research. In the *American Journal of Primatology* from 1983 through 1987, for example, only about one of every ten articles devoted to behavioral or ecological issues made clear a connection to humans.

This is not to gainsay the advances in pure primatology. The contributions to the primate literature have gone a long way to establish the field as a respectable and respected academic discipline. Nonetheless, as primatology has grown, the raison d'etre that undergirded the work of Carpenter, Hall, and the other pioneers has receded from sight, and the potential value of the primatological data for human affairs has been ignored.

The "blame" for this circumstance—Zuckerman would say "credit"—can, it seems, be laid primarily at the feet of sociologists, anthropologists, psychologists, and political scientists. With some significant exceptions, the promise of primatology to provide insight into human affairs has been regarded by the practitioners of the human disciplines as not only false but harmful. The storm of controversy that accompanied the publication of *Sociobiology: The New*

Synthesis (Wilson, 1975) exhibited in acute form the jealousy with which the human disciplines protect their areas of inquiry. Although some of the issues that surround the "sociobiology controversy" are peculiar to the approach and what might be counted as the excesses of Wilson, Barash, and others, the refusal of the human disciplines to entertain the idea that principles of animal (including primate) behavior might illumine some persistent issues of human sociality reflects a general disposition to discount explanations and accounts of human behavior that smack of what social scientists call "biology."

This fundamental reluctance among the human disciplines to give serious consideration to behavioral data from animals—Mario von Cranach remarks wistfully in his introduction to *Methods of Inference from Animal to Human Behaviour* (1976) that "social scientists proved not to be particularly interested in the problem" (p. 1)—has more than anything else encapsulated primatology and prevented a thorough and cooperative exploration of the primate–human connection. The resistance of the human disciplines to "biological" accounts of human behavior, including those that lie hidden in the primatological data, is not entirely irrational. The histories of anthropology, psychology, sociology, and political science can to a large extent be understood as efforts to establish each discipline as a separate entity, not as a subfield of history or philosophy or biology. Although these wars for intellectual independence have been declared over (and the human disciplines have won), they have left as their heritage a wariness on the part of social scientists willingly to embrace other disciplines and approaches as potential sources of knowledge about human life and behavior.

The independence of the human disciplines has rendered most of their practitioners incapable of evaluating critically data drawn from primates or even from studies of human population biology. Undergraduate and graduate education in most of the human disciplines, with the possible exceptions of anthropology and comparative psychology, is bereft of coursework in evolution, genetics, and ethology, the ideas and techniques that form the foundation of primatological enquiry. Animals have been banished from the human disciplines, not only as models of human behavior but as objects of worthwhile study. Indeed, what most social scientists know about animals and animal behavior, including that of monkeys and apes, they have gleaned from Marlin Perkins and the National Geographic.

The exception to this general expulsion of animals from the fields of the human disciplines is, of course, the white rat. Although a host of accounts involving cost, ease of handling, reproductive speed, moral considerations, etc., can be given to explain the ubiquitous presence of the rat, it is startling that students of humanity esteem it more highly than the baboon or chimpanzee. This, however, may not be so odd as it first seems. The laboratory rat exists largely as a human creation, a tool not altogether different from a computer simulation program. It is not the rat that is studied; it is learning, or fear, or some other formal behavior. The rat itself is held at arm's length (always figuratively, sometimes literally). There is nothing very human-like about the rat, and except for a few wags who persist in the old saw that psychology is the science of college sophomores and white rats, there is little if any danger that the independence of the human disciplines will be compromised.

The same is quite obviously not true of the chimpanzee, the baboon, and other monkeys and apes. The power of these creatures to captivate and charm is no less real for the human disciplines, despite the austere posture of a scientific outlook. Harlow's experiments were evocative not because of their careful crafting (though that was clearly important) but rather because of the all-too-human response of his subjects. Belloc's observation that except for respectable clothes and a whiskered countenance the baboon would be like Mister So-and-so hides beneath its whimsy a specter that haunts the human disciplines. The chimpanzee, the orang-utan, the gorilla, and to a lesser extent the other apes and monkeys repeatedly raise the suspicion that they are too much like us to be ignored.

The discomfort produced among the human disciplines by this suspicion has been allayed by a fervent attachment to human uniqueness. We're not *just* animals, the argument goes, we have—and here follow a series of human qualities not found in the simian mimics of Mister So-and-so—speech, culture, rationality, souls, tool use, self-awareness, and the list continues. The upshot of all of this is that social scientists, at the end of a line that runs from Aristotle to Descartes to Kant to Marx to Zuckerman, find themselves defending as unbridgeable a gap between humans and the (other) primate species. In this, the human disciplines resemble alcoholics recovering from their addiction. The substance, in this case monkeys and apes, is still so enticing that unless it is sworn off completely it is likely to possess them against their will.

The exhibition in the human disciplines of what William King Gregory (1929) termed "pithecophobia," the dread of apes as relatives or ancestors, is then one significant barrier to a more productive engagement with primatology (for a recent discussion of pithecophobia in anthropology, see Lewin, 1987). It is not the only one. For most of the twentieth century, and certainly since 1940, social scientists have focused on the development of what Merton (1967) called "theories of the middle range." This focus on "theories that lie between the minor but necessary working hypotheses that evolve in abundance during day-to-day research and the all-inclusive systematic efforts to develop a unified theory that will explain all the observed uniformities of social behavior, social organization, and social change" (p. 39) stands in marked contrast to the grand schemes of Comte, Spencer, Marx, Tylor, and Morgan. The nineteenth century's search for the origins of human institutions (family, politics, and others), and the device of evolutionary schemes of human society and culture, efforts that owed much to Darwin, are now regarded as a sterile, angels-on-the-head-of-a-pin exercise (see Tiryakian, 1976). Few in the human disciplines bemoan this shift in orientation. The grand theorizing of the past remains of interest only to historians of social thought. For contemporary scholars, the ideas, assumptions, and aspirations of their Victorian forebears are both quaint and irrelevant. Who, indeed, now reads Spencer? (This famous rhetorical question, posed by Crane Brinton in 1933 and repeated by Talcott Parsons in 1937 is the widely accepted epitaph of the evolutionary dreams of the nineteenth century.)

The movement of the human disciplines away from an evolutionary focus on the development and uniformities of human culture and toward a concentration on the nature and function of current differences in human societies has placed them on a trajectory that runs away from, not toward, primatological data. Clearly, the questions that primatologists are most likely to be able to help in answering are those that concern the origins, evolution, and uniformity of human behavior and culture. Just as clearly, those sorts of questions are not the ones that move late twentieth century social science. There is no little irony in this. When the questions of human nature and the evolution of society were the rage, the primatological data that may have shed considerable light were scanty and unreliable. Now that the data are available, the questions are passé.

Needless to say, it is the premise of this volume that neither the "pithecophobia" of the human disciplines nor the apparent mismatch between primatological data and the lines of inquiry in contemporary social science are insurmountable barriers. We, both editors and authors, believe that studies of monkeys and apes have something to teach us about ourselves and that what they teach is worthy of note. This is not to say that the lines from chimpanzee or baboon or rhesus to human are straight and easily drawn. They are not. Those lines are, however, worth tracing, because in doing so we may learn more about who we are (and are not), about where we came from (and where we did not), and about what we can hope for (and what we can not).

PLAN OF THE BOOK

We asked several prominent primatologists to assess just what studies of nonhuman primates had revealed about human behavior. At the outset, we considered asking them to adhere to some common structure so that each chapter would bear a formal resemblance to the others. In the end, however, we chose to allow our contributors to make their own decisions about how best to approach the issues and how best to organize their analyses. Not surprisingly, their choices reveal a variety of approaches. In some chapters, the data from humans and the data from nonhumans are reported separately, with parallels drawn between the two bodies of knowledge. In other chapters, data from nonhumans are used as a lens to focus humans in a comparative framework and as the ground for explanations of human behavior. In addition, the chapters employ different levels and types of analysis. Some invoke proximate explanations of behavior almost exclusively; others seek to develop ultimate explanations embedded in evolutionary scenarios; still others employ methods that are largely historical and interpretive.

In some respects, this variability is disconcerting. The chapters are difficult to compare and, after reading them, the question, "What have we learned about humans from studying nonhuman primates," can, it seems, only be answered, "It depends." In other ways, however, the unevenness in approach is reflective of the realities of comparative primatology. The task of extracting from data on monkeys and apes those findings that shed light on human behavior is a formidable one. Moreover, there is little agreement on how (or even whether) such comparisons should be made. That the authors here have chosen to make different assumptions, to select different methods, and to settle for different levels of analysis is a response to both the difficulty of the assignments we gave them and the largely uncharted world of human-nonhuman comparison.

We believe that the variability of chapters has enriched the volume, although it has made the task of organizing them in some logical order more difficult. In general, we have organized the chapters along two dimensions. First, we attempted to keep together those chapters that deal with behavior in a sex-specific way. Accordingly, the chapters on maternal behavior and paternal behavior are sequential as are the chapters on female sexuality and male sexuality. Second, we attempted to place those chapters with relatively more unified discussions of human and nonhuman primate behavior early in the volume, saving until later those chapters that are more segmented. Although some bumpy transitions between chapters remain, we believe the book can be read as a whole.

Following this introductory essay, then, are six chapters that deal with basic areas of behavior and their development: maternal behavior, paternal behavior, ontogeny of behavior, female sexuality, male sexuality, and aggression. Two chapters that are more interpretive, one on kinship, and one on primatological discourse as a way of learning about ourselves, complete the text proper. The book concludes with a brief, annotated list of suggestions for further reading.

Chapter 2, by anthropologist Nancy Nicolson, contains an analysis of maternal behavior. After outlining the basic patterns of primate mothering, she describes several variables that affect the development of maternal nurturance. She then discusses the extreme interindividual variation found in the maternal behaviors of primates and the contributions of several factors (ecology, infant age and needs, social context, etc.) on mother-infant interactions. Nicolson's chapter ends with a discussion of how one determines what constitutes "normal" maternal behavior.

Chapter 3 investigates the connections between the paternal behaviors of nonhuman primates and those of men. Written by anthropologists David Taub and Patrick Mehlman, the chapter provides analyses of both direct and indirect paternalistic investment patterns. Taub and Mehlman identify several factors that influence paternalistic investment in humans and nonhumans alike, among the most important of which are social structure and mating system. The chapter ends with a discussion of cross-cultural and historical variation in men's paternalistic behaviors.

Anthropologist Janice Chism describes the ontogeny of behavior in Chapter 4. After examining the implications of an evolutionary biological perspective for understanding human development, she discusses several topics that historically have received attention from primatologists: the relative influence of genetic and environmental factors, the stages of development, the development of attachment, and the ontogeny of social behavior.

In Chapter 5, anthropologist Linda Wolfe presents an analysis of female sexual behavior that focuses primarily on "ultimate" (i.e., evolutionary) questions including the problem of the evolution of women's apparent lack of estrous cycling, and the evolution of the female orgasm. Wolfe also attempts to place women's homosexual behavior in a comparative perspective among the primates. She concludes her chapter with speculations about mating patterns among early hominids.

In Chapter 6, psychologists Ronald Nadler and Charles Phoenix examine male sexual behavior. Their discussion centers on "proximate" problems of male sexuality, and the ways primate data have helped us understand the human situation. For example, they explain how comparative studies have shed light on hormonal influences on the development of male sexual behavior and its display during adulthood (and especially in old age). Nadler and Phoenix end their chapter with speculations about the evolution of hominid sexual patterns drawn from data on the behavior and sexual anatomy of apes.

Aggression, power relations, and politics are the topics of anthropologist Bernard Chapais's essay in Chapter 7. Chapais provides a detailed framework for the analysis of power relations in primates and then discusses the unique human elaborations in this area of behavior. As Chapais explains, power relations among humans changed significantly with the advent of language and technology, but the changes came with a long legacy of structured aggressive behavior from our nonhuman forebears.

In Chapter 8, anthropologist Donald Sade focuses on the importance of kinship for both nonhuman primate societies and human cultures. Sade argues that, although genealogy has clear and direct affects on behavior (and thus some forms of status) among the nonhuman primates, the situation among humans has been significantly altered by the evolution of symbolized status defined by cultural rules. According to Sade, it is the definition of symbolized status that provides the key to differences between human and nonhuman primate societies. Therefore, what Earl Count (1958) termed an "eidolon" should be the focus of anthropological research. Sade provides several suggestions about avenues such research might profitably follow.

Sociologist C. B. Peters brings his knowledge of human behavior and the primate literature to bear on the question of culture in Chapter 9. After a short historical review of the ideas of humanity and culture and their relationships to nonhuman primates, Peters reflects on the potential of primatological literature to illuminate human culture. Peters argues that through an analysis of primatology as a form of discourse we can gain a new perspective on the meaning of humanity.

Of course, these chapters are not an exhaustive look at the results of comparative primatology. They are only a start, but they do make interesting reading. So let us return to the question posed in Belloc's poem at the head of this chapter: How like Mister So-and-so? Some answers wait ahead.

REFERENCES

Carpenter, C. R. (1942a). Characteristics of social behavior in nonhuman primates. *Transactions of the New York Academy of Science 4*:248–258.

———— (1942b). Societies of monkeys and apes. *Biological Symposium 8:*177–204.

———— (1952). Social behavior of nonhuman primates. *Colloque Internationaux du Centre National de la Recherche Scientifique 34:*227–246.

———— (1964). *Naturalistic Behavior of Nonhuman Primates.* University Park, PA: The Pennsylvania State University Press.

Cranach, M. von (1976). *Methods of Inference from Animal to Human Behaviour.* Chicago: Aldine.

Count, E. W. (1958). The biological basis of human sociality. *American Anthropologist 60:*1049–1085.

Darwin, C. (1872). *The Expression of the Emotions in Man and Animals.* Chicago: The University of Chicago Press [1965].

Gilmore, H. A. (1981). From Radcliffe-Brown to sociobiology: some aspects of the rise of primatology within physical anthropology. *American Journal of Physical Anthropology 56:*387–392.

Gingerich, P. D. (1984). Primate evolution: evidence from the fossil record, comparative morphology, and molecular biology. *Yearbook of Physical Anthropology 27:*57–72.

Gregory, W. K. (1929). *Our Face from Fish to Man.* New York: G. P. Putnam's Sons.

Hall, K.R.L. (1963). Some problems in the analysis and comparison of monkey and ape behavior. Pp. 273–300 in S. L. Washburn, ed., *Classification and Human Evolution.* Chicago: Aldine.

Hinde, R. A. (1987). Can nonhuman primates help us understand human behavior? Pp. 413–420 in B. Smuts, D. Cheney, R. Seyfarth, R. Wrangham, and T. Struhsaker, eds., *Primate Societies.* Chicago: The University of Chicago Press.

Hooton, E. (1955). The importance of primate studies in anthropology, Pp. 1–10 in J. A. Gavan, ed., *The Non-Human Primates and Human Evolution.* Detroit: Wayne University Press.

Lewin, R. (1987). *Bones of Contention.* New York: Simon and Schuster.

Mayer, E. (1982). *The Growth of Biological Thought: Diversity, Evolution, and Inheritance.* Cambridge, MA: Harvard University Press.

Merton, R. K. (1967). *On Theoretical Sociology.* New York: Free Press.

Midgley, M. (1980). *Beast and Man.* New York: New American Library.

Napier, J. R., and P. H. Napier (1985). *The Natural History of the Primates.* Cambridge, MA: The MIT Press.

Southwick, C. H., ed. (1963). *Primate Social Behavior.* Princeton, NJ: D. Van Nostrand.

Symons, D. (1979). *The Evolution of Human Sexuality.* New York: Oxford University Press.

Tiryakian, E. (1976). Biosocial man, sic et non. *American Journal of Sociology 82:*701–706.

Washburn, S. L., and I. DeVore (1961). The social life of baboons. *Scientific American 204*(6):62–71.

Wilson, E. O. (1975). *Sociobiology: The New Synthesis.* Cambridge, MA: Harvard University Press.

Yerkes, R. M. (1932). Yale laboratories of comparative psychobiology. *Comparative Psychology Monographs 8*(3).

Zuckerman, S. (1932). *The Social Life of Monkeys and Apes,* 1st edition. New York: Harcourt, Brace.

———— (1981). *The Social Life of Monkeys and Apes,* 2nd edition. London: Routledge and Kegan Paul.

Maternal Behavior in Human and Nonhuman Primates

NANCY A. NICOLSON

Of all aspects of animal behavior, a mother's care for her young is probably the most widely admired. The immediate responsiveness of a mother to her newborn infant and her willingness to take great risks in protecting and defending it have been taken as evidence for a "maternal instinct." Early naturalists were particularly impressed by the maternal behavior of monkeys: the mother's positioning of the newborn on the breast, cradling, embracing, and gazing at the infant all look like human behaviors. Marais, the South African writer who made the first detailed observations of a free-ranging primate troop at the turn of the century, concluded that "in the baboon mother-love reaches a higher stage of development than in any other animal in our country" (Marais, 1975:56). When Yerkes and his contemporaries initiated systematic observations of primates in captivity, accurate descriptions of maternal behavior confirmed the intensely interactive nature of early mother—infant relations (Bingham, 1927; Tinklepaugh and Hartman, 1932; Yerkes and Tomilin, 1935). In the 1960s, the first scientific field observations of primate maternal behavior were published (De Vore, 1963; Jay, 1963; Goodall, 1967); around the same time, long-term descriptive and experimental studies of mother–infant relations were being undertaken in England (e.g., Rowell et al., 1964; Spencer-Booth et al., 1965) and the United States (e.g., Harlow et al., 1963; Jensen, 1965; Kaufman and Rosenblum, 1969). In the 1970s and 1980s, a rapid growth of research on both free-ranging groups and captive primates living in large, seminatural enclosures provided a wealth of information on interspecific differences and socioecological influences on maternal behavior.

In humans, there are enormous cultural influences on maternal behavior, as revealed in historical (e.g., Aries, 1962; DeMause, 1974) as well as cross-cultural studies (e.g., Minturn and Lambert, 1964; Whiting and Whiting, 1975; Le Vine 1977). Even within a culture, maternal behavior is highly variable.

Sources of these individual differences in human maternal behavior are often regarded as more interesting than any patterns we might share with our closest nonhuman relatives. What, then, can primate studies contribute to our understanding of human motherhood? In this review, we will discuss how data from nonhuman primates can help us answer the following questions.

1. Are there universal patterns of maternal care among the nonhuman primates, and to what extent are these shared by humans?
2. How does the capacity to behave nurturantly toward infants develop?
3. How is maternal behavior initiated in the postpartum period?
4. What is the nature of maternal attachment?
5. What factors contribute to the enormous observable variability in maternal behavior?

From the outset, it is useful to distinguish two different types of data that are relevant to understanding maternal behavior. On the one hand, in studies of normal ontogeny, rearing effects, sex differences, and hormonal influences, we usually are measuring interest in or nurturance toward infants *in general* (that is, to infants other than offspring). Studies of the quality of maternal behavior and the growth and gradual waning of maternal attachment are based, on the other hand, on observations of a mother's interactions with her own offspring. Failure to attend to this distinction has caused much confusion; as we will later see, responsiveness to infants is not equivalent to maternal attachment.

PRIMATE MATERNAL CARE: BASIC PATTERNS

Maternal behavior is critical for infant survival in mammals, and for this reason we can assume that it has been subject to intense selection throughout evolutionary history. There are a number of fish, reptile, and insect species that display maternal care, and the majority of bird species are characterized by heavy participation of both parents in infant care. But, with the evolution of lactation, the behavior and physiology of mammalian mothers and their young were inextricably linked together in a unique adaptive complex (Pond, 1977). The primates share this general mammalian heritage. In addition, monkeys and apes, relative to other mammals of comparable size, produce fewer young at a time and have both longer intervals between births and a longer period of infant dependence. Intensive and prolonged parental care increases the chances that offspring will survive and reproduce.

In most species, females perform the greatest share of infant caretaking. One of a mother's most crucial tasks is to provide for her infant's nutritional needs. Communal hunting or regular provisioning with solid food are not found among nonhuman primates; instead, the mother produces milk until her infant can feed itself. In comparison to that of other mammals, primate milk is high in carbohydrates and low in fat and protein. Primate infants suckle slowly, in short but frequent episodes. Weaning from the breast is a gradual process, which is often completed only when the mother gives birth to her next infant. In many of the smaller, seasonally breeding species such as the vervet monkeys (*Cercopithecus aethiops:* Whitten, 1982), infants are nursed on average for almost 1 year, whereas chimpanzee *(Pan troglodytes)* infants continue nursing for up to 4–5 years (Clark, 1977).

In contrast to those mammalian species whose young are cached in nests or dens, primate neonates are in constant contact with their mothers. While the locomotor system is altricial, behavioral adaptations such as the clinging/grasp reflex and infant vocalizations serve to maintain contact. Through extensive body contact, the mother also plays a major role in infant thermoregulation. Even after the infant is old enough to locomote skillfully on its own, the mother carries it during long journeys, over difficult passages, and away from predators. She also defends her infant from aggression. A mother performs a large share of the grooming her infant receives, keeping its skin free of dirt and parasites. Although direct teaching is rare among nonhuman primates, mothers

Fig. 2–1 Japanese macaque mother simultaneously nurses and grooms an older infant. (Photo: James Loy.)

socialize their infants through modeling appropriate social behaviors and feeding techniques, sharing feeding sites or food items, and actively encouraging independence.

These basic features of primate maternal care have been described in greater detail in a number of reviews (e.g., McKenna, 1979a, 1981; Nash and Wheeler, 1982; Higley and Suomi, 1986). Bowlby (1969) was among the first to argue that the high levels of mother–infant contact observed in the higher primates must also have been adaptive, because of the continued risk of predation, in the environments in which humans evolved. Blurton Jones (1972) examined this hypothesis in more detail, comparing human milk composition and infant sucking rate data with data from a variety of other mammals. Human milk is relatively low in protein and fat, and babies suck slowly, leading to the conclusion that human mothers and infants are adapted, as are other primate species, for frequent feeding and high levels of contact. Indeed, in cultures in which mother and infant are in close contact and nursing occurs on demand, intervals between suckling episodes are short (averaging 15 minutes among !Kung hunter-gatherers; Konner and Worthman, 1980).

A growing awareness of these patterns has led to some marked changes in clinical perspectives and recommendations on infant care practices. For example, there has been movement away from strict 4 hour feeding schedules, and mothers have been reassured that breast-feeding "on demand" will not lead to overfeeding. From an evolutionary perspective, infant crying is now understood as an active attempt to cope with distress by initiating contact with the care giver. Not only does prompt maternal responding not "spoil" an infant, but failure to do so may lead to even more frequent crying (Bell and Ainsworth, 1972).

A recent study on the effects of infant carrying is another example of how primate studies are providing a new framework for human clinical research. Human infants have retained the grasp reflex, but do require additional support to cling to their hairless, bipedal mothers. In traditional cultures, mothers use simple leather or cloth slings to support babies on their hips or backs, often with skin-to-skin contact. With slight repositioning, nursing is easily accomplished. Whether through more frequent nursing, better communication of infant needs to the mother, stimulation of a quiet alert state, or simply the reassuring effects of contact itself, frequent carrying may explain why infants in non-Western cultures are reported to cry less than those in cultures with less mother–infant contact. To test whether increasing the amount of infant carrying time could reduce the frequency of crying in the first 12 weeks of life, Hunziker and Barr (1986) randomly assigned mothers on a Canadian maternity ward to an extra carrying and a control group, asking them to record contact, carrying, feeding, and infant crying episodes each day. When infants were carried for just an extra 2 hours each day, crying, especially in the evening hours, was significantly reduced. The authors concluded that this simple intervention can be more effective than pharmacological or other available means of treating a common, developmentally normal but nevertheless distressing problem for many parents of newborns and may have prophylactic potential in more serious cases of infant colic.

INTERSPECIFIC DIFFERENCES

Early studies of primate maternal behavior focused almost exclusively on the rhesus macaque *(Macaca mulatta)*. It soon became evident, however, that even closely related species display different forms of social behavior, including mother–infant interactions. One of the best known examples comes from the work of Kaufman and Rosenblum (1969) on pigtail *(Macaca nemestrina)* and bonnet macaques *(Macaca radiata)*. In identical laboratory environments, these two species displayed marked differences in patterns of mother–infant contact, maternal rejection, and infant response to separation. With the burgeoning of primate research, especially field studies, in the 1960s and early 1970s, many more species were investigated, including baboons (*Papio* sp.), langurs (*Presbytis* sp.), and the great apes. For practical reasons, much more fieldwork has been done with the larger, more terrestrial species. Although laboratory studies often leave us wondering about the adaptive significance observed interspecific differences might have in the natural environment, they have enriched our knowledge of maternal behavior in rare, small, or arboreal species, most notably a number of New World species (e.g., squirrel monkeys [*Saimiri sciureus*], marmosets, tamarins). From this body of data, we can detect both broad phylogenetic trends and a number of patterns related to species' differences in social organization and ecology. With respect to the former, the time course of maternal care is closely linked to body size, brain size, and rate of development, along with other aspects of species' life histories (Harvey et al., 1987).

Phylogenetic differences in maternal play provide a more concrete behavioral example. Only in our closest relatives, the great apes, do we find extensive mother–infant play and "teaching," with striking similarities to these interactions in humans (Higley and Suomi, 1986). Although it is true that the social organizations of orang-utans *(Pongo pygmaeus)* and chimpanzees limit opportunities for peer play, which could have led to selection for maternal playfulness, both the occurrence of mother–infant play in gorillas *(Gorilla gorilla)* and the improbability that hominid social organization resembled that of the living great apes argue for a more general anthropoid adaptation.

For insights into the evolution and proximate causes of paternal behavior, which is prominent in many human cultures, we can better turn to our phylogenetically more distant relatives, the South American marmosets and tamarins (Callitrichidae and Callimiconidae families). Among these species, the combination of monogamy with high costs of reproduction in females, who typically give birth to twins, has led to the evolution of extensive paternal care. The relevance of studies of primate paternal care for an understanding of both paternal and maternal behavior in humans has been summarized by Lamb and Goldberg (1982) and Taub and Mehlman (Chapter 3, this volume).

THE DEVELOPMENT OF NURTURANCE TOWARD INFANTS

Biological Bases

Early studies of captive rhesus monkeys documented that females other than the mother, including adolescents who had not yet had infants of their own, were extremely interested in and nurturant toward infants, embracing, carrying, and grooming them (Rowell et al., 1964). Field studies have reported the same phenomenon. In most species, juvenile and adolescent females are even more likely than adult females to show interest in and to act nurturantly toward the infants of other females (see Nicolson, 1987:Table 27–3). In contrast, immature males seldom interact with unrelated infants except in play (Walters, 1987). Even with no previous exposure to infants, young captive rhesus females show more interest in neonates than do young males (Chamove et al., 1967, Gibber and Goy 1985; but see also Brandt and Mitchell, 1973). Similarly, squirrel monkey females of all ages are more likely to retrieve unrelated infants than are males (Rosenblum, 1972). These findings suggest that females may be biologically predisposed, from a very young age, to behave nurturantly toward infants. In fact, it is more accurate to speak of an inhibition of attraction toward infants in males. In an experimental study of rhesus monkeys, distinct female and male response patterns to infants could be identified; although prenatal androgen treatment was not effective in inducing the male pattern in genetic females, castration of males did lead to an increase in the female pattern, pointing to a likely role for postnatal (particularly pubertal) androgens in the sexual differentiation of infant-directed behavior (Gibber and Goy, 1985).

As was mentioned above, an increasing body of observations has documented extensive paternal behavior in a number of species, most notably the tiny South American marmosets and tamarins. Not only do fathers carry infants more frequently than mothers but sex differences in infant carrying by immature siblings are also absent (Goldizen, 1987). Androgens do not, therefore, place absolute limits on the kind or amount of infant care males can show. Furthermore, in those species with the most pronounced sexual division of infant-care activities, such as rhesus macaques or squirrel monkeys, males who typically behave indifferently toward infants can, under certain circumstances, display the full range of nurturant responses. For example, when rhesus male and female pairs were tested with an unfamiliar infant, only the female showed caretaking behaviors, and some males actively avoided the infant. When paired alone with the test infant, however, both males and females responded nurturantly (Gibber, 1981).

This finding is especially interesting in light of attempts to demonstrate sex differences in responsiveness to infants in humans at various stages in the life cycle. The results of these studies have been summarized by Lamb and Goldberg (1982). In general, sex differences are more likely to be found on behavioral or self-report measures than on psychophysiological responses to infant stimuli. Moreover, sex differences are not found consistently at all ages but are most pronounced at those stages at which pressures to show traditionally sex-stereotyped behaviors are greatest, for example, in teenagers and new parents. When asked to rate their attraction to pictures of primate infants, women in another study expressed a greater attraction and men a lower attraction to infants when ratings were disclosed in same or mixed-sex groups (Berman et al., 1975). In this case, the results can be explained in terms of social pressures on both men and women to display culturally sex-typed responses. This same interpretation cannot be applied to the primate experiment above, forcing us to look still further to discover how social factors can amplify sex differences in the absence of culturally based normative expectations. More information on how these social factors operate in primates could give us new insights into how sex-role stereotypes "evolve" in human groups. It is possible, for example, that immature males and females selectively focus attention on same-sexed adult models or on specific types of social interactions (Pereira and Altmann, 1985). Additional well-designed field and experimental studies in species with and without adult male participation in infant care are needed.

The Role of Experience

In the natural environment, an innate attraction to infants would increase the likelihood that immature females will seek out opportunities to observe, handle, or even care for infants. "Play mothering" was originally described in vervet monkeys by Lancaster (1971), who suggested that hands-on experience was important in the development of maternal behavior. Reviews of allomaternal behavior (that is, interactions with infants by individuals other than the mother) in primates have produced a variety of explanations for its form, function, and evolution (Hrdy, 1976; Quiatt, 1979; McKenna, 1979b; Nicolson, 1987). Here we limit the discussion to the "learning-to-mother" hypothesis, which asserts that allomaternal behavior among immatures is selectively advantageous because it increases the likelihood of successful reproduction as an adult. Although there are a few species in which older siblings make a positive contribution to infant care, in other species immatures also interact with unrelated infants, and doubts have been raised about whether these interactions help either the infant or its mother. Allomaternal care is often clumsy and incompetent and sometimes is downright nasty. After holding a squealing infant for a few minutes, a young allomother may experience a rapid waning of interest in her charge and roughly reject or abandon it, as reported for wild langurs (*Presbytis entellus*: Hrdy, 1977). Not surprisingly, mothers tend to be selective in entrusting their infants to allomothers. In species such as the macaques, baboons, and chimpanzees, mothers of young infants actively resist allomothering initiatives. Good allomaternal performance is reinforced, however, in that mothers are less likely to retrieve their infants as long as they are content (Hrdy, 1976, 1977).

From captive studies linking a lack of previous experience with infants with decreased parenting success (for example, in the callitrichids; Epple, 1978; Snowdon et al., 1985), we can infer that learning processes prior to adulthood are important. On the other hand, observations of inadequate mothering in wild group-living primates (see below) indicate that early experience does not guarantee good maternal behavior, especially in first-time mothers. As Chism (Chapter 4, this volume) suggests, the species with higher firstborn infant mortality may be those in which immature females have the fewest opportunities to handle young infants.

Abnormal Development

Laboratory observations and experiments over the last 30 years have demonstrated the crucial role of experience in the ontogeny of primate maternal behavior. Harlow's well-known rearing experiments with rhesus infants produced marked long-term behavioral pathology. With regard to maternal behavior, "motherless mothers" (rhesus females reared either in isolation, with wire or cloth-covered surrogate mothers, or with peers only) were grossly inadequate. These females refused to allow their infants to cling, pushing them away, biting them, or crushing them to the floor. Violent attacks were common, and in several cases infants had to be removed from the cage to save their lives. Even when maternal behavior approached the normal range, responses to infants were inconsistent and seemingly arbitrary. In social preference tests, "motherless mothers" displayed no preference for neonates over older stimulus monkeys, suggesting that infants are not as attractive to isolation-reared females as they are to normally reared females. Moreover, after living together for 8 months, motherless mothers still showed no preference for their own infants, spending more time in proximity to unrelated infants in the test apparatus (Sackett and Ruppenthal, 1974). The most abusive mothers were those reared in total social isolation. Exposure to peers in infancy resulted in better but still not completely appropriate maternal behavior (Harlow et al., 1966; Arling and Harlow, 1967; Ruppenthal et al., 1976; Suomi and Ripp, 1983).

A very important finding of longitudinal studies of the motherless mothers was the marked improvement they displayed with successive births. Females that had spent just 2 days or more with previous infants usually exhibited adequate maternal behavior. Females with less neonatal contact, on the other hand, rarely showed any improvement in maternal care even after as many as five births. Mothers that were simply indifferent to their first offspring were much more likely to improve with later births than initially abusive mothers (Ruppenthal et al., 1976).

Total deprivation from an adult caregiver is an extreme intervention and one that fortunately has few parallels in human rearing experience. More relevant to the human case are studies of the effects of early or repeated separations from the mother. For example, chimpanzee females that had been taken from their mothers before they were 18 months old showed inadequate care to their first infants, avoiding, ignoring, or failing to hold the newborn properly. Here again, practice helped. Maternal behavior improved with successive offspring, and by the third birth most of the mothers were rated as highly competent (Rogers and Davenport, 1970). In gorillas, as in chimpanzees, early separation from the mother often leads to indifference towards and neglect of offspring, although infant abuse is rare (Nadler, 1983). Laboratory rhesus monkeys reared with their mothers for 4 months and then caged alone, with opportunities for visual and auditory but no physical contact with other monkeys, also showed deficits in maternal behavior toward their firstborn infants. Females that had not observed other mothers caring for infants neglected their own offspring; in contrast, females that had been able to watch mother–infant interactions became adequate mothers themselves, although they initially held their infants in atypical positions (Dienske et al., 1980). It thus appears that observation of maternal care can compensate to some extent for limited direct experience.

In addition to maternal deprivation, any early experiences that impair the development of psychosocial competence can clearly impinge on later maternal behavior as well as on other forms of social interactions. In one experiment, for example, rhesus monkeys were repeatedly separated from the peer groups in which they were raised; in several cases, this intervention resulted in depression, and these depressed females were later more likely to abuse their own offspring (Suomi, 1978; Suomi and Ripp 1983).

INADEQUATE MATERNAL BEHAVIOR IN WILD PRIMATES

Although the most dramatic pathologies of maternal behavior are seen in captivity, not all wild primate females are perfect mothers. Paralleling laboratory findings, loss of the mother in infancy can contribute to later maternal incompetence. Japanese macaque *(Macaca fuscata)* females who survived being orphaned in late infancy later showed deficits in care toward their firstborn infants, resulting in high infant mortality. Here, too, improvement was seen with subsequent births (Hasegawa and Hiraiwa, 1980). Even in wild females with normal rearing histories and prior opportunities to interact with infants, first-time mothers are often clumsy and occasionally refuse to carry or nurse their infants. Needless to say, if maternal behavior does not improve very rapidly, the infant will die, as has been observed in a number of free-ranging species (e.g., chimpanzees: Goodall, 1986; savannah baboons: personal observations). In most cases, maternal skills do improve dramatically in the postpartum period; nevertheless, primiparas have higher infant mortality, particularly in the early months, as reported in a variety of free-ranging and captive groups (e.g., captive squirrel monkeys: Taub, 1980; rhesus macaques: Wilson et al., 1978; wild howler monkeys [*Alouatta palliata*]: Glander, 1980; olive baboons [*Papio anubis*]: Nicolson, 1982). This highlights the fact that the development of maternal behavior in primates is a continuous process from infancy at least through the first experience of motherhood.

In humans, there is evidence for a higher risk of child neglect and abuse among adolescent mothers (Field et al., 1980). It is tempting to speculate that the higher mortality observed in firstborn infants in many primate species is a result of their mothers' social immaturity, but the exact cause of infant death is rarely known. It does appear that the youngest mothers in wild groups are more likely to neglect or, alternatively, to overprotect their infants, and infants of low-ranking primiparas are especially vulnerable to kidnapping or mistreatment by other group members.

For primatologists, adolescent motherhood represents a contradiction in terms, since females are considered adult as soon as they reproduce. There may, however, be circumstances under which physical and social development are not in synchrony, for example, when reproductive maturity is accelerated by nutritional supplementation. Altmann (1985) summarizes the available data and suggests how these could provide new perspectives on the culture-bound topic of adolescent motherhood in humans.

Pathological development of maternal behavior in nonhuman primates has attracted much interest from researchers and clinicians attempting to understand the etiology of human child abuse; this has led to a productive exchange of data and ideas (see, e.g., Horenstein, 1977; Nadler, 1980; and the volumes edited by Reite and Caine, 1983; Hausfater and Hrdy, 1984; Gelles and Lancaster, 1987). Under optimal laboratory rearing and housing conditions and in the natural environment, however, primate maternal behavior varies along a continuum from protective to permissive to rejecting, but physical abuse resulting in bodily harm is extremely rare. Humans are thus unique among the primates in that mothers may sometimes deliberately kill their own infants (for discussions of the social and ecological circumstances that may contribute to abuse, neglect, and infanticide by human mothers, see Hausfater and Hrdy [1984], DeVries [1987], and Scheper-Hughes [1987]).

INITIATION OF MATERNAL BEHAVIOR POSTPARTUM

Primate studies could potentially contribute much to an understanding of the physiological basis of maternal responsiveness in humans. In comparison to more extensively studied species such as the rat, nonhuman primate maternal physiology closely resembles that of humans. Maternal behavior is, moreover, much more complex and variable in humans and other primates than in rodents. Beyond what can be learned directly from human research, primate studies give us the opportunity to manipulate hormone levels and perform other experiments that would be ethically impossible in humans and to investigate maternal responsiveness in the absence of cultural expectations and sex-role stereotypes about infant care and nurturance.

The available data from nonhuman primates are, however, surprisingly scarce and inconclusive. In general, hormones appear to play a much smaller role in the initiation and maintenance of primate maternal behavior than has been found in a number of other mammalian species (Rosenblatt and Siegel, 1981; Capitanio et al., 1985). In rats and sheep, for example, the hormonal events of pregnancy and parturition are sufficient to initiate maternal behavior in females with no previous exposure to infants. By mimicking these hormonal changes experimentally, it is even possible to induce maternal responses toward unfamiliar infants in virgin females.

Major increases in hormone concentrations, especially in estradiol and progesterone, occur during pregnancy, followed by sharp decreases at parturition and more gradual declines over the next week in humans (Fleming and Corter, 1988) and other primates (Pryce et al., 1988). It is reasonable to ask whether hormones contribute to changes in maternal responsiveness in the pre- or post-partum period. Observations of kidnapping or adoption of neonates by primate females who are in late pregnancy or who have recently lost an infant (Thierry and Anderson, 1986) do suggest that hormonal conditions associated with late pregnancy, parturition, or lactation may heighten responsiveness to young infants.

Only a few studies have systematically examined the effect of reproductive state on responsiveness to infants. For example, in an experiment with squirrel monkeys, pregnant females nearing term were more likely to retrieve a test infant than females earlier in pregnancy, adolescent females, or females with young (Rosenblum, 1972). Similarly, Hrdy (1977) observed a prepartum increase in allomothering attempts by wild Hanuman langurs. When nulliparous rhesus females were exposed to young infants at intervals throughout pregnancy, however, no increase in initiations of contact with the stimulus infants was observed. Because the same females subsequently displayed adequate maternal behavior postpartum, it appeared that hormonal events following parturition may have played a role in inducing maternal responsiveness (Gibber, 1986). It should be noted that the females in this experiment had been reared without opportunities to interact with infants, which may have contributed to the lack of responsiveness during their first pregnancy.

A preference experiment conducted by Sacket and Ruppenthal (1974) gives perhaps the clearest results. Pigtail macaque females were separated from their own infants immediately after giving birth; on the same day and again at intervals up to 6 months later, their preference for an unrelated neonate relative to three other classes of stimulus monkeys was assessed in a choice apparatus. The females showed a significant preference for the neonate on the day of birth, at 2 weeks, and at 1 month, with the most stable and persistent preferences at 2 weeks. These data are particularly striking in light of the fact that these females had had no further contact with their own infant or with the stimulus infants in the period between parturition and testing.

Taken together, these studies point to an increase in the salience of infant stimuli in the puerperium that is most likely hormonally mediated. Unfortunately, they do not give any direct evidence on which hormone or hormones are responsible. Estradiol/estrogen is a likely candidate for involvement in changes observed in late pregnancy. Recently, Pryce et al. (1988) reported an association between prepartum estradiol levels and the quality of subsequent maternal behavior in red-bellied tamarin monkeys *(Saguinus labiatus)*. Good mothers (those that displayed appropriate infant care and whose infants survived their first week) had stable urinary estradiol levels in the last week of pregnancy; estradiol dropped significantly over the same period in poor mothers, which refused to carry or nurse their infants, resulting in high neonatal mortality. This difference was limited to those mothers with no prepubertal experience with infants.

Following parturition, activation of the noradrenergic, oxytocin, and β-endorphin peptidergic systems may prove to be important in enhancing maternal sensitivity to sensory signals from the neonate (Keverne, 1988). The possibilities that the smell or taste of birth fluids (Lundblad and Hodgen, 1980) or ingestion of the placenta (see Higley and Suomi, 1986) facilitates maternal behavior have not been fully evaluated, but these seem unlikely to play a major role in either humans or other primates.

A recent review of the human data (Fleming and Corter, 1988) concludes that physiological and physical events during the puerperium may indeed facilitate the onset of maternal responsiveness in first-time mothers by "sensitizing" the mother to stimuli from the infant, but that these effects are very slight in comparison to those in nonprimates. Although women do report an increasing level of attachment toward their own infant over the course of pregnancy through the first 2–3 days postpartum, there is no evidence for a heightened responsiveness toward babies in general nor any discernible relationship between levels of circulating hormones (estradiol, progesterone, testosterone, cortisol, prolactin, β-endorphin) and changes in self-reported attraction to infants (Fleming and Corter, 1988). The correlational approach has admitted limitations: even in rats, in which hormonal influences on maternal behavior have been experimentally demonstrated, no significant correlations between maternal responsiveness and hormone levels in intact pregnant and postpartum females have been found. For this reason, experiments with nonhuman primates that more closely parallel the work on rodents could prove informative.

In summary, it appears that previous experience, whether by increasing positive motivation or by decreasing inhibition, is sufficient to ensure adequate responsiveness to the infant postpartum; when such experience is lacking, however, as it is in first-time mothers, hormonal changes may be important for the initiation of maternal behavior. It is safe to assume that, over the course of human evolution, a certain percentage of first births have not been planned or even desired, increasing the adaptive advantage of physiological mechanisms that heighten the motivation for maternal behavior (Konner, 1982).

A number of years ago, Klaus and Kennell (1976) proposed that there was a "sensitive period" for maternal attachment in the first few hours after birth; if the mother did not have the opportunity to contact her baby during this period, the process of "bonding" would be inhibited or delayed, possibly leading to long-term disturbances in mother–infant relations. This argument had a major impact on maternity ward practices. Although most clinicians, human ethologists, and primatologists would agree that increased early contact, when possible, is a good thing, major objections were raised to Klaus and Kennell's proposal (see detailed critiques by Leiderman, 1978; Lamb, 1983; Sluckin, 1986; DeVries, 1987). Very briefly, much of the rationale for postulating a sensitive period was based on imprinting-type phenomena in rodents, goats, and other nonprimates. In addition, the implication that postpartum contact is essential for the normal development of attachment is at best misleading and at worst detrimental in cases when mother and infant must be separated for medical reasons. In response to such criticisms, the authors themselves have adopted a compromise position (Klaus and Kennell, 1983).

Much of the confusion about "bonding" could have been avoided had closer attention been paid to nonhuman primate patterns, as described above. Again, although responsiveness to neonatal stimuli may increase around the time of giving birth, responsiveness in primates is not at all restricted to the first hours or even days postpartum. Nor is there is any evidence from primates for the stereotypic responses to infants seen in rodents or for maternal imprinting as seen in goats. Finally, there is no evidence that early interactions (as long as they are sufficient to ensure survival) are crucial for the development of emotional bonds or attachment.

THE NATURE OF MATERNAL ATTACHMENT

In primates, the nature of maternal attachment can be assessed in a number of ways. Here, we consider first the specificity of a mother's interactions with her own offspring, second the dynamic properties of the relationship over time, and third the evidence for emotional bonds as revealed in maternal response to separation or loss.

Specificity

It is likely that primate mothers learn to recognize their own infants on the basis of visual, auditory, and possibly olfactory cues within the first few days after birth. Even in groups with many similarly aged neonates and early breaks in mother–infant contact, mix-ups rarely if ever occur. Although we do not know exactly how fast this recognition process takes place, there is good evidence that mothers can discriminate their own infant's vocalizations within the first month (for example, in squirrel monkeys; Kaplan et al., 1978). What is more important than recognition, however, is the fact that mothers preferentially interact with their own young. Certain forms of care, such as nursing, carrying, and defense from predators, are limited to the mother's own infant. Females also preferentially groom their own offspring and intervene on their behalf during fights, although they may extend these forms of care to other infants, especially if they are relatives. Affiliative responses toward infants such as embracing or touching are much less specific, but these kinds of interactions cost the actor little and are at the same time of little benefit to the infant. In general, the more time, energy, or risk a given form of care entails, the more likely that a female will direct it disproportionately to her own offspring.

Aggressive acts, on the other hand, are far more likely to be directed toward unrelated infants. Although females do not harm their own infants, they can under certain circumstances be extremely aggressive, abusive, or even infanticidal toward the offspring of other females. In many species of primates, experienced mothers have been observed to threaten, bite, hit, throw, drag, sit on, or attack unrelated infants (Hrdy, 1976). In captive bonnet macaques, aggression from adult females appears to increase mortality in infants of low-ranking mothers (Silk, 1980). In functional terms, aggression toward the offspring of other females appears to be an aspect of female reproductive strategy through which the perpetrator may reduce resource competition for herself and her own offspring. That aggression toward infants is most often displayed by females who are themselves experienced, successful mothers underscores the fact that infants (including neonates) do not invariably release an instinctive nurturant response.

Even after weaning, the relationship between a mother and her offspring continues to differ both qualitatively and quantitatively from relationships among other social partners. Vervet monkey mothers, for example, respond more strongly to their juvenile offspring's screams than do other adult females (Cheney and Seyfarth, 1980) and give more alarm calls in response to an artificial predator when their own juvenile offspring are present (Cheney and Seyfarth, 1985). Mothers continue to perform a large share of the grooming their juvenile, adolescent, and even adult offspring receive (chimpanzees: Pusey, 1983; rhesus macaques: Missakian, 1974). For a yellow baboon *(Papio cynocephalus)* troop, Walters (1981) compared the relationships of 3-year-old females with all adult females and found that the juveniles' mothers, if still living, could easily be identified on the basis of grooming patterns and interventions during fights. If the mother had died and her daughter had formed strong affiliative ties with another adult female, this relationship resembled a mother–offspring relationship but could still be distinguished by multivariate techniques. Actual adoption by an adult female is exceedingly rare among wild primates and, in the case of unweaned infants, is unlikely to be successful (Thierry and Anderson, 1986). If an infant does survive its mother's death, a relationship very similar to that between mother and offspring may gradually develop, usually with a sibling or other kin, as has been carefully documented in a case study of freeranging rhesus monkeys (Berman, 1983). Nonetheless, adoptive relationships, in general, appear to be less reciprocal and less stable over time than true mother–offspring relationships (Dohlinow and DeMay, 1982; Thierry and Anderson, 1986).

In summary, a wide variety of individuals can and do respond nurturantly toward infants. What is special about the interactions between a mother and her offspring is that these occur in the context of a long-term, reciprocal *relationship* (Hinde, 1976) in which both biological predispositions and intense mutual exchange contribute to the formation of emotional bonds. In nonhuman primates as well as in humans, these bonds extend well beyond the period of physical dependency.

Dynamics of Mother–Infant Relationships

Harlow et al. (1963) described three stages of what they called "mother love," a stage of care and comfort, a stage of ambivalence, and a stage of relative separation. Since then, a great deal of effort has been focused on describing further how the primate mother–infant relationship changes over time. Hinde and his colleagues have developed quantitative measures for assessing maternal and infant contributions to changes in the relationship over time (see, e.g., Hinde and Atkinson, 1970); these methods have been widely applied in studies of both captive and wild primates as well as in ethological studies of human infant development (e.g., Konner, 1977; Elias et al., 1986).

In general, primate mother–infant contact declines steadily over time, with the primary responsibility for the maintenance of contact and proximity shifting at some point from mother to infant. The timing of this process depends on a host of factors, including both species and environmental characteristics. A controversy has developed over the relative roles of mother and infant in the development of independence and, more specifically, over how maternal rejection might influence this process (see Higley and Suomi, 1986). Maternal rejection may have the immediate effect of increasing infant contact-seeking behaviors, but in the long run rejection seems to accelerate the development of independence.

Maternal Response to Separation or Loss

Another approach to investigating the nature of emotional bonds between primate mothers and infants is to observe responses to physical separation. Although by far the most attention has been focused on infant responses to separation, there is at least some information available on mothers. The immediate maternal response to separation from her infant is to vocalize intensely and show increased activity (Kaplan, 1970). Among free-living primates, active searching ensues if a young infant appears to be lost (personal observations).

Hormonal responses to mother–infant separation in primates have been reviewed by Levine and Wiener (1988). In squirrel monkeys, both mothers and infants displayed significant elevations in plasma cortisol in response to a brief (30 minute) separation. Moreover, this response did not habituate; after six repeated separations of 1 hour duration, mothers as well as infants continued to display cortisol increases (Coe et al., 1983). Mothers did show a reduction in the cortisol response to separation if they were left in the familiar home environment during infant removal (Coe et al., 1985). Rhesus mothers also showed cortisol elevations during the first day after separation, even if they remained in contact with the social group (Levine et al., 1985). Infant responses (both behavioral and hormonal) are usually greater and more persistent over time than mothers' responses, sometimes resulting in marked depressive symptoms. However, the often overlooked maternal reaction to infant separation or loss may also yield insights into the reciprocal nature of attachment in humans as well as other primates.

When a dependent primate infant dies, its mother generally carries the body for several days, often until the corpse is decomposed and barely recognizable as an infant. The mother grooms the corpse and threatens others away. A wild baboon female who, while feeding, misplaced the body of her infant who had died the day before, continued distress vocalization and searching behavior for the rest of the day (personal observation). These observations strongly suggest that carrying dead infants is a specific maternal grief response. The carrying behavior could have evolved to prevent too early abandonment of sick infants; at the proximate level, contact with the dead infant may alleviate maternal distress by helping the mother adjust to the finality of the loss. Unfortunately, no data are available on the hormonal aspects of this response. Based on the results of experimental separations, maternal cortisol levels could be expected to rise in reaction to infant death, and, if carrying the dead infant is a form of active coping, it may allow a quicker termination of the physiological stress response.

TOWARD AN UNDERSTANDING OF INDIVIDUAL DIFFERENCES IN MATERNAL BEHAVIOR

Evolutionary Perspectives

Over the last 2 decades, parental behavior has attracted increasing attention from evolutionary theorists attempting to explain the great diversity of life history strategies among contemporary animal species. A broad distinction can be made between species that produce large numbers of offspring, show little or no parental care, and have high preadult mortality and species that produce few offspring, invest much care in each, and have a larger percentage of offspring surviving to reproduce in turn. Thus species such as the Pacific salmon produce many offspring in one breeding season and promptly die, whereas a chimpanzee female produces only about four or five offspring during her comparatively long lifetime but invests a great deal of care in each. Parental care that both benefits current offspring and entails costs to future reproduction is referred to as "parental investment" (Trivers, 1972). The most obvious example of parental investment among humans and other primates is lactation. The longer and more frequently a female nurses her current infant, the longer the period of postpartum infertility and the delay in further reproduction (humans: McNeilly, 1979; wild baboons: Nicolson, 1982).

Parental investment theory has generated a number of predictions and hypotheses that can be investigated in human and nonhuman primates. Areas that have received particular attention include sex differences in parental investment, differential investment in sons and daughters, and parent–offspring conflict. The first topic is discussed in depth by Taub and Mehlman (Chapter 3, this volume); the others are summarized very briefly below.

In general, parental investment theory predicts that mothers will provide equal care to sons and daughters (Trivers, 1972). Exceptions to this rule should occur only when offspring of one sex are more costly to produce or to rear or when the anticipated returns in terms of reproductive success differ (Trivers and Willard, 1973). In nonhuman primates, mothers in a number of matrilineally organized primate species appear to reject male infants earlier, at least in captive or provisioned groups (macaques: Berman, 1984; Simpson and Simpson, 1985; vervets: Fairbanks and McGuire, 1985). Few differences in maternal care given to males and females have been found among wild primates, however, although numerous studies have addressed this issue.

In contrast, sex-biased parental investment is common in many human societies. Not only are female infants more often the target of infanticide but they are often nursed less frequently, weaned earlier, or fed less than males. Such practices can be interpreted in light of the costs and benefits of producing sons vs. daughters under various economic, social, and ecological conditions. (For a thorough discussion of both theory and data, see the recent review by Hrdy [1987].)

Parent–offspring conflict theory, as formulated by Trivers (1974), calls attention to the different adaptive strategies a mother and her offspring are expected to pursue. Current evolutionary theory identifies the individual as the unit of selection; it follows that the genetic self-interests of a mother and her young are not identical. Although each primate infant, after a lengthy gestation, already represents considerable maternal investment at birth, the amount and duration of further investment by the mother must be weighed in terms of its costs to future reproduction. The infant, however, being more closely related to itself than to future siblings, can profit from more parental investment than the mother is selected to give. A number of commonly observed features of parent–offspring relations, most notably maternal rejection and infant protest during the weaning process, may be interpretable as behavioral manifestations of the different adaptive strategies pursued by the two actors. (For a thoughtful critique of parent-offspring conflict theory as applied to primate mother-infant relations, see Altmann [1980].)

There are also practical limitations to the application of evolutionary models among primates; adequate life history data are available for only a few free-ranging groups, and estimates of the costs and benefits of parental investment are by necessity indirect. Nevertheless, recent data relating maternal time and energy expenditure to reproductive rate, mortality risk, and other life history parameters—the costs and benefits of motherhood—have provided many new insights (Altmann, 1980, 1983, 1986; Nicolson, 1982; Whitten, 1982; Lee, 1984; Hauser, 1988; Hauser and Fairbanks, 1988).

In summary, an economic analysis that defines the costs and benefits of behaviors in terms of survival and reproductive success can usefully be applied to parent–offspring relations. Different styles of mothering reflect constraints from and facultative adaptations to current social and ecological circumstances. In the next sections, we consider the kinds of social and ecological forces that shape maternal behavior in primates.

The Social Context

The vast majority of primate females live in social groups, in the context of which both competitive and cooperative behaviors have evolved. A female's position in a dominance hierarchy and the presence of closely related kin and other allies within the group can influence the maternal style she displays, especially how much she restricts or encourages infant independence. Group life can be dangerous. Rough treatment of infants by other group members or during encounters with neighboring groups occasionally results in serious injury or death, and a young infant who is kidnapped by an overly zealous allomother may starve before its mother is able to retrieve it (Hrdy, 1976). In species with strong female dominance hierarchies (e.g., macaques, baboons), mothers are more restrictive, rarely allowing their infants to be touched or held by others in the early weeks of life. In contrast, in species with less prominent dominance relations (langurs, colobus monkeys [*Colobus* sp.], howlers), mothers readily allow other females access to even very young infants (Hrdy, 1976; McKenna, 1979b). A female's social rank can thus influence her maternal behavior. Apparently in response to the pressures of aggressive interactions and infant pulling by other females, low-ranking yellow baboon mothers are more restrictive of infant independence than higher-ranking "laissez faire" mothers (Altmann, 1980). Here, variability in maternal behavior can be interpreted as an adaptive strategy. Early infant independence may have benefits, in terms of shorter interbirth intervals (Nicolson, 1982) or better chance of offspring survival if the mother should die. For low-ranking females, however, the risks probably outweigh the potential benefits.

Rank-related effects are often even more pronounced in captivity, especially if there is crowding or stress resulting from unstable social groups. When social groups have been maintained in captivity for several generations, however, close kin networks develop in which mothers may actually be less fearful and restrictive than in free-ranging groups, as Berman (1980) has documented for rhesus monkeys. Focusing on the effects of grandmothers in captive vervet groups, Fairbanks (1988) has reported that young mothers whose own mothers were still present were more relaxed and that infant independence was accelerated. Moreover, with grandmothers in the group, mothers experienced less aggression from nonkin females, were better able to retrieve their neonates from allomothers, and were much more likely to rear their infants successfully (Fairbanks and McGuire, 1986).

In addition to the effects of group members on maternal restrictiveness, the extent to which social activities conflict with infant care can also shape a mother's behavior. Social grooming and interventions in aggressive interactions are important for the maintenance of long-term friendships and alliances, both usually kin-based. If these social relationships are disturbed—for example, when the mother is temporarily removed from the social group during a mother–infant separation experiment—reunion may produce conflict in mother–infant relations; the mother has to respond to increased demands from her infant at the same time she must reintegrate herself into the group (Hinde and Davies, 1972). Similarly, when a female is trying to establish her position in a recently formed captive group, she may be less attentive to her infant than normal or may even abuse it.

In contrast, captive mothers isolated with their infants are generally less restrictive and more rejecting than group-living mothers. Not only is there less danger from conspecifics but infants approach and seek contact more or, in the absence of peers, attempt to play with their mothers more often leading to negative responses (Hinde and Spencer-Booth, 1967). Interestingly, rhesus females caged alone with their infants, in contrast to group-living females, failed to show a preference for their own infants in a choice experiment and were less attracted to infants in general (Sackett and Ruppenthal, 1974). Nadler (1980) has reported that captive gorilla mothers are much more likely to display adequate maternal behavior if they give birth while housed in a social group. He suggests that group life stimulates good mothering by increasing the necessity for a mother to protect and restrict her infant from contact with other group members.

For human mothers, social relationships more often represent a source of support than a danger. In contemporary societies, however, changes in social structure, subsistence, and housing patterns have resulted in the relative social isolation of mothers of small children. If we view this situation as very roughly analogous to the experience of captive primate mother–infant pairs housed alone, it seems likely that maternal irritation could increase when a mother and young children spend long periods of time together in close quarters. Social isolation may indeed be a contributing factor in human child abuse (Horenstein, 1977).

The Ecology of Motherhood

In addition to infant care, a primate mother in the natural environment has to make a living, obtaining energy and nutrients sufficient for her own needs as well as those of her growing infant. Foraging and feeding take up a large percentage of the day, varying in relation to such factors as food quality, abundance, pattern of distribution, and seasonality. To what extent does motherhood impose extra demands on an adult female's time budget? How are infant care activities integrated with maintenance activities, and what changes occur over time as the infant grows and becomes increasingly able to contribute to its own maintenance? When time or resources are in short supply, are there quantitative or qualitative changes in patterns of mother–infant interaction? Although subsistence patterns in contemporary cultures have diverged in major ways from those of nonhuman primates, these questions have clear parallels among humans.

Lactation entails a large increase in a primate female's energy requirements. Additional costs are incurred in carrying the growing infant until it is capable of efficient independent locomotion. In meeting these energetic demands, mothers either must increase food intake by spending more time feeding each day or must lose weight. Based on time budgets and the estimated energetic costs of lactation and infant carriage, Altmann (1980, 1983) has modeled how wild yellow baboon females cope with the demands of motherhood. Adult females without dependent young spend an average of 43% of the day feeding and 23% of the day walking, with the remainder available for rest and social interaction. Mothers spend an increasing proportion of the day feeding as their infants grow older, peaking at almost 60% when infants are 5–6 months old, by which time infants are contributing substantially to their own nutrition and can also walk considerable distances. Inevitably, an increase in time allocated to feeding must be compensated for by a reduction of time spent in another activity. If we assume that rest is an obligatory category of behavior, then the increased energetic demands of motherhood must lead to decreased time available for socializing.

Gelada baboons *(Theropithecus gelada)* live in a habitat where food is relatively abundant but is subject to large seasonal fluctuations. In this species, mothers also spend more time feeding as their infants grow older, increasing from 35% to 70% of the day over the first 4 months (Dunbar and Dunbar, 1988). The rate at which feeding time increases is affected by seasonal differences in food quality and the development of independent feeding by infants. In this case, however, extra feeding time is gained at the expense of rest time. Dunbar and Dunbar suggest that the maintenance of social relationships is highly important to a female's future reproductive success, and social time is thus conserved as long as possible. When social time is eventually constrained by increases in feeding time, gelada mothers continue to devote the same amount of time to interactions with their closest companions but narrow the number of social partners.

Coordination of Maternal and Infant Activities

At birth, a primate infant typically weighs less than 10% of its mother's body weight (Whitten, 1982) and can soon cling so effectively to her that infant transport and even nursing can be accomplished without interruption of the mother's other activities. Until many months have passed, infants are most likely to be in contact when their mothers are moving (Altmann, 1980; Nicolson, 1982), especially when long distances are covered (Altmann, 1980; Johnson, 1986). Even in the early months, infant contact imposes some costs. For example, baboon mothers were more likely to sit than to stand while feeding during their infants' first 2 months (Altmann, 1980). Maternal feeding efficiency is likely to be impaired as a result. Wild vervet mothers did, indeed, have a lower rate of food intake when their infants were in as opposed to out of contact (Whitten, 1982). As infants grow older, heavier, and the more active, mothers begin to restrict contact and nursing to times when these activities do not interfere with their own feeding. Over time, contact occurs disproportionately when mothers are resting or involved in social grooming, as has been observed in wild baboons (Altmann, 1980; Nicolson, 1982) and rhesus monkeys (Johnson, 1986).

The relationship between mother–infant contact and maternal feeding has also been studied experimentally. Bonnet macaque infants spent slightly less time in contact with their mothers in captive groups maintained under a "high foraging demand" condition, in which mothers had to spend more time procuring food, in comparison to "low foraging demand" groups. The high-demand mothers also displayed more rejection of contact-seeking, and their infants were more likely to be out of contact when the mothers were feeding than under the low-demand condition (Rosenblum and Sunderland, 1982;

Plimpton and Rosenblum, 1983). When the feeding conditions were varied from time to time in an unpredictable fashion, mothers became more involved in dominance interactions and spent less time in social grooming. They were more likely to break contact and increase distance from their infants. "Variable feeding demand" infants showed increased levels of clinging to their mothers, showed decreased play and exploration, and were most likely to show signs of affective disturbance and depression (Rosenblum and Paully, 1984).

In summary, primate mothers are constrained in their daily activities by both increased energetic demands and the incompatibility of certain forms of infant care with maintenance and social activities. Behavioral solutions to these problems can be observed in the conservation of those activities most crucial to survival and future reproduction and in temporal restructuring of maintenance and social activity, including mother–infant interaction. It has become clear that ecological factors exert a major influence on maternal behavior, mother–infant relations, and infant development. In relatively poor habitats, for example, interbirth intervals and the period of infant dependence are often longer than in rich habitats. Infant play and other social interactions may be less frequent under these circumstances (Lee, 1984) and mother–infant conflict more pronounced (Hauser and Fairbanks, 1988).

Although human mothers, particularly those who also work outside the home, are well aware of the challenge of juggling maintenance and social activities within a framework of limited time and energy, there appear to be relatively few empirical data available. In an analysis of time-use data from industrialized countries, Stone (1978) found that women who were both employed and had children got fewer hours of sleep than other women or men. Whereas primate mothers must invest more time in feeding and foraging activities, in humans the costs of motherhood are more likely translated into increases in time devoted to family maintenance activities, such as laundry and food preparation, in addition to direct child care demands. It should be noted that such "costs" are unlikely to have a direct impact on future reproduction, but they certainly have an impact on the perceived quality of life.

Although recently popularized devices such as carrying slings and backpacks (simple versions of which undoubtedly have an evolutionary history dating back to the emergence of human bipedalism) make it possible for women to maintain contact with their infants while doing something else, many activities are incompatible with simultaneous infant care, especially as infants grow and become more active. The primate pattern of shifting the infant from front to back carriage and the eventual necessity of weaning a heavy toddler from riding to walking, even before full locomotor competence has been achieved, will be recognizable to many Western parents as well as to mothers and child nurses in cultures where carrying infants is the norm rather than the exception. Among !Kung hunter-gatherers, the energetic costs of carrying an infant while gathering food are so high that birth intervals of less than the average 4 years not only are considered undesirable by mothers but also appear to be maladaptive in terms of reproductive success (Lee, 1972, 1979; Blurton Jones and Sibley, 1978).

Cross-cultural surveys have shown an association between women's workloads and patterns of maternal care. On the one hand, the kinds of work women perform have been shaped in part by perceived compatibility with child-rearing demands. On the other hand, in cultures with the heaviest workloads for women, child care is delegated earlier and more frequently to other caretakers, including grandparents and siblings (Whiting and Whiting, 1975). Among Western mothers, constraints on the daily time budget have led to concerns about "quality time." It is better to structure time use so that interaction takes place in less frequent sessions when the mother's attention can be focused solely on the child, instead of the mother frequently interrupting her activities to respond? This is a common problem when mother and child spend the entire day together or when the mother's daily schedule is overly full. Studies of children in daycare facilities (Kagan et al., 1978), for example, support the notion that quality is more important than quantity in the development of infant attachment to the mother. Very little is known, however, about effects on *mothers,* especially how time use and the integration of child care, maintenance, and social activities influence mood, attitudes, and behavior. Based on results from repeated time sampling of mothers's reported activities and self-esteem, Wells (1988) concluded that increased demands on mothers' time, attention, and energy may affect mothers' subjective experiences and, in turn, their behavior and interactions with their children. More data of this sort are needed in order to understand the ecology of human families (see Bronfenbrenner, 1986).

FACULTATIVE ADAPTATIONS IN MATERNAL CARE

Dangerous Environments

As was noted above, low-ranking females in species with strong female dominance hierarchies have been reported to be more restrictive or protective of their infants (Altmann, 1980). Mothers appear to be extremely good at adjusting the care they provide in relation to the level of social or environmental risk. Predation, for example, has been postulated as a major selective pressure in the evolution of infant attachment behaviors (Bowlby, 1969) and has also led to adaptations on the part of mothers. In habitats where predators are abundant and mortality from predation high, extra vigilance on the part of both mothers and infants is required, especially during the stage of infant development when young are frequently at some distance from the mother. Observations of wild primates do indicate that mothers are sensitive to the presence of predators. For example, in rhesus monkeys living at a site where humans and dogs posed a considerable danger, mother–infant contact was greater than at safe sites (Johnson and Southwick, 1984). Long after a baboon mother has weaned her infant from riding, she will still carry it when she flees from a predator (Nicolson, 1982). In an ingenious playback experiment, vervet monkey mothers were found to be most responsive, in terms of scanning and protective behaviors, to predator alarm calls during the stage of infant development when mortality risk was highest (Hauser, 1988).

There is, in short, no single formula for ideal mothering; the best strategy will depend on the risks and opportunities provided by the particular social and ecological context. Similar conclusions can be drawn from a recent study of human parental attitudes and behavior. Baldwin et al. (1989) report that U.S. mothers in high-risk inner-city environments who were more restrictive or authoritarian had children with a better cognitive outcome than permissive mothers. In low-risk environments, however, the pattern was reversed: restrictive parenting tended to be slightly detrimental for the development of competence.

Maternal Assessment of Offspring Needs

Primate mothers also appear to be highly sensitive to changes in infants' abilities and needs. Wild chimpanzee mothers are most likely to share food items such as hard-shelled fruits that their infants are incapable of preparing themselves; when offspring skills improve, maternal generosity declines (Silk, 1978). As another example, mothers go to extraordinary lengths to nurture and protect handicapped, sick, or wounded infants (see e.g., Rosenblum and Youngstein, 1974; Berkson, 1974). In both field and captive studies, mothers increase contact with infants during periods of ill health (wild baboons: Altmann, 1980), permit them to nurse more frequently (Nicolson, 1982), assist them in clinging (squirrel monkeys: Rumbaugh, 1965) and carry them more frequently. It thus appears that nonhuman primate mothers are capable of adjusting the kind and amount of care they provide in relation to the costs and benefits of investment at each stage of development. There is every reason to believe that human mothers do the same.

DETERMINING THE NORMAL RANGE OF MATERNAL BEHAVIOR

From studies that provide quantitative data, it is clear that there is an enormous range in normal human maternal behavior, even within the same culture. In a study of mother–infant relations over the first 2 years of life in a Boston middle-class sample, large and consistent differences related to the mother's infant care ideals were found in the number of breast feeding episodes per day and time spent in contact, yet no measurable effects on infant social, cognitive, or language development in the first 2 years of life could be demonstrated (Elias et al., 1986). Similarly, few differences have been found between home– and day-care-reared infants, despite differences in the frequency of mother–infant interaction (Clarke-Stewart and Fein, 1983). These findings underscore the resilient nature of child development. Not only is there no single ideal pattern of mothering, but wide ranges in patterns of mother–infant interaction are tolerated. For normal development, it is essential only that mothering be "good enough," in the words of a noted child psychiatrist

(Winnicott, 1958). But how can we learn, except through potentially biased retrospective accounts, what constitutes "good enough"? Here, ethological descriptions of mother–child interaction may be useful in establishing a better idea of the normal behavioral ranges. Two examples from primate studies will illustrate this point.

The first case involves a wild-born Japanese macaque female living in a stable social group in captivity (Troisi et al., 1982; Troisi and D'Amato, 1984). Although she exhibited competent social behavior in the group, this female abused all of her three successive infants so severely that none survived longer than a few months. Insight into this rare example of maternal abusiveness under normal rearing and housing conditions was gained through detailed, quantitative analysis of mother–infant contact and interaction patterns. Surprisingly, the behavior of the abusive mother deviated most dramatically from the behavior of other mothers in the group with respect to the very high levels of mother–infant contact, possessiveness, and warmth and low level of rejections she displayed. Infant attempts to break contact or play with peers often instigated episodes of abuse; other triggers included infant emotional disturbance or failure to respond to the mother's signals, resulting in a vicious circle in which abuse led to infant distress and distress to more abuse. The authors concluded that this anxiously attached, ambivalent mothering style was related to an underlying emotional disturbance.

The second study points to a possible causal relationship between deviant patterns of mother–infant interaction and subsequent infant illness in wild chimpanzees (van de Rijt-Plooij and Plooij, 1988). From detailed longitudinal observations of five chimpanzee mothers with offspring under 2 years of age, the relative roles of mother and infant in maintaining close proximity could be determined using an index developed by Hinde and coworkers (Hinde and Atkinson, 1970; Hinde, 1976). Normally, an infant's role gradually increases over time until, by about 1 year of age, the chimpanzee offspring is relatively more responsible for maintaining proximity than its mother. Two of the mother–infant pairs in this study diverged from this developmental pattern, as reflected both in Hinde's index and in qualitative aspects, indicating stress or conflict in mother–infant relations. Thereafter, both infants became ill. The authors argued that there are "limits of tolerance" in the normal range of mother–infant interactions, which, if exceeded, can have negative consequences for infant health and development (van de Rijt-Plooij and Plooij, 1988). In the cases they observed, self-correcting processes appeared to operate, since proximity measures returned to the normal range following the illnesses.

With respect to human studies, a plea must be made for greater emphasis on the contributions of both mother and infant to the relationship (Hinde, 1987). Attachment theorists, for example, have classified babies as "securely attached," "anxiously attached," or "angrily attached" to the mother on the basis of the behavior of the infant only. A recent reinterpretation of individual differences in the Strange Situation paradigm shows how ethological and evolutionary perspectives can indeed shed new light on mother-infant attachment in humans (Lamb et al., 1985).

CONCLUSIONS

Primate maternal care is characterized by high levels of physical contact and frequent nursing throughout the day and night. Human milk composition, infant clinging and grasping reflexes, and patterns of mother–infant interaction in extant human hunter-gatherer groups all suggest that human mothers and infants are evolutionarily adapted to higher levels of contact and nursing frequency than are currently observed in Western cultures. Although we still do not know whether deviations from the basic primate pattern have any consequences for infant development, the tolerable range appears to be large. Thus, although primate studies should not be used to set criteria for optimal levels or forms of mother–infant contact, they have been beneficial in "naturalizing" infant care recommendations.

Across the order, primate maternal behavior varies along phylogenetic lines and in relation to species differences in ecology and social organization. The implication for understanding human maternal behavior is that there is no single primate species that can serve as the best model in all respects. With expansion in the number of primate species studied and a better understanding of their socioecology from field observations, we now understand the limitations and the dangers of extrapolating findings from a single species to humans.

Both field and laboratory studies have confirmed the overriding importance of experience for the development of normal maternal behavior in primates. A female who has had a prolonged and positive relationship in infancy with her own mother, contact with peers, and an opportunity to handle or observe infants is very likely to become a good mother herself. Maternal behavior appears to be highly overdetermined; adequate maternal behavior can often develop even if rearing conditions were less than optimal, and, furthermore, poor maternal performance very often improves with practice. Primate studies have shifted away from extreme interventions such as total social or maternal deprivation toward short-term separation studies or observations of spontaneously occurring deficits in maternal behavior, which may have more direct relevance to the human case. Infant neglect and abuse by mothers are rare among wild primates. Not all females are equally competent in handing their newborns, however, and, perhaps as a result, infant mortality is higher for first-time mothers in a number of species. In species in which females are the primary parental caregivers, juvenile and adolescent females show stronger attraction to infants than do males and frequently direct affiliative and nurturant behaviors toward them. Nurturance toward infants may be inhibited in males of these species by the actions of prenatal or postnatal androgens, or both. Sex differences in responsiveness to infants are amplified in mixed-sex social groups, indicating that these differences are relative and that males are capable of displaying appropriate caretaking responses even in those species in which male involvement with infants is normally minimal.

Primate maternal responsiveness in the postpartum period is strongly related to the mother's own rearing history, including the adequacy of maternal care she received, her previous opportunities to interact with infants, and the social

environment in which the birth takes place. Hormones may act in late pregnancy and immediately postpartum to increase the mother's sensitivity to
infant stimuli or otherwise facilitate the initiation of maternal behavior, thus
promoting infant survival during the puerperium, a period of mutual adjust-
ment and rapid learning. Although it is probably unlikely for practical reasons
that such subtle hormonal effects can be measured in human mothers, well-
designed experiments in primates could be helpful in clarifying the role of physiological mechanisms in the initiation of maternal behavior.

Even if hormones can be shown to increase the mother's attraction to infant
stimuli in the postpartum period, there is no evidence to support the notion
that specific forms of mother–infant interaction during this "sensitive period"
are crucial for the development of attachment. As in humans, the specificity of
a primate mother's responsiveness to her own infant develops over a period of
several days, and emotional bonds continue to develop over longer periods of
time.

Within a species, primate mothers show a great deal of individual variability
in maternal behavior, some of which can be related to social, ecological, and
life-historical factors. An evolutionary perspective, which seeks to identify the
relative costs and benefits of maternal investment from the point of view of the
mother as well as the infant, allows a better understanding of the differences
both between individual dyads and within individual dyads over the course of
infant dependency. The interactions between the two partners in this relationship reflect conflicts and compromises between different strategies of mother
and offspring for maximizing survival and future reproduction.

Much attention has been focused on the effects of different mothering styles
on infants, but the consequences for mothers are rarely discussed. Primate
studies have shown that rearing an infant is an energetically costly business that
requires a restructuring of daily activities and time use. We can assume that
mothering has evolved to be an intrinsically pleasurable activity, with emotional rewards that maintain a high level of motivation in spite of the hardships
involved. What we do not know is how infant care and mutually rewarding
interactions between mother and infant can best be integrated into a woman's
daily time budget and her longer term commitments, goals, and aspirations.
Here we can learn little from primate studies and much from women in our
own culture and in other cultures.

REFERENCES

Altmann, J. (1980). *Baboon Mothers and Infants.* Cambridge, MA: Harvard University
        Press.
——— (1983). Costs of reproduction in baboons *(Papio cynocephalus).* Pp. 67–88 in W.
        P. Aspey and S. I. Lustick, eds., *Behavioral Energetics: The Costs of Survival in
        Vertebrates.* Columbus, OH: Ohio State University Press.
——— (1985). Adolescent pregnancies in non-human primates: an ecological and dev-
        elopmental perspective. Pp. 247–262 in J. Lancaster and B. Hamburg, eds.,

School-Age Pregnancy and Parenthood: Biosocial Dimensions. New York: Aldine.

———(1986). Parent-offspring interactions in anthropoid primates: an evolutionary perspective. Pp. 161–178 in M. H. Nitecki and J. A. Kitchell, eds., *Evolution of Animal Behavior: Paleontological and Field Approaches.* Oxford, England: Oxford University Press.

Aries, P. (1962). *Centuries of Childhood: A Social History of Family Life.* New York: Vintage.

Arling, G. L., and H. F. Harlow (1967). Effects of social deprivation on maternal behavior of rhesus monkeys. *Journal of Comparative Physiological Psychology 64:*371–377.

Baldwin, A. L., C. Baldwin, A. Sameroff, and R. Seifer (1989). Protective factors in adolescent development. Paper presented at the Society for Research in Child Development meetings, Kansas City.

Bell, S. M., and M. S. Ainsworth (1972). Infant crying and maternal responsiveness. *Child Development 43:*1171–1190.

Berkson, G. (1974). Social responses of animals to infants with defects. Pp. 233–249 in M. Lewis and L. A. Rosenblum, eds., *The Effect of the Infant on Its Caregiver.* New York: John Wiley & Sons.

Berman, C. M. (1980). Mother-infant relationships among free-ranging rhesus monkeys on Cayo Santiago: a comparison with captive pairs. *Animal Behaviour 28:*860–873.

———(1983). Effects of being orphaned: a detailed case study of an infant rhesus. Pp. 79–81 in R. A. Hinde, ed., *Primate Social Relationships: An Integrated Approach.* Oxford, England: Blackwell.

———(1984). Variation in mother-infant relationships: traditional and nontraditional factors. Pp. 17–36 in M. F. Small, ed., *Female Primates: Studies by Women Primatologists.* New York: Alan R. Liss, Inc.

Berman, P., P. Abplanalp, P. Cooper, P. Mansfield, P., and S. Shields (1975). Sex differences in attraction to infants: when do they occur? *Sex Roles 1:*311–315.

Bingham, H. C. (1927). Parental play of chimpanzees. *Journal of Mammalogy 8:*77–89.

Blurton Jones, N. G. (1972). Comparative aspects of mother-child contact. Pp. 305–328 in N. G. Blurton Jones, ed., *Ethological Studies of Child Behaviour.* Cambridge, England: Cambridge University Press.

Blurton Jones, N. G., and R. M. Sibley (1978). Testing adaptiveness of culturally determined behaviour: do Bushman women maximize their reproductive success by spacing births widely and foraging seldom? Pp. 135–157 in N. Blurton Jones and V. Reynolds, eds., *Human Behaviour and Adaptation.* London: Taylor & Francis.

Bowlby, J. (1969). *Attachment and Loss. Vol. 1: Attachment.* New York: Basic Books.

Brandt, E. M. and G. Mitchell (1973). Pairing preadolescents with infants *(Macaca mulatta). Developmental Psychology 8:*222–228.

Bronfenbrenner, U. (1986). Ecology of the family as a context for human development. *Developmental Psychology 22:*723–742.

Capitanio, J. P., M. Weissberg, and M. Reite (1985). Biology of maternal behavior: recent findings and implications. Pp. 51–92 in M. Reite and T. Field, eds., *The Psychobiology of Attachment and Separation.* Orlando, FL: Academic Press.

Chamove, A., H. Harlow, and G. Mitchell (1967). Sex differences in the infant-directed behavior of preadolescent rhesus monkeys. *Child Development 38:*329–335.

Cheney, D. L., and R. M. Seyfarth (1980). Vocal recognition in free-ranging vervet monkeys. *Animal Behaviour 28*:362–367.

—— (1985). Vervet monkey alarm calls: manipulation through shared information? *Behaviour 94*:150–166.

Clark, C. B. (1977). A preliminary report on weaning among chimpanzees of the Gombe National Park, Tanzania. Pp. 235–260 in S. Chevalier-Skolnikoff and F. E. Poirier, eds., *Primate Bio-Social Development: Biological, Social, and Ecological Determinants.* New York: Garland.

Clarke-Stewart, K. A., and G. G. Fein (1983). Early childhood programs. Pp. 917–999 in M. M. Haith and J. J. Campos, eds., *Infancy and Developmental Psychobiology,* Vol. II of P. H. Mussen, ed., *Handbook of Child Psychology,* 4th edition. New York: John Wiley & Sons.

Coe, C. L., S. G. Wiener, and S. Levine (1983). Psychoendocrine responses of mother and infant monkeys to disturbance and separation. Pp. 189–214 in L. A. Rosenblum and H. Moltz, eds., *Symbiosis in Parent–Offspring Interactions.* New York: Plenum.

Coe, C. L., S. G. Wiener, L. T. Rosenberg, and S. Levine (1985). Endocrine and immune responses to separation and maternal loss in nonhuman primates. Pp. 163–199 in M. Reite and T. Field, eds., *The Psychobiology of Attachment and Separation.* Orlando, FL: Academic Press.

DeMause, L. (1974). *The History of Childhood.* New York: The Psychohistory Press.

DeVore, I. (1963). Mother–infant relations in free-ranging baboons. Pp. 305–335 in H. L. Rheingold, ed., *Maternal Behavior in Mammals.* New York: John Wiley & Sons.

DeVries, M. W. (1987). Alternatives to mother-infant attachment in the neonatal period. Pp. 109–130 in C. Super and S. Harkness, eds., *The Role of Culture in Developmental Disorder.* New York: Academic Press.

Dienske, H., W. van Vreeswijk, and H. Koning (1980). Adequate mothering by partially isolated rhesus monkeys after observation of maternal care. *Journal of Abnormal Psychology 89*:489–492.

Dohlinow, P., and M. G. DeMay (1982). Adoption: the importance of infant choice. *Journal of Human Evolution 11*:391–420.

Dunbar, R.I.M., and P. Dunbar (1988). Maternal time budgets of gelada baboons. *Animal Behaviour 36*:970–980.

Elias, M. F., N. A. Nicolson, and M. Konner (1986). Two subcultures of maternal care in the United States. Pp. 37–50 in D. M. Taub and F. A. King. eds., *Current Perspectives in Primate Social Dynamics.* New York: Van Nostrand Reinhold.

Epple, G. (1978). Reproductive and social behaviour of marmosets and tamarins with special reference to captive breeding. Pp. 50–62 in N. Gengozian and F. Deinhardt, eds., *Marmosets in Experimental Medicine.* Basel: Karger.

Fairbanks, L. A. (1988). Vervet monkey grandmothers: effects on mother-infant relationships. *Behaviour 104*:176–188.

Fairbanks, L. A., and M. T. McGuire (1985). Relationships of vervet monkeys with sons and daughters from one to three years of age. *Animal Behaviour 33*:40–50.

—— (1986). Age, reproductive value and dominance-related behavior in vervet monkey females: cross-generational influences on social relationships and reproduction. *Animal Behaviour 34*:1710–1721.

Field, T., S. Widmayer, S. Stringer, and E. Ignatoff (1980). Teenage, lower-class, Black mothers and their preterm infants: an intervention and developmental follow-up. *Child Development 51*:426–436.

Fleming, A. S., and C. Corter (1988). Factors influencing maternal responsiveness in humans: usefulness of an animal model. *Psychoneuroendocrinology 13*:189–212.

Gelles, R., and J. Lancaster (eds.) (1987). *Child Abuse and Neglect: Biosocial Dimensions.* New York: Aldine.

Gibber, J. R. (1981). Infant-directed behaviors in male and female rhesus monkeys. PhD Dissertation, University of Wisconsin.

—— (1986). Infant-directed behavior of rhesus monkeys during their first pregnancy and parturition. *Folia Primatologica 46*:118–124.

Gibber, J. R., and R. W. Goy (1985). Infant-directed behavior in young rhesus monkeys: sex differences and effects of prenatal androgens. *American Journal of Primatology 8*:225–237.

Glander, K. E. (1980). Reproduction and population growth in free-ranging mantled howling monkeys. *American Journal of Physical Anthropology 53*:25–36.

Goldizen, A. W. (1987). Tamarins and marmosets: communal care of offspring. Pp. 34–43 in B. Smuts, et al., eds., *Primate Societies.* Chicago: University of Chicago Press.

Goodall, J. (van Lawick-Goodall) (1967). Mother-offspring relations in free-ranging chimpanzees. Pp. 365–436 in D. Morris, ed., *Primate Ethology.* Chicago: Aldine.

—— (1986). *The Chimpanzees of Gombe.* Cambridge, MA: Harvard University Press.

Harlow, H. F., M. K. Harlow, R. O. Dodsworth, and G. L. Arling (1966). Maternal behavior of rhesus monkeys deprived of mothering and peer associations in infancy. *Proceedings of the American Philosophical Society 110*:58–66.

Harlow, H. F., M. K. Harlow, and E. W. Hansen (1963). The maternal affectional system of rhesus monkeys. Pp. 254–281 in H. L. Rheingold, ed., *Maternal Behavior in Mammals.* New York: John Wiley & Sons.

Harvey, P. H., R. D. Martin, and T. H. Clutton-Brock (1987). Life histories in comparative perspective. Pp. 181–196 in B. Smuts et al., eds., *Primate Societies.* Chicago: University of Chicago Press.

Hasegawa, T., and M. Hiraiwa (1980). Social interactions of orphans observed in a free-ranging troop of Japanese monkeys. *Folia Primatologica 33*:129–158.

Hauser, M. D. (1988). Variation in maternal responsiveness in free-ranging vervet monkeys: A response to infant mortality risk? *American Naturalist 131*:573–587.

Hauser, M. D., and L. A. Fairbanks (1988). Mother-offspring conflict in vervet monkeys: variation in response to ecological conditions. *Animal Behaviour 36*:802–813.

Hausfater, G., and S. B. Hrdy (eds.) (1984). *Infanticide: Comparative and Evolutionary Perspectives.* New York: Aldine.

Higley, J. D., and S. J. Suomi (1986). Parental behaviour in non-human primates. Pp. 152–207 in W. Sluckin and M. Herbert, eds., *Parental Behaviour.* Oxford, England: Basil Blackwell.

Hinde, R. A. (1976). Interactions, relationships, and social structure. *Man 11*:1–17.

—— (1987). Can nonhuman primates help us understand human behavior? Pp. 413–420 in B. Smuts, D. Cheney, R. Seyfarth, R. Wrangham, and T. Struthsaker, eds., *Primate Societies.* Chicago: University of Chicago Press.

Hinde, R. A., and S. Atkinson (1970). Assessing the role of social partners in maintaining mutual proximity, as exemplified by mother–infant relations in rhesus monkeys. *Animal Behaviour 18*:169–176.

Hinde, R. A., and L. Davies (1972). Removing infant rhesus from mother for 13 days compared with removing mother from infant. *Journal of Childhood Psychological Psychiatry 13*:227–237.

Hinde, R. A. and Y. Spencer-Booth (1967). The effect of social companions on mother–
    infant relations in rhesus monkeys. Pp. 343–364 in D. Morris, ed., *Primate Ethol-
    ogy.* Chicago: Aldine.
Horenstein, D. (1977). The dynamics and treatment of child abuse: can primate research
    provide the answers? *Journal of Clinical Psychology 33:*563–565.
Hrdy, S. B. (1976). The care and exploitation of nonhuman primates by conspecifics
    other than the mother. *Advances in the Study of Behavior 6:*101–158.
——— (1977). *The Langurs of Abu.* Cambridge, MA: Harvard University Press.
——— (1987). Sex-biased parental investment among primates and other mammals: a
    critical evaluation of the Trivers-Willard hypothesis. Pp. 97–147 in R. Gelles and
    J. Lancaster, eds., *Child Abuse and Neglect: Biosocial Dimensions.* New York:
    Aldine.
Hunziker, U. A., and R. G. Barr (1986). Increased carrying reduces infant crying: a ran-
    domized controlled trial. *Pediatrics 77:*641–648.
Jay, P. (1963) Mother-infant relations in langurs. Pp. 282–304 in H. L. Rheingold, ed.,
    Maternal Behavior in Mammals. New York: John Wiley & Sons.
Jensen, G. D. (1965). Mother-infant relationship in the monkey *Macaca nemestrina:*
    development of specificity of maternal response to own infant. *Journal of Com-
    parative Physiological Psychology 59:*305–308.
Johnson, R. L. (1986). Mother-infant contact and maternal maintenance activities
    among free-ranging rhesus monkeys. *Primates 27:*191–203.
Johnson, R. L., and C. H. Southwick (1984). Structural diversity and mother-infant rela-
    tions among rhesus monkeys in India and Nepal. *Folia Primatologica 43:*198–
    215.
Kagan, J., R. B. Kearsley, and P. Zelazo (1978). *Infancy: Its Place in Human Develop-
    ment.* Cambridge, MA: Harvard University Press.
Kaplan, J. N. (1970). The effects of separation and reunion on the behavior of mother
    and infant squirrel monkeys. *Developmental Psychobiology 3:*43–52.
Kaplan, J. N., A. Winship-Ball, and L. Sim (1978). Maternal discrimination of infant
    vocalizations in the squirrel monkey. *Primates 19:*187–193.
Kaufman, I. C., and L. A. Rosenblum (1969). The waning of the mother-infant bond in
    two species of macaque. Pp. 41–59 in B. M. Foss, ed., *Determinants of Infant
    Behavior,* Vol. IV. London: Methuen.
Keverne, E. B. (1988). Central mechanisms underlying the neural and neuroendocrine
    determinants of maternal behaviour. *Psychoneuroendocrinology 13:*127–141.
Klaus, M. H., and J. H. Kennell (1976). *Maternal-Infant Bonding.* St. Louis: C.V.
    Mosby.
——— (1983). Parent-to-infant bonding: setting the record straight. *Journal of Pediatrics
    102:*575–576.
Konner, M. (1977). Aspects of the developmental ethology of a foraging people. Pp.
    285–304 in N. G. Blurton Jones, ed., *Ethological Studies of Child Behaviour.*
    Cambridge, England: Cambridge University Press.
——— (1982). *The Tangled Wing: Biological Constraints on the Human Spirit.* New
    York: Holt, Rinehart and Winston.
Konner, M., and C. Worthman (1980). Nursing frequency, gonadal function, and birth
    spacing among !Kung hunter-gatherers. *Science 207:*788–791.
Lamb, M. E. (1983). Early mother-neonate contact and the mother–child relationship.
    *Journal of Childhood Psychological Psychiatry 24:*487–494.
Lamb, M. E., and W. A. Goldberg (1982). The father-child relationship: a synthesis of
    biological, evolutionary, and social perspectives. Pp. 55–73 in L. W. Hoffman, R.

J. Gandelman, and H. Schiffman, eds., *Parenting: Its Causes and Consequences.* Hillsdale, NJ: Lawrence Erlbaum Associates.

Lamb, M. E., R. A. Thompson, W. Gardner, and E. L. Charnov (1985). *Infant-Mother Attachment: The Origins and Developmental Significance of Individual Differences in Strange Situation Behavior.* Hillsdale, NJ: Lawrence Erlbaum Associates.

Lancaster, J. B. (1971). Play-mothering: the relations between juvenile females and young infants among free-ranging vervet monkeys *(Cercopithecus aethiops). Folia Primatologica 15:*161–182.

Lee, P. C. (1984). Ecological constraints on the social development of vervet monkeys. *Behaviour 91:*245–262.

Lee, R. B. (1972). The !Kung Bushmen of Botswana. Pp. 327–368 in M. G. Bicchieri, ed., *Hunters and Gatherers Today.* New York: Holt, Rinehart and Winston.

———— (1979). *The !Kung San: Men, Women, and Work in a Foraging Society.* Cambridge, England: Cambridge University Press.

Leiderman, P. H. (1978). The critical period hypothesis revisited: mother to infant social bonding in the neonatal period. Pp. 43–77 in F. D. Horowitz, ed., *Early Developmental Hazards: Predictions and Precautions.* Boulder, CO: Westview Press.

LeVine, R. (1977). Child rearing as cultural adaptation. Pp. 15–27 in P. H. Leiderman, S. Tulkin and A. Rosenfeld, eds., *Culture and Infancy.* New York: Academic Press.

Levine, S., D. F. Johnson, and C. A. Gonzalez (1985). Behavioral and hormonal responses to separation in infant rhesus monkeys and mothers. *Behavioral Neuroscience 99:*399–410.

Levine, S., and S. G. Wiener (1988). Psychoendocrine aspects of mother-infant relationships in nonhuman primates. *Psychoneuroendocrinology 13:*143–154.

Lundblad, E. G., and G. D. Hodgen (1980). Induction of maternal-infant bonding in rhesus and cynomolgus monkeys after caesarian delivery. *Laboratory Animal Science 30:*913.

Marais, E. N. (1975). *My Friends the Baboons.* London: Blond & Briggs [originally published 1939].

McKenna, J. J. (1979a). Aspects of infant socialization, attachment, and maternal caregiving patterns among primates: a cross-disciplinary view. *Yearbook of Physical Anthropology 22:*250–286.

———— (1979b). The evolution of allomothering behavior among colobine monkeys: Function and opportunism in evolution. *American Anthropologist 81:*818–840.

———— (1981). Primate infant caregiving behavior. Pp. 389–416 in D. Gubernick and P. Klopfer, eds., *Parental Care in Mammals.* New York: Plenum.

McNeilly, A. S. (1979). Effects of lactation on fertility. *British Medical Bulletin 35:*151–154.

Minturn, L., and W. W. Lambert (1964). *Mothers of Six Cultures: Antecedents of Child Rearing.* New York: John Wiley & Sons.

Missakian, E. A. (1974). Mother-offspring grooming relations in rhesus monkeys *(Macaca mulatta)* on Cayo Santiago. *Primates 13:*169–180.

Nadler, R. D. (1980). Child abuse: evidence from nonhuman primates. *Developmental Psychobiology 13:*507–512.

———— (1983). Experiential influences on infant abuse of gorillas and some other nonhuman primates. Pp. 139–150 in M. Reite and N. Caine, eds., *Child Abuse: The Nonhuman Primate Data.* New York: Alan R. Liss, Inc.

Nash, L. T., and R. L. Wheeler (1982). Mother-infant relationships in non-human primates. Pp. 27–61 in H. Fitzgerald, J. Mullins, and P. Gage, eds., *Child Nurturance, Vol. 3, Studies of Development in Nonhuman Primates.* New York: Plenum.

Nicolson, N. (1982). Weaning and the development of independence in wild baboons. PhD Dissertation, Harvard University.

—— (1987). Infants, mothers, and other females. Pp. 330–342 in B. Smuts, D. Cheney, R. Seyfarth, R. Wrangham, and T. Struhsaker, eds., *Primate Societies.* Chicago: University of Chicago Press.

Pereira, M. E., and J. Altmann (1985). Development of social behavior in free-living nonhuman primates. Pp. 217–309 in E. S. Watts ed., *Nonhuman Primate Models for Human Growth and Development.* New York: Alan R. Liss, Inc.

Plimpton, E., and L. Rosenblum (1983). The ecological context of infant maltreatment in primates. Pp. 103–117 in M. Reite and N. Caine, eds., *Child Abuse: The Nonhuman Primate Data.* New York: Alan R. Liss, Inc.

Pond, C. (1977). The significance of lactation in the evolution of mammals. *Evolution 31:*177–199.

Pusey, A. E. (1983). Mother-offspring relationships in chimpanzees after weaning. *Animal Behaviour 32:*363–377.

Pryce, C. R., D. H. Abbott, J. H. Hodges, and R. D. Martin (1988). Maternal behavior is related to prepartum urinary estradiol levels in red-bellied tamarin monkeys. *Physiology and Behavior 44:*717–726.

Quiatt, D. D. (1979). Aunts and mothers: adaptive implications of allomaternal behavior among nonhuman primates. *American Anthropologist 81:*311–319.

Reite, M., and N. Caine (eds.) (1983). *Child Abuse: The Nonhuman Primate Data.* New York: Alan R. Liss, Inc.

Rogers, C. M., and R. K. Davenport (1970). Chimpanzee maternal behavior. Pp. 361–368 in G. H. Bourne, ed., *The Chimpanzee,* Vol. 3. Basel: Karger.

Rosenblatt, J. S., and H. I. Siegel (1981). Factors governing the onset and maintenance of maternal behavior among nonprimate mammals: the role of hormonal and nonhormonal factors. Pp. 13–76 in D. J. Gubernick and P. H. Klopfer, eds., *Parental Care in Mammals.* New York: Plenum Press.

Rosenblum, L. A. (1972). Sex and age differences in response to infant squirrel monkeys. *Brain, Behavior, and Evolution 5:*30–40.

Rosenblum, L. A., and G. S. Paully (1984). The effects of varying environmental demands on maternal and infant behavior. *Child Development 55:*305–314.

Rosenblum, L. A., and G. Sunderland (1982). Feeding ecology and mother-infant relations. Pp. 75–110 in L. W. Hoffman, R. J. Gandelman, and H. Schiffman, eds., *Parenting: Its Causes and Consequences.* Hillsdale, NJ: Lawrence Erlbaum Associates.

Rosenblum, L. A., and K. P. Youngstein (1974). Developmental changes in compensatory dyadic response in mother and infant monkeys. Pp. 141–161 in M. Lewis and L. A. Rosenblum, eds., *The Effect of the Infant on Its Caregiver.* New York: John Wiley, & Sons.

Rowell, T. E., R. A. Hinde, and Y. Spencer-Booth (1964). "Aunt"–infant interactions in captive rhesus monkeys. *Animal Behaviour 12:*219–226.

Rumbaugh, D. (1965). Maternal care in relation to infant behavior in the squirrel monkey. *Psychological Reports 16:*171–176.

Ruppenthal, G., G. L. Arling, H. F. Harlow, G. P. Sackett, and S. J. Suomi (1976). A ten-year perspective of motherless-mother monkey behavior. *Journal of Abnormal Psychology 85:*341–349.

Sackett, G. P., and G. Ruppenthal (1974). Some factors influencing the attraction of adult female macaque monkeys to neonates. Pp. 163–185 in M. Lewis and L. A. Rosenblum, eds., *The Effect of the Infant on Its Caregiver.* New York: John Wiley & Sons.

Scheper-Hughes, N. (ed.) (1987). *Child Survival: Anthropological Perspectives on the Treatment and Maltreatment of Children.* Dordrecht, Holland: D. Reidel.

Silk, J. B. (1978). Patterns of food sharing among mother and infant chimpanzees at Gombe National Park, Tanzania. *Folia Primatologica 29:*129–141.

———— (1980). Kidnapping and female competition among captive bonnet macaques. *Primates 21:*100–110.

Simpson, A. E., and M.J.A. Simpson (1985). Short-term consequences of different breeding histories for captive rhesus macaque mothers and young. *Behavioral Ecology and Sociobiology 18:*83–89.

Sluckin, W. (1986). Human mother-to-infant bonds. Pp. 208–227 in W. Sluckin and M. Herbert, eds., *Parental Behaviour.* Oxford, England: Basil Blackwell.

Snowdon, C. T., A. Savage, and P. B. McConnell (1985). A breeding colony of cotton-top tamarins *(Saguinus oedipus). Laboratory Animal Science 35:*477–480.

Spencer-Booth, Y., R. A. Hinde, and M. Bruce (1965). Social companions and the mother-infant relationship in rhesus monkeys. *Nature 208:*301–308.

Stone, P. J. (1978). Women's time patterns in eleven countries. Pp. 113–150 in W. Michelson, ed., *Public Policy in Temporal Perspective.* The Hague: Mouton.

Suomi, S. J. (1978). Maternal behavior by socially incompetent monkeys: neglect and abuse of offspring. *Journal of Pediatric Psychology 3:*28–34.

Suomi, S. J., and C. Ripp (1983). A history of motherless mother monkey mothering at the University of Wisconsin primate laboratory. Pp. 49–78 in M. Reite and N. Caine, eds., *Child Abuse: The Nonhuman Primate Data.* New York: Alan R. Liss, Inc.

Taub, D. M. (1980). Age at first pregnancy and reproductive outcome among colony-born squirrel monkeys *(Saimiri sciureus* Brazilian). *Folia Primatologica 33:*262–272.

Thierry, B., and J. R. Anderson (1986). Adoption in anthropoid primates. *International Journal of Primatology 7:*191–216.

Tinklepaugh, O. L., and C. G. Hartman (1932). Behavior and maternal care of the new-born monkey *(Macaca mulatta–Macaca rhesus). Journal of Genetic Psychology 40:*257–286.

Trivers, R. L. (1972). Parental investment and sexual selection. Pp. 136–179 in B. Campbell, ed., *Sexual Selection and the Descent of Man.* Chicago: Aldine.

———— (1974). Parent-off spring conflict. *American Zoologist 14:*249–264.

Trivers, R. L., and D. E. Willard (1973). Natural selection of parental ability to vary the sex ratio of offspring. *Science 179:*90–91.

Troisi, A., and F. R. D'Amato (1984). Ambivalence in monkey mothering: infant abuse combined with maternal possessiveness. *Journal of Nervous and Mental Disease 172:*105–108.

Troisi, A., F. R. D'Amato, R. Fuccillo, and S. Scucchi (1982). Infant abuse by a wild-born group-living Japanese macaque mother. *Journal of Abnormal Psychology 91:*451–456.

Van De Rijt-Plooij, H., and F. Plooij (1988). Mother–infant relations, conflict, stress and illness among free-ranging chimpanzees. *Developmental Medicine and Child Neurology 30:*306–315.

Walters, J. R. (1981). Inferring kinship from behavior: maternity determinations in yellow baboons. *Animal Behaviour 29:*126–136.

——— (1987). Transition to adulthood. Pp. 358–369 in B. Smuts, et al., eds., *Primate Societies.* Chicago: University of Chicago Press.

Wells, A. J. (1988). Variations in mothers' self-esteem in daily life. *Journal of Personality and Social Psychology 55:* 661–668.

Whiting, B., and J. Whiting (1975). *Children of Six Cultures: A Psycho-Cultural Analysis.* Cambridge, MA: Harvard University Press.

Whitten, P. L. (1982). Female reproductive strategies among vervet monkeys. PhD Dissertation, Harvard University.

Wilson, M. E., T. P. Gordon, and I. S. Bernstein (1978). Timing of births and reproductive success in rhesus monkey social groups. *Journal of Medical Primatology 7:*202–212.

Winnicott, D. W. (1958). *Collected Papers: Through Paediatrics to Psycho-Analysis.* London: Tavistock [republished by Hogarth Press, 1982].

Yerkes, R. M., and M. I. Tomilin (1935). Mother–infant relations in chimpanzees. *Comparative Psychology 20:*321–349.

Primate Paternalistic Investment: A Cross-Species View

DAVID TAUB
AND PATRICK MEHLMAN

The keystone of mammalian evolution is intrauterine gestation and the attendant physiological adaptations of females that allow them to provide all nourishment to neonates after parturition. These adaptations have had revolutionary consequences on the evolution of diverse social systems and mating patterns (Trivers, 1972; Brown, 1975; Maynard-Smith, 1977; Wittenberger and Tilson, 1980; Williams, 1966; Gubernick and Klopfer, 1981) as well as intensifying the mother–offspring bond (Trivers, 1972), which is the central configuration of all mammalian social systems. So pervasive and crucial has been the bond between mammalian mother and child that in both the scientific and the lay literature we have equated the term "parental behavior" with mothering (hence the term "biparental care" when males play a role in parenting). Consequently, most studies of both human and nonhuman infant development among mammals focus on the mother's relationship with its infant(s) (see Chapters 2 and 4, this volume).

However maternocentric investigations of mammalian social behavior have been historically, a new interest in the male's role in infant development and socialization was kindled with the publication of Trivers' (1972) provocative analysis of parental investment and sexual selection. Perhaps nowhere has this interest been expressed more than in sociological and psychological studies of the role of human fathers in child rearing, especially in Western industrialized societies (Lamb, 1981). This new interset in mammalian "paternal" behavior has also occurred among anthropologists and psychologists, and they have begun to focus increased attention on the behavior of male nonhuman primates and their role in the care and rearing of offspring (Taub, 1985; Snowdon and Suomi, 1982). Stimulated by a tremendous growth in the study of primate social biology in general and by early studies showing a pronounced role for males in infant development (e.g., Itani, 1959), the past two and one-half decades of primate studies have witnessed the development of a large body of scientific data documenting a plethora of structurally diverse and functionally disparate "paternal" interactions. We are discovering from these data that paternal behaviors among primates are characterized by fundamental differences in structure, in contexts of occurrence, in the rates of exhibition, in the distribution among males, in various biosocial characteristics of the males (such as rank, age, tenure of group residence, kinship), and in the form of the mating system and/or social structure characterizing the species.

As data on male–infant interactions among primates began to accumulate in the late 1960s, the first attempts to summarize and synthesize these data also appeared (Mitchell, 1969) and since then a number of reviews have been published (Hrdy, 1976; Redican and Taub, 1981; Taub and Redican, 1984; Whitten, 1987; Mehlman, 1988; Taub, 1990). By the late 1970s, as more empirical data became available, it seemed as though primate paternalistic behavior could be classed into two essentially separate functional categories correlated with the mating system and kinship: caretaking or true paternal investment in monogamous species and exploitation in nonmonogamous species. It is clear today, however, that the diversity of primate paternalistic behaviors does not lend itself to so straightforward a functional interpretation (see Taub, 1990). We have discovered that nonhuman primate males engage infants in a multitude of ways, with many structural and functional explanations; too many for a single chapter to deal effectively with all of them. Therefore, in this chapter we focus on three major themes. First, we limit our discussion to what we call "paternalistic investment" among primates (defined below). Second, we characterize the distribution, structure, and function of paternalistic investment among the nonhuman primates in order to define from a comparative point of view the nature of human paternalistic behavior. Third, we attempt to assess the degree to which the study of nonhuman primates can provide an understanding of human paternalistic investment.

DEFINITION OF PATERNAL INVESTMENT

Trivers was the first to operationalize parental investment as "any investment by the parent in an individual offspring that increases the offspring's chance of surviving (and hence reproductive success) at the cost of the parent's ability to invest in other offspring" (1972:139). Maynard-Smith (1977), however, pointed out that investment may occur when the parent incurs a cost, but that cost may not necessarily occur at the expense of investing in other offspring. S. A. Altmann et al. (1977) made the same observation and suggested the terms "depreciable" and "nondepreciable" investment, the former following Trivers' original formulation. In their scheme, nondepreciable investments would be those investments that a parent could make in offspring that would *not* decrease the potential for investing in other offspring (e.g., behaviors that are directed simultaneously to several offspring).

Kleiman and Malcolm (1981) proposed another definition of parental investment (for males) that includes both nondepreciable and depreciable patterns. Concentrating on benefits to the offspring, they defined male parental investment as "any increase in a prereproductive mammal's fitness attributable to the presence or action of a male" (1981:348). We prefer a broader definition of parental investment, much like that of Kleiman and Malcolm, but we do not see any theoretical reason to limit the definition to prereproductive offspring. Thus we define parental investment as any investment (with an associated cost) by a parent that increases the offspring's fitness during all prereproductive and reproductive phases of the offspring's life cycle.

Since this chapter is concerned with male parental investment, we would prefer to use the term *paternal* investment or *paternalism,* i.e., those behaviors characteristic of fathers. However, since *paternalism* is by strict definition limited to interactions between a male and its offspring, we follow Redican (1976), and generally use the term *paternalistic* investment, since we often do not know true paternity relationships, and behaviors directed at close kin (e.g., siblings or siblings' offspring) are often structurally identical to fathers' behaviors toward their offspring. Paternalistic investment is further composed of two broad classes or axes: "direct" and "indirect" investment patterns (Kleiman and Malcolm, 1981). Direct paternalistic investment (hereafter referred to as DPI) includes all behavioral interactions occurring directly between a male and one or more offspring or kin-related individuals. Indirect paternalistic investment (hereafter referred to as IPI) in contrast, is oriented towards one or more offspring or kin-related individuals but occurs without direct behavioral interactions.

DIRECT PATERNALISTIC INVESTMENT IN NONHUMAN PRIMATES

Theories of kinship selection (Trivers, 1972; Kurland and Gaulin, 1984) predict a close, positive correlation between a monogamous mating system and high levels of male parental investment. As we shall see, this appears to be a fairly basic "rule" (not without its exceptions) of mammalian social structure. Therefore, it is not too surprising to find some degree of DPI among the monogamously (and polyandrously) mating New World primates (Vogt, 1984). Monogamous New World primates are unsurpassed among nonhuman primates in the extent of DPI, and this phenomenon is particularly well developed among four genera of marmosets and tamarins *(Callithrix, Saguinus, Cebuella, Leontopithecus)* of the family Callitricidae and among the cebid monkeys *Callicebus* and *Aotus.* In these species, males not only share all caring duties except nursing but in addition they can make significant contributions to infant care (Higley and Suomi, 1986; Vogt, 1984; Kleiman, 1985; see Taub, 1990 for details). In most species, adult males share food with infants (Brown and Mack, 1978; Goldizen, 1987), carry and groom them, promote the infants' emerging independence (Dixson, 1983), and protect them from other group members and potential predators (Wolters, 1978). Results of several studies of three species of wild and captive callitrichids have indicated that males do substantially more infant carrying than do the mothers (*Saguinus fuscicollis:* Epple, 1975; Vogt et al., 1978; Goldizen and Terborgh, 1986; Ingram, 1978; *Leontopithecus rosalia:* Hoage, 1978; *Callithrix jacchus:* Box, 1975), whereas other studies have found the opposite (Box, 1975, 1977; Izawa, 1978).

Sources of variability in DPI among the males of the monogamous callitrichids seem to revolve around its timing of onset, who participates, and to what degree the "father" actually participates. In some species, males begin caretaking at parturition (Epple and Katz, 1983, 1984; Stevenson, 1970, 1978); in others, males may not begin DPI for several days or even weeks (Pook, 1978; Hoage, 1978; Vogt, 1984). Both the number of potential care givers (see below) and the extent to which the mother promotes or allows other care givers to approach and obtain the infant (Stevenson, 1978; Fragaszy et al., 1982) strongly effect the timing and extent of DPI. Male DPI is also strongly correlated with group size and the phenomenon of "helpers at the nest." In all of these species, juveniles and subadults remain in the group (i.e., family) until they reach sexual maturity, so nonbreeding relatives are available to assume caretaking duties. The norm is that all group members, especially older, sexually immature siblings, share infant caretaking duties with the "father." For example, in eight captive groups of *S. fuscicollis* (Cebul and Epple, 1984), the dominant male accounted for 95% of all carrying, with two male helpers in the group, but, in another group, the male accounted for only 38% of carrying. So strong was the tendency of male *S. fuscicollis* in this study to care for infants that nonrelated "step fathers" (males introduced into the groups prior to parturition) were among the most active and vigorous caretakers. McGrew (1988) found that, the more helpers available, the less "babysitting" was done by the father, and, in some other studies, siblings contributed more than fathers (Cebul and Epple, 1984; Ingram, 1977; Vogt et al., 1978; Cleveland and Snowdon, 1982). "Helping" by these relatives may be at least partly explained by a combination of nepotistic gains through enhanced "inclusive fitness" (Hamilton, 1964) and the benefits of becoming experienced at infant care. Ingram (1978) found that the probability of a male showing care for infants was directly related to whether the male had had prior experience with infants. Thus it seems likely that care given by these "helpers" not only provided additional support for the infant but also was critical for them to exhibit adequate DPI as adults.

Many species of birds have evolved communal rearing of and investment in offspring as a reproductive strategy; perhaps some species of primates have as well. Recent work by Goldizen and others has shown that some callitrichids have a polyandrous mating system, and others have even more complex social systems, with multiple breeding females (Goldizen and Terborgh, 1986; Dawson, 1978; Soini, 1982; Rylands, 1981; Garber et al, 1984; Goldizen, 1987). Studies of marked groups of saddle-backed tamarins *(S. fuscicollis)* in Peru over 5 years show marked changes in group composition, with frequent intergroup transfers of individuals. Monogamous groups were observed only 17% of the time, whereas units containing one female and more than one adult male were observed 62% of the time. Groups containing two females and more than one male (12%), two females and one male (6%), and only males (3%) were also seen. Furthermore, groups changed composition over time, being monogamous at one time only to become polyandrous sometime later. Remarkable as this flexible social system now appears to be, even more remarkable is the fact that in every case *all* adult group members participated in all aspects of infant care. Because twins are common in these species and the weights of the neonates represent upwards of 25% of maternal weight (increasing to 50% by weaning; Leutenegger, 1980), more than a single care giver (especially maternal) appears to be necessary for successful reproduction. It is not coincidental that in 5 years of study no single pair without "helpers" was seen to produce young in wild saddlebacked tamarins. Groups without nonreproductive helpers typically accept another male as a second breeder and helper; by sharing the probability of siring infants and then helping to care for them cooperatively, males enhance their mutual reproductive success. Marmosets and tamarins have solved the reproductive dilemma of multiple births and large infants with a flexible social and mating system that recruits additional caretakers.

The monogamously mating cebid monkeys, *Aotus* and *Callicebus,* also show extensive male care (Fragaszy et al., 1982; Wright, 1984; Robinson et al., 1987). *Aotus* males are the primary carrier, even from the first day of life (Dixson and Fleming, 1981; Taub, unpublished observations), and may account for upwards of 80% of carrying in the first month of life. Wright's study of wild *Aotus* has confirmed that the "father" accounts for a majority of carrying duties for the first 6 months of life or until weaning. In both genera males not only carry the infant but will share food and play with and groom it. Although not as common in these cebid species as in marmosets and tamarins, sibling "helpers" will carry infants, but only for the first few months (Wright, 1984; Dixson and Fleming, 1981).

All the "lesser apes," gibbons and siamangs (*Hylobates* sp.), are monogamously mating, arboreal, territorial, and monomorphic (as are the callitrichids) (Chivers, 1972, 1974; Chivers and Raemaekers, 1980; Tenaza, 1975; Tilson, 1981), but DPI in infants occurs only in the siamang (Chivers, 1974). It is most curious that males in none of the other seven hylobatid species show DPI and this is a major exception to what appears to be a general rule of a correlation between a monogamous mating system and male DPI. Unlike the New World species, however, siamang males carry, groom, and sleep with their offspring only after they are older. Thus the mother is the primary care giver in the first year and one-half of the infant's life (although on occasion males carry infants as young as 6–8 months of age; Chivers, 1974). Later there is a shift to the male in the second year; during this time the father becomes the primary care giver until the offspring are weaned and independent by age 3 years (Chivers and Raemaekers, 1980). Males may spend up to three-fourths of their social activity time carrying and grooming infants during their period of DPI, but typically they spend considerably less (Chivers, 1974). In sharp contrast to the "helpers at the nest" phenomenon in the New World monogamous species, maturing hylobatids are actively driven from their natal group, sometimes very aggressively (Tilson, 1981; Fox, 1972), and it is the father that takes the most active

role. The difference in the degree of DPI between monogamously mating callitrichids and hylobatids appears to be related to the differential reproductive
burden on the mother (Leutenegger, 1980): Hylobatids bear singleton young,
callitrichids bear twins routinely; the callitrichid offspring-to-mother weight
ratio is five times greater than for hylobatid mothers; interbirth interval is considerably longer for hylobatids.

There are many types of affiliative male–infant interactions among the nonmonogamously mating species, especially baboons (*Papio* sp.) and macaques
(*Macaca* sp.). DPI in these so-called "affiliative" species (Whitten, 1987) tends
to be (1) much less frequent (see below), (2) irregular and opportunistic, and
(3) shown by only some males (but it is probably a regular component of the
social repertory in that all males at some point in their life probably engage in
some degree of interaction with infants). The most critical evidence that is typically missing or is conjectural among these nonmonogamously mating species
is, of course, the biological relationship between males and their infant partners. In many cases, however, there is strong circumstantial evidence that the
participants are genetically related (e.g., Taub, 1985; Anderson, N.D.). An
important characteristic of DPI in these species is its reduced rate of occurrence
compared with DPI among the monogamously mating species. For example,
rates for all types of baboon male–infant interactions range from a low of one
interaction (triadic) for every 53 observation hours (Packer, 1980) to a high of
one episode for every 5 observation hours (Busse, 1984); adjusted for the number of males in the group, the rates range from one for every 1,250 to one for
every 100 observation hours. For all baboon studies that provide quantified
data (Taub, 1985), the rate of DPI is about one episode for every 19 observation
hours; likewise, when adjusted for the number of possible male actors, the average is about one episode for every 344 observation hours. Estrada and Sandoval
(1977) found that stumptail macaque males *(Macaca arctoides)* interacted with
infants at rates of 1.4 episodes per hour total and 2.7 episodes per hour when
adjusted for the number of males present. Taub (1980, 1984) found the rate of
dyadic interactions of Barbary macaques *(M. sylvanus)* to be one every 23 minutes (one per hour for triadic interactions); in Deag's (1980) study the figure
was one every hour (one every 7.5 hours and for triadic interactions). Typically,
then, rates of DPI among the males of these polygynously mating species tend
to be absolutely and relatively low (with the possible exception of Barbary and
stumptail macaques) compared with monogamous species. However, intensive
and long-term affiliations among males and infants likely to be related have
been reported for many primates, and we describe below DPI in some of these
species.

The Barbary macaque is a remarkable species in that all males engage infants
regularly and in every conceivable way, from birth through the first year of life
(Taub, 1984; Mehlman, 1986). Taub (1978) observed over 2,231 episodes of
male–infant interactions in a group containing seven adult, four subadult and
five juvenile males. All males showed highly specific preferences for particular
infants for interaction (e.g., among the adult males, five had one "primary"
infant, one had two, and one had three, whereas all four subadult males pre-

ferred only one "primary" infant). There were significant differences among males in the degree to which each was involved with infants in general and with specific infants in particular. Subadult males were the most intensively involved with infants; the four in Taub's study group accounted for 44% of all male–infant interaction vs. 46% for the seven adult males. All males showed essentially the same types of paternal behaviors, but they appeared to operate functionally at multiple levels. Taub (1980, 1984) argued that subadult males may be investing in younger siblings; some adult males are investing in both actual and probable offspring, some in offspring of their relatives, and some in nonrelatives. Males also may be interacting with infants to induce females to copulate with them during the mating season.

Although early studies of stumptail macaques suggested that males were uninterested in infants (Bertrand, 1969), Gouzoules found that adult and subadult males "showed interest in, and interacted with infants almost as much as females did" (1975:413). However, rather than kinship, male and maternal social ranks appeared to be most important in influencing the patterns of male–infant interactions. Other studies (Hendy-Neely and Rhine, 1977; Rhine and Hendy-Neely, 1978; Smith and Peffer-Smith, 1984) also found a correlation with both male age and social status, but stumptail males in some other studies probably exhibited DPI toward their relatives (Estrada et al., 1977; Estrada and Sandoval, 1977; Estrada, 1984; Smith and Peffer-Smith, 1984). In one study, males showed "substantial amounts of care behavior to their infant siblings" (Estrada and Sandoval, 1977:803); for example, three juvenile males with sibling infants showed a significant preference for interactions with them over other infants (67% and 52%). In the Smith and Peffer-Smith study, males interacted at a higher rate with immatures to whom they were known to be related than they did with nonrelated individuals. Studies of Japanese macaques (*Macaca fuscata:* Itani, 1959; Hasegawa and Hiraiwa, 1980; Hiraiwa, 1981; Alexander, 1970; Gouzoules, 1984) have shown that males may interact frequently with infants, but these same studies failed to find a correlation with kinship.

Several types of male-infant interactions have been reported for savannah baboons. A review of these studies indicates that males may both affiliatively care for infants and exploit them by using the youngsters to ameliorate intermale conflicts ("agonistic buffering"). The same infant may be used in both situations, and occasionally affiliation and agonistic buffering occur simultaneously (Collins, 1986; Ransom and Ransom, 1971; Altmann, 1980; Stein, 1984; Smuts, 1982, 1985; Strum, 1984; Packer, 1980; Popp, 1978).

The baboon studies fall into three distinct groups as they relate to the degree of relatedness between the "caretaking" male and infant (and the "opponent" in triadic interactions). At one end, Popp (1978) has asserted that the actor is unlikely to have sired the infant he carries, but the male to whom he directs his interactions ("the opponent") has a high probability of being the infant's father. Busse (Busse and Hamilton, 1981; Busse, 1984) has asserted that males carry their own infants to protect them against the possibility of injury by a nonrelated opponent. For example, in 111 of 112 triadic interactions, the actor

had a high probability of being the sire of the offspring with whom he interacted, but the recipient could not have been the sire because he was not a resident in the social group at the time of the infant's conception. Other baboon
studies fall in between concerning the degree of relatedness between male and
infant (but most of them assert some form of exploitation, use, or "agonistic
buffering" to be operating [Ransom and Ransom, 1971; Altmann, 1980;
Packer, 1980; Smuts, 1982; Strum, 1984; Nicholson, 1982; Collins, 1981;
Klein, 1983; Gilmore, 1977, cited in Strum, 1984; Stein, 1981; Stein and Stacey, 1981]). Smuts (1982) and Strum (1984) suggest that because actors sometimes could and sometimes could not have been sires, "special relationships"
or "affiliations" between the males and the infants were more important than
probable kinship.

Taub (1985) has shown that there is strong circumstantial evidence suggesting a close kinship relationship between "affiliated" males and infants in most
baboon studies. Packer's (1980) data reveal that "potential" fathers showed
more male–infant behavior than did males who could not have been fathers,
and subadult males preferentially carried their siblings. Stein (1981, 1984)
noted that most male–infant interactions were accounted for by one or two
adult males interacting with specific infants with whom each had a "preferred
relationship," i.e., males who had a high likelihood of siring the infants. In
Klein's (1983) study, adult males acted more affiliatively toward mates and
infants they were likely to have fathered than toward other infants and non-
mates. Data on four immigrant males and their activities with infants conceived after their entrance into the troop, vs. infants born earlier, show a strong
relationship between probable paternity and social attraction to and interactions with probable offspring. Collins (1981, 1986) too found that, because
interactions between males and infants involved only resident and subadult
males, the associations were restricted to possible fathers. Similarly, Anderson
(N.D.) found among chacma baboons that males probably sired the infants
with whom they interacted.

In summary, at present there is no definitive, independent data available on
paternity certainty among these various primate populations. Clearly, baboon
and macaque males can form strong, affiliative, mutually preferred attachments to infants (Taub, 1984, 1990). Although in some cases males have a low
probability of having sired the infants they carry, a large number of instances
of male–infant interactions seem to cluster around specific males and particular infants with whom they are probably related (the Popp study is exceptional
here). Given the strong relationship between male–infant associations and
probable paternity in many studies, it is likely that many adult male baboons
care for (and exploit) their own offspring.

Most nonhuman primate species show no DPI at all; it does not appear to
be part of their normal, species-typical repertoire. However, on a sporadic and
idiosyncratic basis, positive, affiliative male–infant interactions do occur in
some of these species. It is quite unknown, however, to what degree any of the
males in these species has any probability of siring their infant partners. Adult
and immature sibling sooty mangabey *(Cercocebus atys)* males sometimes

carry infants (Bernstein, 1976; Chalmers, 1968), although in a captive group one adult male in particular, who was not the dominant male, was observed to carry nine different infants at one time or another. Among gelada baboons *(Theropithecus gelada),* young "follower" males attempting to establish their own "one-male units" (OMU) may sometime groom and carrying an infant in order to establish a relationship with the infant's mother and thus coopt her into his breeding unit (Mori, 1979). If a deposed gelada male remains associated with his former OMU, he may become very solicitous of offspring (since this male was the sole breeder prior to being deposed, these infants are presumably his own) especially if they are threatened by the new leader male (Dunbar and Dunbar, 1975; Dunbar, 1984). Silverback male gorillas *(Gorilla gorilla)* are very solicitous of infants, and they may groom, cuddle, and nest with 3 and 4 year olds, allowing them to play with, climb upon, and tumble over them with impunity (Fossey, 1979, 1983). Since it is generally these dominant males that do the mating, they are quite possibly caring for their own offspring. Adult male black howling monkeys *(Alouatta caraya),* which are monogamous, may develop strong bonds with infants, although they rarely actually carry them (Bolin, 1981). Captive crab-eating macaques *(Macaca fascicularis)* may sometimes carry infants (Auerbach and Taub, 1979), and male squirrel monkeys *(Saimiri sciureus;* a species notorious for a lack of male interest in infants) may huddle, play, and sleep with infants if their mothers are artificially removed from the group (Vaitl, 1977).

"Adoption" of orphaned infants, sometimes by siblings and sometimes by unrelated males, has also been reported for several species, including rhesus monkeys (*Macaca mulatta;* Berman, 1981; Vessey and Meikle, 1984), Japanese macaques (Hiraiwa, 1981), hamadryas baboons (*Papio hamadryas;* Kummer, 1967), and gibbons (Carpenter, 1940). In these cases, very intense, albeit typically brief and transient, associations develop, and these males assume such otherwise maternal duties as holding, cuddling, comforting, grooming, carrying, and protecting these very vulnerable infants (Fig. 3–1).

As it bears on the evolutionary potential for the expression of DPI, the literature on rhesus monkeys is very provocative. Most studies of rhesus monkeys, whether of wild populations (Lindburg, 1971), provisioned free-ranging colonies (Vessey and Meikle, 1984; Berman, 1982), or captive groups (Rowell, 1974; Spencer-Booth, 1968), indicate that adult males rarely show interest in or interact with infants. However, under the catalytic stimulus of appropriate social conditions, rhesus males may show remarkably intense relations with infants. Postulating that the lack of male interest in infants stemmed more from maternal restrictiveness and few opportunities than it did from a lack of innate motivation, Redican (Redican and Mitchell, 1974; Redican, 1976) removed rhesus mothers and allowed male–infant pairs to live together in small cages. Not only were males tolerant of these infants (as they would be among their wild counterparts) but they became quite accomplished surrogate "mothers." These males exhibited levels of carrying and grooming equal to those of mothers, and they actively protected the infants from sources of danger. Play within these pairs was far more intense and reciprocal than that within

Fig. 3–1 Adult male rhesus monkey cuddles an older infant in a ventroventral position. (Photo: James Loy.)

mother–infant control pairs. Suomi (1977, 1979), in a series of similar investigations, allowed infants to be raised in "nuclear" families. Infants raised under such conditions preferred their mothers to other adult females but their father to other adult males (Suomi, 1979). Suomi has interpreted this as consistent with Redican's view that maternal restrictiveness holds male–infant associations in check.

Lack of opportunity and/or maternal restrictiveness may not explain all cases of male disinterest in infants, however, and there seems to be fundamental species differences in this potential or at least in the variables that promote or discourage its exhibition. For example, it seems to be a species-typical phenomenon that langur (*Presbytis* sp.) females are extremely nonrestrictive concerning access to neonates, but male langurs are nevertheless characteristically uninvolved with infants. We have also noted elsewhere that males may show an extreme form of caretaking by adopting orphaned infants; apparently the stimulus of a vulnerable, abandoned infant in the absence of its mother is sufficient to elicit the expression of this "paternal" potential, at least on a short-term basis, in some species where males typically do not exhibit such paternalistic behavior.

INDIRECT PATERNALISTIC INVESTMENT IN NONHUMAN PRIMATES

IPI has not been systematically examined in nonhuman primates, principally because adequate methodologies have not been developed for (1) quantitatively determining the long-term differential benefits to offspring receiving IPI, (2) determining the relative costs and benefits to males performing IPI, and (3) routinely determining paternity (Dunbar, 1988; Altmann et al., 1988). Because of these methodological difficulties, and the concomitant lack of attention to IPI, primatologists have not as yet formulated or classified the potential types of IPI that may exist for nonhuman primates. In this section, we first propose a heuristic classification of the potential types of indirect benefits that could be received by offspring and close kin as the result of male behaviors that do not involve direct interaction with infants. Kleiman and Malcolm (1981) have proposed a useful typology of IPI for mammals; their categories, with the addition of two others—resource inheritance and protection against infanticidal conspecifics—provide a useful catalogue for examining the potential benefits that might accrue to kin as a result of male activities (Table 3–1). Second, we differentiate between those benefits resulting from kin selection (i.e., male strategies related to parental effort) and those occurring secondarily as the result of male strategies related to copulatory effort. Third, we propose that the type and amount (both absolute and relative) of benefits accruing to offspring (and other kin) in any given species are theoretically related to a number of species-specific social and biological features, permitting a ranking of species on a relative scale with respect to the amount of IPI they can potentially exhibit.

Perhaps the most easily identifiable type of potential IPI among the nonhuman primates is surveillance and defense against predators (Table 3–1). Adult males of almost all nonsolitary primate species have been observed to perform sentinel behavior, active defense against predators, alarm calling, and other antipredator behaviors (Nagel and Kummer, 1974; Hrdy, 1976; Alley, 1986; Cheney and Wrangham, 1987). Sentinel behavior, a kind of vigilant surveillance of the habitat often from strategic vantage points, occurs in many species, e.g., patas (*Erythrocebus patas;* Hall, 1965), macaques, and Hanuman Langurs

Table 3–1 Types of indirect paternalistic investment in nonhuman primates and humans

General primate traits	Uniquely human traits
1. Surveillance and defense against predators	5. Food and resource provisioning to females
2. Defense and maintenance of resources within home range or territory	6. Aid in shelter construction
3. Inheritance of resources within home range or territory	
4. Protection against infanticidal conspecifics	

(*Presbytis entellus;* Hrdy, 1977). More active defense, such as lunging toward, mobbing, and physical assaults on predators, has been observed among baboons (DeVore and Washburn, 1963; Stoltz and Saayman, 1970), rhesus monkeys (Lindburg, 1977), gorillas (Schaller, 1963), and common chimpanzees (*Pan troglodytes;* Kortlandt, 1972). Patas males regularly perform diversionary displays rather than directly confronting predators, a tactic probably designed to lead predators away from other group members (Hall, 1965). This also occurs on rare occasions in Barbary macaques (Mehlman, personal observations).

Another type of potential IPI can result from males defending and maintaining resources within a territory or home range (Table 3–1). In many mammals and birds (Kleiman and Malcolm, 1981), spacing mechanisms such as long calls and active exclusion of conspecifics from a territory are thought to convey longer term benefits (e.g., food and other resources) to offspring that remain in the parental territory until or beyond adulthood. In nonhuman primates, if this type of IPI exists, it is most clearly expressed when males participate in agonistic intergroup encounters, as is common among most species of New and Old World monkeys (reviewed in Wrangham, 1980; Cheney, 1983, 1987), hylobatids (Carpenter, 1940), chimpanzees (Goodall et al., 1979; Wrangham, 1977), and gorillas (Harcourt, 1978; Fossey, 1979). A rarer form of benefit develops among some nonhuman primate species when renewable resources within a defended territory (or partially defended home range) are inherited by offspring (Table 3–1; e.g., cotton-topped tamarins, *Saguinus oedipus:* McGrew and McLuckie, 1986).

Adult male protection of offspring against infanticidal conspecifics is another potential type of male behavior that may benefit offspring and close kin (Table 3–1). For example, in groups with a unimale structure, a resident male who successfully defends his group against intrusion by extragroup males is at the same time indirectly preventing any potential infanticide by the latter. This is in contrast to *direct* defense and protective behaviors of infants (classified as DPI) that have been documented both in species in which males are thought regularly to practice infanticide as a reproductive strategy (redtail monkeys, *Cercopithecus ascanius;* blue monkeys, *Cercopithecus mitus;* red colobus, *Colobus badius;* Leland, et al., 1984) as well as in some species (mainly baboons; e.g., Busse and Hamilton, 1981) in which male infanticide appears to be infrequent.

A major difficulty in determining the degree to which any of these potential benefits to offspring (and close kin) actually represent IPI is our inability to assess what evolutionary selection pressures canalized and shaped these behavioral configurations. For example, regarding antipredator strategies, there is a divergence of opinion whether they evolved by kin selection (and thus would truly represent IPI) or were produced by individual selection (Maynard Smith, 1965; Trivers, 1971). Moreover, it may be the case that antipredator behaviors function in different ways in species with monogamous mating or unimale-multifemale social organizations vs. species organized into multimale-multifemale groups. In the former, male antipredator behavior may function as a male reproductive strategy of parental investment; i.e., paternity certainty is high in groups with this social structure, and protection from predators increases offspring survival. For species with multimale systems, where paternity certainty is lower, it is not clear whether antipredator behaviors evolved as a type of generalized kin selection (i.e., to protect many members with various degrees of genetic relatedness and thus increase inclusive fitness) or if they function specifically to increase the fitness of offspring. Another possibility for the origin of these behaviors in multimale groups is that they function as displays of innate male fitness (however defined) and enhance access to reproductively active females through the operation of female choice. At present, the consensus is that, at the very least, in primate groups with only one resident male (i.e., "harem" polygyny and monogamy), male alarm calls (and thus, by analogy, most antipredator behaviors) have evolved as protective mechanisms for increasing offspring fitness (see, e.g., Cheney and Wrangham, 1987; Dunbar, 1988). In these cases, antipredator behaviors by definition would qualify as IPI.

It is even more difficult to determine the nature of the benefits accruing to offspring (and kin) through other types of IPI (Table 3–1) vs. the benefits obtained by the males performing the behaviors. For example, other factors, such as copulatory effort and competitive restriction of access to females, probably have created the selection pressures that canalized male agonistic behavior during intergroup encounters. By "defending" the group and its range from extragroup conspecifics, males may be primarily ensuring their access to reproductive females rather than ensuring access to ecologically based resources for their offspring. Dunbar (1988) recently suggested that among nonhuman primates, including monogamous species, most male xenophobia is directly related to a reproductive strategy of protecting access to females rather than resource defense. This same argument could be applied to protection against infanticide, since in each case a reproductive strategy focused on copulatory effort may have provided the most important selective forces directing the behavior patterns' evolution. This is not to deny that multiple selective pressures may canalize a behavior pattern or that a single, major selective force may have an impact on multiple behavioral phenomena. Specifically, although xenophobic males may be investing in copulatory effort by attempting to restrict access to females, simultaneously they may be conveying secondary benefits to immatures by excluding infanticidal competitors or by securing resources within a territory or a core area in a home range. If so, xenophobic behaviors were not shaped by parental effort and should be excluded from the class of IPI. Thus, of the general primate traits in Table 3–1, only antipredator behaviors can presently be classified as true IPI; for the rest, more data are needed to clarify their relationship to parental efforts.

Two classes of behavior, food provisioning of females and male aid in shelter construction (Table 3–1), qualify as IPI because they are oriented towards offspring and kin-related individuals and appear to have evolved as forms of parental effort. These two types, however, do not occur among any of the nonhuman primates, although their absence is not surprising given their rather limited occurrence in all mammalian orders. For example, male food provisioning of females has evolved only in canids (Malcolm and Kleiman, 1981; Moehl-man, 1986), and appears to be linked to conditions in which nutritional and energetic benefits to immatures may accrue from transporting small, highly nutritious food packets (e.g., meat) to mothers and other caretakers stationed in shelters. Among mammals, male aid in shelter construction occurs only among the mustelids and one species of viverrids (Malcolm and Kleiman, 1981), and shelter construction among these mammalian families has probably evolved as a protection against severe climatic conditions as well as against predators.

Male aid in shelter construction is absent among the nonhuman primates, probably as a result of their arboreal heritage, their relative restriction to trop-ical and semitropical climatic zones, their small litter sizes, the persistent trans-port of infants by adults, and the concomitant highly developed grasping reflex of infants. In those species that do construct shelters (e.g., nests among the great apes), they function primarily for comfort during resting and sleeping, not as protective caches for infants. Similar reasoning could explain the absence of male provisioning of females among nonhuman primates, an adaptation that appears to be linked to the coevolution of carnivory and regurgitation, shelter use, and a strategy of at least one parent or caretaker remaining at the shelter to protect immatures. Nonhuman primates have not evolved carnivory as a principle feeding adaptation, and shelter construction is virtually absent. Although shelter use is found among some prosimians for caching their off-spring, no adults remain to protect them, and thus there is no opportunity for male provisioning of mothers or helpers at the nest.

Given the above findings, what then is the frequency and occurrence of IPI in nonhuman primates? First, the potential amount or relative significance of IPI in any given primate species is influenced by a number of factors. The fore-most is a social organization that includes the presence of one or more males (Table 3–2). Thus the solitary life-style of orang-utans (*Pongo pygmaeus;* Rod-man, 1973; Galdikas, 1978) and special cases of multimale, polygamously mat-ing species in which males are driven from breeding groups during the non-mating season (squirrel monkeys: Baldwin and Baldwin, 1972; talapoins, *Miopithecus talapoin:* Gautier-Hion, 1970) preclude most opportunities for the expression of IPI. These cases rank low on a relative scale of potential IPI (Table 3–2).

Another factor is adult male tenure in the group. The longer a single male retains residence in a social group containing his offspring, the greater his potential for IPI. In many forest guenons (*Cercopithecus* sp.) and colobines (subfamily Colobinae) with unimale-multifemale social organizations, extra-group males compete with and oust resident males at relatively high frequen-cies (see, e.g., Struhsaker and Leland, 1987; Cords, 1987). These adult males have shorter group tenures than, for example, hylobatid males that reside in monogamous family groups. Thus in those social systems with long periods of male tenure (e.g., the monogamous systems and their derivatives), the potential for IPI would be higher than with unimale social units (Table 3–2).

A third factor bearing on the existence of true IPI and the degree to which

Table 3–2 Potential for indirect paternalistic investment correlated with mating system/social organization and other related variables (see text). Types of IPI correspond to Table 3–1. Numbered mating system/social organization types are fully identified in notes to table.

Mating system/social organ.	Types of IPI	Length of male tenure	Territoriality	Male infanticide	Examples	Relative IPI
(1) Monogamy/facultative polygyny	1,2,3,4,5,6	long	moderate	low to moderate	humans	very high
(2) Monogamy/facultative polyandry	1,2,3?	long	high	absent	marmosets/tamarins	high
(3) Monogamy	1,2	long	high	absent	gibbons, titi, & owl monkeys	high
(4) Unimale polygyny & fem. dispersal	1,2,3,4	long	low	moderate	gorillas	moderate
(5) Unimale polygyny	1,2,4	short	moderate	high	langurs/guenons	low to moderate
(6) Multimale polygamy	1,2,4	medium	low	low	macaques/baboons	moderate
(7) Multimale polygamy & fem. dispersal	1,2,3,4	long	moderate	moderate	chimpanzees	low
(8) Multimale polygamy with male periph.	1?	—	low	absent	talapoins/squirrel monkeys	absent
(9) Solitary	none?	—	low	absent	orangutans	absent

Mating system/social organization types: (1) monogamy/facultative polygyny; (2) monogamy/facultative polyandry; (3) monogamy; (4) unimale polygyny and female dispersal; (5) unimale polygyny; (6) multimale polygamy; (7) multimale polygamy and female dispersal; (8) multimale polygamy with males driven to group's periphery during nonmating season; (9) solitary.

male activities might convey benefits to offspring and kin-related individuals is the number of competing and potentially reproductively active males in a social group. Thus, in groups with multiple males (and thus many potential sires for any given infant), any male's behaviors could simultaneously convey benefits to most or all of the immatures in the group, many of which may not be his offspring or close kin. Thus the potential for IPI among "average" males in multimale groups would be less than that of males in unimale polygynous groups or monogamously pair-bonded family groups. In the latter systems, males would have a high probability that any energy/time expenditures and/or risks incurred in performing IPI would benefit only their own offspring. In the former system, however, males with only a small number of offspring and close kin (or none of either) would have little to gain relative to parental investment by performing antipredator behaviors. Thus, we might predict differential performance of IPI by males in multimale groups: males with high reproductive success may perform more IPI than, for example, recent immigrant males (see Dunbar, 1988, and Cheney and Wrangham, 1987, for some suggested empirical tests).

A fourth factor influencing the potential significance of IPI is the degree of range defense and/or true territoriality displayed by the species. In those species in which males actively defend a range and exclude conspecifics from food, water, sleeping sites, or other resources, males might effectively preserve those resources for the exclusive use of the female(s) he mates with and any of his offspring. The potential benefits (although secondarily derived) to offspring and kin-related individuals from true territorial behavior in males generally would be higher than those of males in multimale groups living in overlapping home ranges that can only be partially defended. These same arguments would of course apply not only to the possibilities for resource defense and any immediate benefits to offspring and close kin but also to the possibility for inheritance of resources.

What then can we say about the general nature of IPI among the nonhuman primates? First, without more empirical data, only one type of true IPI can be identified among the nonhuman primates: antipredator behaviors. Second, it is clear that in those species where males exhibit the greatest amounts of DPI, there is also the greatest potential for male activities to convey indirect benefits to offspring and close kin, despite the fact that many of these male behaviors may have evolved as part of a reproductive strategy focused on copulatory effort. Thus, in species with monogamous systems and evolutionary derivations of monogamous systems (siamangs, *Hylobates syndactylus;* titi monkeys, *Callicebus* sp.; owl monkeys, *Aotus trivirgatus;* marmosets and tamarins, family Callitrichidae), all the factors that increase the potential for males to convey indirect benefits to offspring and close kin converge. In these species, males are always present in the group, defend territories, enjoy relatively long tenure, and have relatively high paternity certainty.

GENERAL PATTERNS OF PATERNALISTIC INVESTMENT AMONG NONHUMAN PRIMATES

In this section, we summarize several features that are diagnostic and characteristic of male paternalistic investment among nonhuman primates.

Frequency

Measured ordinally, DPI among nonhuman primates is a relatively infrequent phenomenon. If we use such indices as the number of taxa in which such interactions occur, the number of males among all potential actors who actually engage in some form of infant interaction, or the rate or frequency at which those few participating males actually do associate with infants, then we are led to the conclusion that DPI is not a common occurrence among nonhuman primate males.

Similar conclusions result from a survey of IPI among nonhuman primates. Of the many potential types of IPI, it appears that only antipredator behavior qualifies as true IPI. Moreover, although most males of nonsolitary species do perform antipredator behaviors, it is only for species with monogamous and unimale mating structures that we can confidently state that offspring and close kin might directly accrue benefits. It is tempting to speculate that, although the males of most primate species do not engage in DPI, they at least perform some IPI by virtue of their antipredator behaviors. Without the empirical evidence, however, that offspring or close kin are benefiting from these behaviors, especially in multimale systems, this speculation is premature.

Interspecific Variability and Diversity

Although DPI may not be a frequent event, males of many different primate species do interact with immature conspecifics in a wide diversity of ways. This may be seen by contrasting the paternalistic behavior of siamangs, which consists mostly of carrying, and that of the versatile Barbary macaque, who carries, holds, grooms, monitors, protects, and engages in ritualized triadic interactions with infants. Thus a characteristic feature of DPI across taxa (i.e., interspecifically) is structural (and functional) heterogeneity. Different interpretations for superficially similar paternalistic behaviors are not mutually incompatible; evidence for one does not ipso facto exclude another. This becomes clear when we recognize that among baboons, for example, a male may both care for and exploit an infant, even his own. Paternalistic behaviors may differ in a number of fundamental, especially functional, ways depending on the social context, the value of the resource being contested, the selective advantages of the behaviors, and to whom the advantages accrue. We believe that this is why the heterogeneous male–infant behaviors can appear to be so functionally contradictory among the nonhuman primates. Similarly, the two types of indirect male behaviors that potentially confer benefits to offspring and kin-related individ-

uals, namely, range defense and indirect protection against infanticidal conspecifics, are expressions most likely not of parental effort but rather of copulatory effort, and any benefits conveyed to offspring and close kin are only
secondary and fortuitous in nature.

Intraspecific Plasticity and Malleability

There is some intraspecific variability in the expression of DPI among nonhuman primates. This variability demonstrates both behavioral plasticity and
the importance of ecological, social, demographic, and/or social features in
shaping the exact nature of its expression. For example, supernumerary male
saddleback tamarins may become incorporated into a group, thereby creating
a new social structure and mating system (i.e., converting monogamy to polyandry), and the exhibition of DPI is affected proportionately. Male callitricids
may also show varying amounts of DPI depending on how many "helpers" are
available to assist in caretaking duties.

These examples highlight the difference between a particular species' propensity or potential for DPI and the expression of that potential as mediated by
social and ecological factors. That males of such usually nonpaternalistic species as rhesus macaques can develop intensive and long-term relationships with
infants given the appropriate social opportunity and environmental context
gives clear evidence of the within-species malleability of paternalistic behavior.
The influence of socioecological factors on this plasticity is also probably
responsible for the sporadic and opportunistic expresson of "paternalistic"
behavior in those species where DPI occasionally appears but is not a regular
or species-typical behavior. Much less is known about the possible intraspecific
variability of IPI among primate males, but we suspect a similar phenomenon
of potential vs. expressed behaviors. For example, among vervet monkeys (Cercopithecus aethiops), there is some empirical evidence that individuals are quite
capable of withholding alarm calls when it may be advantageous to do so (Cheney and Seyfarth, 1986).

Social Structure, Mating Systems, and Paternalistic Investment

DPI can be found in species that are characterized by every type of social system exhibited by primates (except solitary living). Thus DPI may be shown by
males living in unimale-unifemale groups, in groups characterized by multiple
males and females, and even in groups featuring multiple males bonded to a
single female. Similarly, DPI is found in species exhibiting every type of mating
system: monogamy, polyandry, and polygyny. Likewise, among all types of
nonsolitary social structures, there exists the potential for indirect benefits
derived by offspring and close kin as the result of male activities as well as the
possibility for true IPI in the form of antipredator behaviors.

There appears, however, to be a general, positive correlation between the
most intensive and well developed forms of DPI and IPI (as well as secondary
benefits to offspring and close kin) and a monogamous mating system. This

correlation between a high degree of paternalistic investment and monogamous mating systems (and mating systems derived from monogamy, as in the
facultatively polyandrous callitricids) bears directly on the notion of the diver-
gent strategies of parental and copulatory effort (and paternity certainty) in
Trivers' (1972) original theoretical formulation.

PATERNALISTIC INVESTMENT IN HUMANS

Detailed observational studies of human father–child relationships began only
in the 1970s, with most studies focusing on attachment and the influence of
the father on child development rather than on the quantitative distribution
and types of direct interactional behaviors and cost/benefit analyses of paternalistic investment. As a consequence of the nature of the early studies of
human paternal behavior and also the relative newness of primatology as a
discipline, the developing methodologies in the two areas have not yet provided
us with databases that lend themselves to comprehensive comparisons. Despite
these weaknesses, however, a comparative examination of paternalistic investment in nonhuman primates and humans reveals that certain features of
human paternalistic behavior are unique with respect to our phylogenetic
relatives.

Direct Paternalistic Investment in Humans

One type of measurement common to both human and nonhuman primate
studies is that of proximity, i.e., direct interaction time spent between males
and offspring (or close kin). Katz and Konner (1981) conducted a cross-cultural survey of 80 preindustrial societies in search of correlates between family
structure, marriage styles, economic subsistence, and patterns of parenting.
They reported that, in 59% of societies, fathers are rarely or never with their
infants. Moreover, in those 4% of societies with the highest degree of paternal
proximity (e.g., the !Kung San bushmen), fathers account for only 6% of the
direct parental care given to infants, and they spend on average only 14% of
their time directly interacting with infants. Thus, if we attempt to rank human
males on a scale of male–infant interaction time, as Whitten (1987) has done,
they fall somewhere in the middle of the continuum of this measure among all
primates (i.e., they spend less time with infants than marmosets, tamarins, owl
and titi monkeys, and Barbary macaques and can be ranked close to gorilla
males).

Structurally, human males exhibit many types of interactions with infants
that also characterize DPI among the nonhuman primates: human males carry,
caretake, clean, groom, and play with infants. There may be, however, a quantitative structural difference between DPI in human and nonhuman primate
males. Data from many studies of Westernized cultures indicate that, in general, fathers are identified with play or playful types of interaction, whereas
mothers are more associated with nurturing and caretaking activities (Kotel-

chuck, 1975). For example, mothers spend an average of almost 2 hours per day feeding infants less than 2 years of age, whereas fathers spend 15 minutes per day (Kotelchuck, 1975). On a proportional basis, mothers allocate about 65% of their caretaking activities to play, whereas fathers allocate over 90% of their interactive time to play (Richards et al., 1975). Moreover, as the child gets older, fathers may play with infants an even greater absolute amount of time than do mothers (Clark-Stewart, 1978). A caveat to all of these findings, however, is that they are based on research from Western industrialized cultures, and, as Katz and Konner (1981) caution, more data from non-Western and preindustrial societies must be collected before we can characterize both the range of variation and any central tendencies for human DPI.

Another way in which human DPI probably shows a quantitative structural difference from that of the nonhuman primates is the degree to which human males share food and cofeed with offspring and close kin. In certain species of nonhuman primates, adult males occasionally share food and cofeed with presumed offspring (many of the marmosets and tamarins: Hershovitz, 1977; Masatata, 1981; titi monkeys: Starin, 1978; Wright, 1984; reviewed in Whitten, 1987). There are a number of human subsistence patterns in which males are the predominant provider, e.g., some hunting and agricultural societies. Although quantitative data on the patterns of distribution of male-derived foodstuffs (i.e., the transfer of food from male directly to offspring and/or close kin) are essentially nonexistent, it is nevertheless clear that, through the evolution of a food provisioning adaptation, human males provide nourishment to offspring on a much more frequent basis than do males of any nonhuman primate species.

These differences between human DPI and nonhuman primate DPI are eclipsed by what may be the major qualitative shift in human DPI compared with that of our phylogenetic relatives. This shift is the result of the profound impact of spoken language and the symbolic transference of critical knowledge on the quality (value of benefits) of direct interactions between males and offspring or close kin. In simplistic terms, a human father, during rather short periods of direct contact with juvenile, subadult, and even adult offspring, can communicate a host of significant benefits without which his offspring may not survive, let alone reproduce. These benefits may include abstract and symbolic information involved in socialization, such as societal rules and traditions, rituals, religious beliefs, etc., or they may be related to subsistence information, such as the location of prey, hunting and farming techniques, ecological knowledge, techniques of warfare, location of hostile conspecifics, etc.

The use of language in human DPI also relates to two additional elements that may characterize father–older infant interactions. First, Parsons (1958) theorized that fathers' interactions may be biased toward competence-directed functions, whereas mother–older infant interactions may be biased towards nurturant or expressive functions. Although many researchers suggest that the empirical evidence from Western cultures supports this notion (Lamb, 1981), it is unknown to what degree, or even whether, this gender bias in interaction styles operates universally. A second element is that, in all human societies, males and females undergo very different socialization paths because of the sexual division of labor. Each sex learns a different set of information regarding gender-based roles. This may explain why cross-culturally human males are more involved in their sons' than in their daughters' socialization processes (Lamb, 1981).

Human DPI is also influenced by two unique features of hominid evolutionary history: the protracted period of juvenile dependence (Lancaster and Lancaster, 1987) and a lengthened subadulthood. Thus direct interactions between human fathers and offspring may take place well beyond the period of infancy. Fathers' direct interactions with postinfant offspring confer important benefits that increase offsprings' survival and reproductive success, but comparative studies of nonhuman and human primate males usually neglect this dimension of potential DPI (e.g., Whitten, 1987). Most human studies focus on interactions between fathers and younger children, and the role of the relations between fathers and adult offspring has never been examined from a paternal investment point of view (Hagestad, 1987). To take an extreme example that highlights the importance in humans of a postinfancy role for males in parental investment, among the Rwala Bedouin (Musil, 1928; cited in Katz and Konner, 1981), children are segregated from their fathers until about 7 years of age, obviously preventing much opportunity for direct interactive investment. During preadolescence and adolescence, however, young boys are trained by their fathers in discipline, obedience, and the techniques of warfare (fathers whip, cut, and stab their sons). Although ethnocentrism may tempt many readers to discount this behavior as paternalistic investment, the circumstances of the Rwala's subsistence and their extensive involvement in warfare indicate that fathers are probably conveying important benefits to their offspring, without which the latter's fitness might be substantially lowered.

Indirect Paternalistic Investment in Humans

Depending on our ability to identify IPI among nonhuman primates, there is probably a structural continuum between some types of IPI performed by nonhuman primate males and similar behaviors exhibited by human males. For example, human males perform antipredator behaviors, perhaps convey benefits to offspring by resource defense and maintenance, and also provide protection against infanticidal conspecifics. Despite these similar features, the human male clearly distinguishes himself from other primate males in performing many derived and specialized types of IPI.

Unlike other primate males, human males have a long history (approximately 1.5 million years of hunting and gathering; Isaac and Crader, 1981; Binford, 1985; Shipman, 1986; Blumenschine, 1987) of being away from their social units for part or all of the diurnal cycle. Rather than being with the group at all times, they left it to hunt and bring back meat, which was then distributed to other group members, including the hunters' mates. By providing direct benefits to their mates, human hunters indirectly provided benefits to presumed offspring. This form of IPI exists today among the remaining hunter-gatherer and intensive agricultural peoples of the world and continues in modified form among all human societies where males act as "breadwinners" (making monetary or nutritional contributions to their mate[s], and thus even more indirectly providing resources for their presumed offspring [reviewed in Lamb et al., 1987]).

Tied to the development of subsistence patterns derived from the domestication of plants and animals and the attendant sexual division of labor is the uniquely human trait of centering group activities at a "home base," a camp that typically contains shelters constructed by group members. Although males do not always participate in all facets of shelter construction, in many societies males play some role, such as providing or preparing raw materials or providing the economic base so that specialists can be hired to construct the shelter. This unique form of IPI clearly provides benefits to presumed offspring or close kin.

In many human societies, when ecological and socioeconomic conditions have allowed the unequal acquisition of status and wealth among males, societal rules have developed regulating how men must distribute their resources (and/or status) to offspring and close kin. This type of IPI, with all the concomitant differential benefits to some offspring and close kin, has no parallel among nonhuman primates. Although there is the possibility that some nonhuman primate offspring may inherit territorial resources secondarily through the activites of their fathers (see above), it is only within some human societies that the succession of property and status appear to be a derivation of other paternalistic investments oriented towards benefiting offspring or close kin. It is important to note here that, in claiming that patrilineal inheritance of status and property is uniquely human, we are not suggesting that the inheritance of status does not operate among the nonhuman primates. For example, the *matrilineal* inheritance of status (dominance rank) is a fundamental principle among many cercopithecines (see Chapters 7 and 8, this volume), but it is only among humans that males transfer material possessions, property, wealth, and status either to their sons in patrilineal societies or to their sister's son(s) in many matrilineal societies (Radcliffe-Brown, 1935).

It is important to emphasize, however, that inheritance of parental status and material possessions is not a characteristic of all human societies, and it may not have been an important aspect of IPI during much of human evolution. Members of the simplest hunter-gatherer societies do not have many material possessions (there is often a great deal of communal ownership), and rules of succession are absent or only poorly developed. In many of these societies, neither status nor property are inherited (Radcliffe-Brown, 1935). It was probably only with the development of sedentary agriculture and the differential accumulation of wealth and property during the Neolithic transition that rules of succession gained enough significance to convey substantial benefits to offspring and close kin (Hoebel, 1972; Lancaster and Lancaster, 1987). It was among these relatively recent societies that succession first began to follow the rule of primogeniture and, in some cases, urogeniture (Hoebel, 1972).

Plasticity in the Expression of Human Paternalistic Investment

A final characteristic of human paternalistic investment, that differentiates it from nonhuman primate systems, is its variability of expression. Witness the major changes that have occurred in modern societies in the last several decades, presumably as a result of changes in the economic base of the family, namely, the large-scale influx of mothers into the work force (over one-half of all American mothers now work outside the home; Glick, 1979). Several studies document increasing levels of fathers' involvement with their children. For example, only 2% of fathers are reported to spend no time with their children vs. 10% (or more) 50 years ago (Caplow and Chadwick, 1979). Recent studies have shown that twice as many children received regular care from fathers during the 1970s as during the 1960s (Daniels and Weingarten, 1981), and additional significant increases in paternal involvement occurred in the 1980s (Juster, 1985). During the 1980s, paternal involvement showed three times the increase of maternal involvement.

Among extant nonindustrialized societies too there is a wide variance in the structure, quality, and nature of parent–offspring relationships. Katz and Konner (1981) have ranked cultures as "close" or "distant" with regard to father–child proximity. In 90% of the 80 cultures they rated, the mother was the primary (or exclusive) care giver; in only 4% was there a regular and close relationship between father and child. In about 40% of these cultures, fathers were "frequently" (i.e., moderately) proximate to their children (Katz and Konner, 1981). Among some societies, especially those that are matrilineally organized, men invest not in their wife's children (putatively their own offspring) but rather in the children of their siblings, especially their sisters' children, a system known as the avunculate (Kurland, 1979; Flinn and Low, 1986). Katz and Konner also examined social and economic variables and how they are related to paternalism. They found that a combination of subsistence patterns, mating, and labor roles rather than any single element correlated with either high or low levels of paternal involvement. Among nonindustrialized populations, "father–infant proximity is greatest where either gathering or horticulture (small scale vegetable gardening and fruit growing as practiced throughout the Pacific Islands) is the primary mode of subsistence and where combinations of polygyny, patrilocal residence, the extended family, or patridominant division of labor are absent" (Katz and Konner, 1981:172). Relative to subsistence patterns, the highest ranked cultures for father–infant proximity are gathering societies such as the !Kung San, among whom subsistence is derived from vegetable foods gathered by the women (60–80%) and from meat obtained by male hunters (20–40%) and each sex works about one-half the week, with much time left for leisure. Also ranked high on a scale of father–infant proximity are societies that practice horticulture in combination with fishing, such as the societies of Polynesia, where each sex contributes about equally to subsistence. When female contribution to subsistence is high, father–infant proximity scores appear high, unless there is a polygynous mating/marriage system. Moreover, all societies with strict monogamy and nuclear families have high father–infant proximity scores, but these are few in number. More commonly, these nonindustrialized cultures have extended families, and here monogamy or low-level polygyny in nonconjunction with nonpatrilocal residence patterns is associated with high father–infant proximity scores. Thus, for many human societies, some generalized rules (more precisely, correlates) regarding the interplay of some socioeconomic variables and the exhibition of paternalism have been generated. It is important to note, however, that it has not been possible to develop cross-phylogenetic rules regarding the ecological determinants of male parental investment.

Historically as well, variation and plasticity have characterized paternalistic investment. We can trace changes in this behavior beginning with the Roman era. As the power of the Roman Empire waned, the general hedonistic character of urban life prompted a breakdown in monogamous family structure, leading to a loss of paternal authority and increases in infanticide, abandonment, and sexual abuse of children. It is at this point in human history that we see the appearance of what is today a uniquely human form of mating: socially imposed monogamy. As an outgrowth of the Roman tradition of a strong, centralized state authority, monogamy was dictated to be the only socially permissible form of marriage through a system of laws and rules imposed on all members of society. This contrasts with what is termed "ecologically imposed monogamy" (the type seen among nonhuman primates), where owing to ecological or economic contingencies (typically living in marginal habitats), males can provide support to the offspring of only a single mate (Flinn and Low, 1986). Throughout Asia, Europe, and North and South America, there has been a spread of socially imposed monogamy within recorded history, so that today it is the predominant form of mating, at least in terms of the actual number of humans exhibiting this system if not in terms of the numbers of cultures.

In spite of socially imposed monogamy, the degree of paternalistic involvement varied in different European historical periods. During the "Dark Ages" of Europe, adults are described as being indifferent, insensitive, and harsh in their treatment of their children (DeMause, 1975). Children were often considered to be property, to be sold at parental will; such sale of children by parents persisted in England through the twelfth century. The Renaissance ushered in an era of the nurturant parent (especially mothers), which contrasted sharply with the harsh and sometimes cruel socialization of children during the previous half millennium. The Industrial Revolution of the eighteenth century witnessed further major changes in relationships between parents and children, as the transformation in social and family relations shifted away from a communal orientation with large extended families toward nuclear (or slightly extended) families and more individualistic expressions of parental behavior. The new emphasis on personal privacy fostered the notion of romantic love with empathetic, reciprocal husband-wife roles and consequently a more empathetic model of child rearing. Ironically, the success of the Industrial Revolution required the male head of household to leave home for work, and this led to a diminished role for him in child care at a time when societal mores were emphasizing a nurturing, caring role for parents. Economic prosperity and the growth of the middle class led to a concentration of female roles on marriage, homemaking, and child bearing and rearing, whereas male roles were oriented toward income generation outside the home, leading to an ever-increasing estrangement for the male from child caretaking. Social changes resulting from World War II and the feminist movement of the past two decades have again changed the family structure and the role that fathers play in child care.

Paternalism among humans shows some continuity with our phylogenetic past. For the most part, females and not males are the primary, and in many cases the exclusive, caretakers. Paternal strategies are linked to such factors as mating systems, economic subsistence patterns, and family (group) structure, and, as these have changed during human evolutionary history, so too have the quality and nature of human paternalism changed. Indirect contributions seem to have eclipsed direct ones in expression and importance. Many human subsistence adaptations have required men to be spatially separated from women and children for extended periods, minimizing the opportunity for direct interaction with infants. Consequently, the provision of resources by males (a pattern nonexistent among nonhuman primates) has become the most important paternal contribution among humans. Finally, in agreement with what appears to be a basic primate potentiality for behavioral plasticity, human paternalistic investment patterns have been and are currently subject to major perturbations in relation to culturally derived and socially imposed regulations.

DISCUSSION

Can the study of paternalistic investment among the nonhuman primates help us analyze and understand human paternalistic behavior? The most facile, tempting, and irreverent response to this question is "no"; there are simply too many unique aspects of human paternalistic investment even to rank human males on the same scale with other primate males. In more common parlance, this endeavor is much like comparing apples and oranges. For example, examine Whitten's (1987) suggestion that human males should be classified as "affiliative" rather than "intensive" with regard to their interactions with infants. As concerns the distribution and frequency of male–infant interaction times, this may indeed be the case, but what this typological perspective overlooks, in evolutionary terms, is the degree to which human males invest (especially indirectly) in offspring and/or close kin measured in terms of the benefits in survival and reproductive success the latter may derive. Thus, to take the example of parent–offspring linguistic communication as a qualitative shift in human DPI (Table 3–3), it is illogical to assume that 1 hour of interaction between marmoset father and offspring is somehow equivalent in terms of paternalistic investment to 1 hour spent between human father and child.

Therefore, with human DPI, the intervening variables of culture and spoken language (with their impact on benefits to offspring) render much of the nonhuman primate data noncomparable. So much of human survival and reproductive success is dependent on technological (i.e., extrasomatic) solutions and

Table 3-3 Characteristic features of human paternalistic investment distinguishing it from that of nonhuman primate species with high levels of paternalistic investment

Quantitative shifts in human DPI patterns

    Increase in play interactions and types of play
    Increase in cofeeding with infants and an increase in provisioning infants
    Increase in paternalistic investments directed toward juvenile, subadult, and adult
       offspring
    Shift toward preferential investment in male offspring

Qualitative shifts in human DPI patterns

    Use of symbolic communication to enhance and condense transfer of information, thus
       increasing DPI per interaction by several orders of magnitude

Qualitative shifts in human IPI patterns

    Food provisioning of females
    Participation in shelter construction
    Inheritance of status, wealth, and possessions

has become singularly influenced by elaborate, symbolic sociocultural rules and constraints that it is difficult to imagine human DPI occurring without recourse to the spoken word. Furthermore, the web of human kinship networks that continue into adulthood and the complex opportunities for male activities (direct and indirect) to benefit offspring are best left for analysis to anthropologists, sociologists, and psychologists; the nonhuman primate data simply do not offer many homologies, analogies, or even guiding principles with which to understand human paternalistic investment. At its most extreme, consider too the manifold instances in which human culture overrides biology, such as when socially imposed rules dictate paternalistic investment in offspring that are not one's own (e.g., adoption). Other animals may be fooled into such investments, as in cuckoldry, but only human males knowingly and readily sacrifice their own reproductive success in a truly altruistic act on behalf of unreciprocating male "competitors."

Although human males exhibit several unique aspects of DPI, they are most extraordinary in the degree to which their parental investment patterns have become indirect. Take the example of male food provisioning of females. It is through this behavior, combined with the protracted period of juvenile dependence (essentially nonexistent among nonhuman primates), that human females are able to support more than one dependent offspring at a time (Lancaster and Lancaster, 1987). This compression of what would otherwise be a very long human birth interval has had revolutionary consequences on human population demography and to a large degree may account for our evolutionary success (Lovejoy, 1981; Lancaster and Lancaster, 1987). Relative to this adaptation of male provisioning of females with dependent young, it is far more satisfactory to turn to data on canids for developing an analogous, referential model. Similarly, for a comparative cost/benefit analysis of human male aid in shelter construction, data from avian studies would likely offer more insight than information from nonhuman primates. With regard to yet another facet of human IPI, inheritance of possessions and wealth, there is simply nothing specific to nonhuman primates that can enable us to understand these uniquely human behaviors.

And, yet, it is precisely because the nonhuman primate data are available to us that we can confidently assert the unique features of human paternalistic behavior, and herein lies one salient reason to argue that nonhuman primate studies do benefit the study of human paternalism. It is only through comparisons with the nonhuman primates that we can determine, in an evolutionary sense, those features of human paternalistic behavior that are shared and "primitive" vs. those that are unique and "derived." During the course of evolution, our primate forebears, and later our hominid grandfathers, developed specialized and unique adaptations with respect to their paternalistic investments. Among humans, these adaptations are, of course, tied into a larger adaptive behavioral complex that bears on every facet of our existence: language, culture, symbolic behavior, tool making and technological innovation, division of labor, base-camp and shelter construction, etc. It is only recently that the data on nonhuman primate parental investment have been incorporated into theories that model the course of hominid social evolution (Lovejoy, 1981; Alexander and Noonan, 1979; Strassman, 1980; Symons, 1979; Parker, 1987; Benshoof and Thornhill, 1979; reviewed in Mehlman, 1988). In each of these evolutionary scenarios, there is an increasing awareness of the important role that male parenting and reproductive strategies have played in the evolution of mating and social systems. Witness the recent publication of an entire volume (Kinzey, 1987) devoted solely to modeling the evolution of human social behavior by drawing on the nonhuman primate data.

For example, among modern hunter-gatherer cultures, which we assume show the type of social organization present during much of hominid prehistory, groups are multimale-multifemale, yet they contain both monogamous and polygamous reproductive subunits. It appears that, when ecological conditions permit, a mild form of polygyny develops, as men with the most wealth and status are capable of supporting more than one mate (Ford and Beach, 1952; Van den Berghe, 1979; Symons, 1979; Flinn and Low, 1986). Even among major industrialized societies that practice socially imposed monogamy, serial replacement among bonded pairs (serial monogamy) is facultatively equivalent to mild polygyny.

Many features of human morphology (e.g., relative testicular size, mild sexual dimorphism wherein body weight of males is 20% greater than that of females and body length is 5–12% greater) suggest that humans were also facultatively polygynous or at least serially monogamous during their evolution (Alexander et al., 1979; Short, 1981; Flinn and Low, 1986). Thus the mating system that appears to have been characteristic of most human evolution is a mild form of polygyny. This suggests that monogamy (with its potential for high levels of DPI) is not our biologically derived condition. Monogamous nonhuman primate males appear to devote most of their reproductive effort to parental care, with little toward copulatory effort, whereas the opposite seems to be true for most polygynous nonhuman primate males. Human males appear to have a unique reproductive strategy (just as their mating system is unique) in that it is "mixed." They devote more of their reproductive effort to copulatory effort (i.e., attempting to mate with several females) than do monogamous nonhuman primate males, while at the same time devoting a substantial amount of energy to parental effort in the form of uniquely derived types of IPI.

It is also the comparative method that allows us to assert confidently that one unique characteristic of human paternalistic behavior is its extreme degree of variability. In the nonhuman primates, we have seen that variation in paternalistic investment occurs mostly at the interspecific level. For example, all male *S. fuscicollis* care for infants in a restricted number of ways, primarily by carrying them. To the reduced degree that it manifests itself, most intraspecific variation among those nonhuman primate species that express the greatest degree of paternalism (i.e., the monogamous/facultative polyandrously mating species) appears to revolve around the number of caretaking males available and thus how much time any given male actually interacts with infants. In fact, across all taxa, the primary manifestation of variation is the limited number of males (of all males available as caretakers) who actually do interact with infants and the variation between them in actual frequency and amount of interaction.

In contrast, intraspecific variation typifies humans. More than any other mammalian species, humans exhibit extreme variance in parental patterns both within and between groups (societies). From ethnographic studies, we have learned that human societies exhibit considerable variability in their mating/marriage patterns and in the attendant roles played by parents (or other proscribed care givers, such as "uncles" in the matrilineal avunculate system). All types of mating strategies (strict monogamy, serial monogamy, polygyny [mild and extensive] and polyandry [rare]) are found in different societies of the single, albeit biologically highly polytypic, human species (indeed it is such behavioral and cultural variability that led some early social scientists to suggest that humans were not a single species). Concordantly, among humans, the paternal patterns associated with these variations in mating and family structure are equally variable cross culturally. Even within a single culture (with the same language, geographic boundaries, social history, and traditions), variations in quality, structure, frequency, or total time of paternal interaction may be extremely pronounced, such as among different socioeconomic classes in the United States and the United Kingdom.

Although humans are characterized by high levels of intraspecific variation in paternalistic investment, an analysis of the nonhuman primate data on interspecific variation in such investment might permit us to generate principles and explanatory models pertinent to humans. An examination of the data has identified at least some of the basic features that positively correlate with mating systems (e.g., monogamy and mild polygyny) in which there are high levels of certainty of paternity: a family structure with few adult males but many other individuals, usually related but nonbreeding, to act as "helpers" and territorial sequestering of subsistence resources.

Analyzing the correlates of paternalistic investment among preindustrial human societies, we find that in fact high levels of such behavior covary with many of the variables that operate among other primates. For example, the highest levels of DPI are found in societies with monogamous or slightly polygynous mating systems. In extended families (i.e., with helpers within the group structure), there are associations between high certainty of paternity and marriage types (monogamy or mild polygyny) and high degrees of paternalism, but only when residence patterns are not patrilocal. DPI is also high when women make a large contribution to subsistence, although this combines with mating patterns to predict high or low paternalism.

CONCLUSION

It has been only 132 years since Darwin (1859) proposed his revolutionary vision of the evolution of life by the action of natural selection and 120 years since his heretical proposition that the same principles applied to human beings (Darwin, 1871). There have been many scientific and social consequences of man's view of himself as conditioned by the development of evolutionary theory, and it has become an article of faith that humans have been subject to the same selective pressures that have shaped the evolution of all animals. This perspective forms the foundation for the use of the "comparative method," which allows us to approach questions concerning how mankind is similar to or different from his evolutionary ancestors and what factors (phylogenetic, ecological, social) are responsible for those similarities or differences. The comparison of the behavior of contemporary, industrialized humans with that of our closest relatives (the nonhuman primates) and with that of nonindustrialized societies such as the !Kung San (thought to represent the modal type of human existence prior to the neolithic revolution) has found its greatest expression and value in the modern disciplines of psychology and anthropology. Analysis of human behavior from the perspective of evolutionary, ecological, and sociobiological theory is in a very preliminary state. Refinements are needed in the empirical, theoretical, and methodological bases for studying the social biology and ecology of nonhuman primates. This should not dissuade us, however, from recognizing the inherent utility of this analytical approach. In its simplest form, the identification of unique features of human mating patterns, parenting styles and family structures can best be accomplished by a search for analogues or homologues among our nearest relatives. Furthermore, the analyses of comparative data from other species support the view that there are ultimate (i.e., evolutionary) underpinnings bearing on humans' proximate behaviors.

REFERENCES

Alexander, B. K. (1970). Parental behavior of adult male Japanese monkeys. *Behaviour* 36:270–285.

Alexander, B. K., and K. M. Noonan (1979). Concealment of ovulation, parental care, and human evolution. Pp. 436–453 in N. A. Chagnon and W. Irons, eds., *Evolutionary Biology and Human Social Behavior: An Anthropological Perspective.* North Scituate, MA: Duxbury Press.

Alexander, B. K., J. L. Hoogland, R. D. Howard, K. M. Noonan, and P. W. Sherman (1979). Sexual dimorphisms and breeding systems in pinnapeds, ungulates, primates, and humans. Pp. 402–435 in N. A. Chagnon and W. Irons, eds., *Evolutionary Biology and Human Social Behavior: An Anthropological Perspective.* North Scituate, MA: Duxbury Press.

Alley, T. (1986). An ecological analysis of the protection of primate infants. Pp. 239–258 in V. McCabe and G. Balzano, eds., *Event Cognition.* Hillsdale, NJ: Erlbaum.

Altmann, J. (1980). *Baboon Mothers and Infants.* Cambridge, MA: Harvard University Press.

Altmann, J., G. Hausfater, and S. A. Altmann (1988). Determinants of reproductive success in savannah baboons *(Papio cynocephalus).* Pp. 403–418 in T. H. Clutton-Brock, ed., *Reproductive Success: Studies of Individual Variation in Contrasting Breeding Systems.* Chicago: University of Chicago Press.

Altmann, S. A., S. S. Wagner, and S. Lenington (1977). Two models for the evolution of polygyny. *Behavioral Ecology and Sociobiology* 2:397–410.

Anderson, C. A. (N.D.). Paternal investment under changing conditions among chacma baboons at Suikerbosrand. *American Journal of Physical Anthropology* (in press).

Auerbach, K. G., and D. M. Taub (1979). Paternal behavior in a captive "harem" group of cynomolgus macaques *(Macaca fascicularis). Laboratory Primate Newsletter* 18(2):7–11.

Baldwin, J. D., and J. I. Baldwin (1972). The ecology and behavior of squirrel monkeys *(Saimiri oerstedi)* in a natural forest in western Panama. *Folia Primatologica* 19:161–184.

Benshoof, L., and R. Thornhill (1979). The evolution of monogamy and concealed ovulation in humans. *Journal of Social and Biological Structures* 2:95–106.

Berman, C. (1981). Effects of being orphaned: a detailed case study of an infant rhesus. Pp. 79–81 in R. A. Hinde, ed., *Primate Social Relationships: An Integrated Approach.* Oxford, England: Blackwell.

——— (1982). The ontogeny of social relationships with group companions among free-ranging infant rhesus monkeys I. Social networks and differentiation. *Animal Behaviour* 30:149–162.

Bernstein, I. (1976). Activity patterns in a sooty managbey group. *Folia Primatologica* 26:185–206.

Bertrand, M. (1969). The behavioral repertoire of the stumptail macaque. *Bibliotheca Primatologica* 11:1–273.

Binford, L. (1985). Human ancestors: changing views of their behavior. *Journal of Anthropological Archaeology* 4:292–327.

Blumenschine, R. (1987). Characteristics of an early hominid scavenging niche. *Current Anthropology* 28:383–407.

Bolin, I. (1981). Male parental behavior in black howler monkeys *(Alouatta palliata pigra)* in Belize and Guatemala. *Primates* 22:349–360.

Box, H. O. (1975). A social developmental study of young monkeys *(Callithrix jacchus)* within a captive family group. *Primates 16:*419–435.

———(1977). Quantitative data on the carrying of young monkeys *(Callithrix jacchus)* by other members of their family groups. *Primates 18:*475–484.

Brown, J. L. (1975). *The Evolution of Behavior.* New York: Norton.

Brown, K., and D. S. Mack (1978). Food sharing among captive *Leontopithecus rosalia. Folia Primatologica 29:*268–290.

Busse, C. B. (1984). Triadic interactions among male and infant chacma baboons. Pp. 186–212 in D. M. Taub, ed., *Primate Paternalism.* New York: Van Nostrand Reinhold.

Busse, C. B., and W. J. Hamilton (1981). Infant carrying by male chacma baboons. *Science 212:*1282–1283.

Caplow, T., and P. Chadwick (1979). Inequality and lifestyles in Middletown 1920–1978. *Social Science Quarterly 60:*367–385.

Carpenter, C. R. (1940). A field study in Siam of the behavior and social relations of the gibbon *(Hylobates lar). Comparative Psychology Monographs 16:*1–212.

Cebul, M. S., and G. Epple (1984). Father-offspring relationships in laboratory families of saddle-back tamarins *(Saguinus fuscicollis).* Pp. 1–19 in D. M. Taub, ed., *Primate Paternalism.* New York: Van Nostrand Reinhold.

Chalmers, N. (1968). The social behavior of free-living mangabeys in Uganda. *Folia Primatologica 8:*263–281.

Cheney, D. L. (1983). Extra-familial alliances among vervet monkeys. Pp. 278–286 in R. A. Hinde, ed., *Primate Social Relationships: An Integrated Approach.* Oxford, England: Blackwell.

———(1987). Interactions and relationships between groups. Pp. 267–281 in B. B. Smuts, D. L. Cheney, R. M. Seyfarth, R. W. Wrangham, and T. T. Struhsaker, eds., *Primate Societies.* Chicago: University of Chicago Press.

Cheney, D. L. and R. M. Seyfarth (1986). The recognition of social alliances among vervet monkeys. *Animal Behaviour 34:*1722–1731.

Cheney, D. L., and R. W. Wrangham (1987). Predation. Pp. 227–239 in B. B. Smuts, D. L. Cheney, R. M. Seyfarth, R. W. Wrangham, and T. T. Struhsaker, eds., *Primate Societies.* Chicago: University of Chicago Press.

Chivers, D. J. (1972). The siamang and the gibbon in the Malay peninsula. *Gibbon and Siamang 1:*103–135.

———(1974). The siamang in Malaya. *Contributions to Primatology 4.* Basel: S. Karger.

Chivers, D. J., and J. J. Raemakers (1980). Long-term changes in behaviour. Pp. 209–260 in D. J. Chivers, ed., *Malayan Forest Primates: Ten Years' Study in a Tropical Rain Forest.* New York: Plenum Press.

Clark-Stewart, K. A. (1978). And Daddy makes three: the father's impact on mother and young child development. *Child Development 49:*466–478.

Cleveland, J., and C. T. Snowdon (1982). The complex vocal repertoire of the adult cotton-top tamarin *(Saguinus oedipus oedipus). Zeitschrift fur Tierpsychologie 58:*231–270.

Collins, D. A. (1981). Social behavior and patterns of mating among adult yellow baboons *(Papio c. cynocephalus* L. 1766). PhD Dissertation, Cambridge University.

———(1986). Relations between adult male and infant baboons. Pp. 205–218 in J. G. Else and P. C. Lee, eds., *Primate Ontogeny, Cognition and Social Behaviour.* Cambridge, England: Cambridge University Press.

Cords, M. (1987). Male-male competition in one-male groups. Pp. 98–111 in B. B.

Smuts, D. L. Cheney, R. M. Seyfarth, R. W. Wrangham, and T. T. Struhsaker, eds., *Primate Societies.* Chicago: University of Chicago Press.

Daniels, P., and K. Weingarten (1981). *Sooner or Later: The Timing of Parenthood in Adult Lives.* New York: Norton.

Darwin, C. (1859). *On the Origin of Species by Means of Natural Selection, or the Preservation of Favoured Races in the Struggle for Life.* London: Murray.

———— (1871). *The Descent of Man and Selection in Relation to Sex.* London: Murray.

Dawson, G. A. (1978). Composition and stability of social groups of the tamarin, *Saguinus oedipus geoffroyi,* in Panama: ecological and behavioral implications. Pp. 23–28 in D. G. Kleiman, ed., *The Biology and Conservation of the Callitrichidae.* Washington, DC: Smithsonian Institution Press.

Deag, J. M. (1974). A study of the social behavior and ecology of the wild Barbary macaque, *Macaca sylvanus.* PhD Dissertation, University of Bristol.

———— (1980). Interactions between males and unweaned Barbary macaques: testing the agonistic buffering hypothesis. *Behaviour 75:54–81.*

DeMause, L. (1975). *The History of Childhood.* New York: Harper.

DeVore, I., and S. L. Washburn (1963). Baboon ecology and human evolution. Pp. 335–367 in F. C. Howell and F. Bourliere, eds., *African Ecology and Human Evolution.* New York: Wenner-Gren Foundation.

Dixson, A. F. (1983). Observations on the evolution and significance of "sexual skin" in female primates. Pp. 63–106 in J. D. Rosenblatt, R. A. Hinde, C. Beer, and M. C. Busnel, eds., *Advances in the Study of Behavior 13.* New York: Academic Press.

Dixson, A. F., and D. Fleming (1981). Parental behavior and infant development in owl monkeys *(Aotus trivirgatus griseimembra). Journal of Zoology 194:25–39.*

Dunbar, R.I.M. (1984). Infant use by male gelada in agonistic contexts: agonistic buffering, progeny protection, or soliciting support? *Primates 25:28–35.*

———— (1988). *Primate Social Systems.* Ithaca, NY: Cornell University Press.

Dunbar, R.I.M. and E. Dunbar (1975). Social dynamics of gelada baboons. *Contributions to Primatology 6.* Basel: Karger.

Epple, G. (1975). Parental behavior in *Saguinus fuscicollis* spp. *(Callitrichidae). Folia Primatologica 24:221–238.*

Epple, G., and Y. Katz (1983). The saddle back tamarin and other tamarins. Pp. 115–148 in J. Hearn, ed., *Reproduction in New World Primates.* Boston: Lancaster.

———— (1984). Social influences on estrogen secretion and ovarian cyclicity in saddle back tamarins *(Saguinus fuscicollis). American Journal of Primatology 6:215–228.*

Estrada, A. (1984). Male-infant interactions among free-ranging stumptail macaques. Pp. 56–87 in D. M. Taub, ed., *Primate Paternalism.* New York: Van Nostrand Reinhold.

Estrada, A., R. Estrada, and R. Ervin (1977). Establishment of a free ranging troop of stumptail macaques *(Macaca arctoides):* social relations I. *Primates 18:647–676.*

Estrada, A., and J. Sandoval (1977). Social relations in a free ranging group of stumptail macaques *(Macaca arctoides):* male care behavior I. *Primates 18:793–813.*

Flinn, M. V., and B. S. Low (1986). Resource distribution, social competition, and mating patterns in human societies. Pp. 217–243 in D. I. Rubenstein and R. W. Wrangham, eds., *Ecological Aspects of Social Evolution: Birds and Mammals.* Princeton, NJ: Princeton University Press.

Ford, C., and F. Beach (1952). *Patterns of Sexual Behavior.* New York: Harper and Row.

Fossey, D. (1979). Development of the mountain gorilla *(Gorilla gorilla beringei):* the first thirty-six months. Pp. 139–186 in D. A. Hamburg and E. R. McCowen, eds., *The Great Apes.* Menlo Park, CA: Benjamin/Cummings.

——— (1983). *Gorillas in the Mist.* Boston: Houghton Mifflin.

Fox, G. (1972). Some comparisons between siamang and gibbon behavior. *Folia Primatologica 18:*122–139.

Fragaszy, D. M., S. Schwartz, and D. Schinosaka (1982). Longitudinal observation of care and development of infant titi monkeys *(Callicebus moloch). American Journal of Primatology 2:*191–200.

Galdikas, B. (1978). Orangutan adaptation at Tanjung Puting Reserve, Central Borneo. PhD Dissertation, University of California, Los Angeles.

Garber, P. A., L. Moya, and C. A. Malaga (1984). A preliminary field study of the moustached tamarin monkey *(Saguinus mystax)* in northeastern Peru: questions concerned with the evolution of a communal breeding system. *Folia Primatologica 42:*17–32.

Gautier-Hion, A. (1970). L'organisation sociale d'un bande de talapoins *(Miopithecus talapoin)* dans le nord-est du Gabon. *Folia Primatologica 12:*116–141.

Gilmore, H. B. (1977). The evolution of agonistic buffering in baboons and macaques. Paper presented at the 46th Annual Meeting of the American Association of Physical Anthropologists, Seattle.

Glick, B. B. (1980). Ontogenetic and psychobiological aspects of the mating activities of male *Macaca radiata.* Pp. 345–370 in D. G. Lindberg, ed., *The Macaques: Studies in Ecology, Behavior, and Evolution.* New York: Van Nostrand Reinhold.

Goldizen, A. W. (1987). Tamarins and marmosets: communal care of offspring. Pp. 34–43 in B. B. Smuts, D. L. Cheney, R. M. Seyfarth, R. W. Wrangham, and T. T. Struhsaker, eds., *Primate Societies.* Chicago: University of Chicago Press.

Goldizen, A. W., and J. Terborgh (1986). Cooperative polyandry and helping behavior in saddle backed tamarins *(Saguinus fuscicollis).* Pp. 191–198 in J. G. Else and P. C. Lee, eds., *Primate Ecology and Conservation,* Vol. 2. Cambridge, England: Cambridge University Press.

Goodall, J., E. Bandora, C. Bergmann, C. Busse, H. Matama, E. Mpongo, A. Pierce, and D. Riss (1979). Intercommunity interactions in the chimpanzee population of the Gombe National Park. Pp. 13–53 in D. A. Hamburg and E. R. McCown, eds., *The Great Apes.* Menlo Park, CA: Benjamin/Cummings.

Gouzoules, H. (1975). Maternal rank and early social interactions of stumptail macaques, *Macaca arctoides. Primates 16:*405–418.

——— (1984). Social relations of males and infants in a troop of Japanese monkeys: a consideration of causal mechanisms. Pp. 127–145 in D. M. Taub, ed., *Primate Paternalism.* New York: Van Nostrand Reinhold.

Gubernick, D. J., and P. H. Klopfer (1981). *Parental Care in Mammals.* New York: Plenum Press.

Hagestad, G. O. (1987). Parent–child relations in later life: trends and gaps in past research. Pp. 405–434 in J. B. Lancaster, J. Altmann, A. S. Rossi, and L. R. Sherrod, eds., *Parenting Across the Life Span: Biosocial Dimensions.* New York: Aldine de Gruyter.

Hall, K.R.L. (1965). Behavior and ecology of the wild patas monkey, *Erythrocebus patas,* in Uganda. *Journal of Zoology, London 148:*15–87.

Hamilton, W. D. (1964). The genetical evolution of social behavior. *Journal of Theoretical Biology 7:*1–52.

Hamilton, W. J. (1984). Significance of paternal investment by primates to the evolution of male-female associations. Pp. 309–335 in D. M. Taub, ed., *Primate Paternalism.* New York: Van Nostrand Reinhold.

Harcourt, A. H. (1979). Social relationships between adult male and female mountain gorillas. *Animal Behaviour 27:*325–342.

Hasegawa, T., and M. Hiraiwa (1980). Social interactions of orphans observed in a free-ranging troop of Japanese monkeys. *Folia Primatologica 33:*129–158.

Hausfater, G. (1984). Infanticide in langurs: strategies, counterstrategies, and parameter values. Pp. 257–282 in G. Hausfater and S. B. Hrdy, eds., *Infanticide: Comparative and Evolutionary Perspectives.* Hawthorne, NY: Aldine.

Hendy-Neely, H., and R. Rhine (1977). Social development of stumptail macaques *(Macaca arctoides):* momentary touching and other interactions with adult males during the infant's first 60 days of life. *Primates 18:*589–600.

Hershkovitz, P. (1977). *The Living New World Monkeys,* Vol. 1 Chicago: University of Chicago Press.

Higley, J. D., and S. J. Suomi (1986). Parental behavior in non-human primates. Pp. 153–207 in W. Sluckin and M. Herbert, eds., *Parental Behavior.* London: Basil Blackwell.

Hiraiwa, M. (1981). Maternal and alloparental care in a troop of free-ranging Japanese monkeys. *Primates 22:*309–329.

Hoage, R. J. (1978). Parental care in *Leontopithecus rosalia rosalia:* sex and age difference in carrying behavior and the role of prior experience. Pp. 293–306 in D. G. Kleiman, ed., *The Biology and Conservation of the Callitrichidae.* Washington, DC: Smithsonian Institution Press.

Hoebel, E. A. (1972). *Anthropology: The Study of Man,* 4th edition. New York: McGraw-Hill Book Company.

Hrdy, S. B. (1976). The care and exploitation of nonhuman primate infants by individuals other than the mother. *Advances in the Study of Behavior 6:*101–158.

——— (1977). *The Langurs of Abu.* Cambridge, MA: Harvard University Press.

Ingram, J. C. (1977). Interactions between parents and infants, and the development of independence in the common marmoset. *Animal Behaviour 25:*811–827.

——— (1978). Parent-infant interactions in the common marmoset *(Callithrix jacchus).* Pp. 281–292 in D. G. Kleiman, ed., *The Biology and Conservation of the Callitrichidae.* Washington, DC: Smithsonian Institution Press.

Isaac, G., and D. Crader (1981). To what extent were early hominids carnivorous? An archaeological perspective. Pp. 37–103 in R. Harding and G. Teleki, eds., *Omnivorous Primates.* New York: Columbia University Press.

Itani, J. (1959). Paternal care in the wild Japanese monkey, *Macaca fuscata fuscata. Primates 2:*61–93.

Izawa, K. (1978). A field study of the black-mantle tamarin. *Primates 19:*241–274.

Katz, M. M. and M. J. Konner (1981). The role of the father: an anthropological perspective. Pp. 155–186 in M. E. Lamb, ed., *The Role of the Father in Child Development,* 2nd edition. New York: John Wiley & Sons.

Kinzey, W. G. (ed.) (1987). *The Evolution of Human Behavior: Primate Models.* Albany, NY: State University of New York Press.

Kleiman, D. G. (1985). Paternal care in New World primates. *American Zoologist 25:*857–859.

Kleiman, D. G., and J. R. Malcolm (1981). The evolution of male parental investment in mammals. Pp. 347–388 in D. J. Gubernick and P. H. Klopfer, eds., *Parental Care in Mammals.* New York: Plenum Press.

Klein, H. D. (1983). Paternal care and kin selection in yellow baboons, *Papio cyno-cephalus*. PhD Dissertation, University of Washington.

Kortlandt, A. (1972). *New Perspectives on Ape and Human Evolution*. Amsterdam: Stichting voor Psychobiologie.

Kotelchuck, M. (1976). The infant's relationship to the father: experimental evidence. Pp. 329–344 in M. E. Lamb, ed., *The Role of the Father in Child Development*. New York: John Wiley & Sons.

Kummer, H. (1967). Tripartite relations in hamadryas baboons. Pp. 63–72 in S. Altmann, ed., *Social Communication among Primates*. Chicago: University of Chicago Press.

Kurland, J. A. (1979). Paternity, mother's brother, and human sociality. Pp. 145–180 in N. A. Chagnon and W. Irons, eds., *Evolutionary Biology and Human Social Behavior: An Anthropological Perspective*. North Scituate, MA: Duxbury Press.

Kurland, J. A., and S.J.C. Gaulin (1984). The evolution of male paternal investment: effects of genetic relatedness and feeding ecology on the allocation of reproductive effort. Pp. 259–308 in D. M. Taub, ed., *Primate Paternalism*. New York: Van Nostrand Reinhold.

Lamb, M. E. (1981). *The Role of the Father in Child Development*, 2nd edition. New York: John Wiley & Sons.

Lamb, M. E., J. H. Pleck, E. L. Charnov, and J. A. Levine (1987). A biosocial perspective on paternal behavior and involvement. Pp. 111–142 in J. B. Lancaster, J. Altmann, A. S. Rossi, and L. R. Sherrod, eds., *Parenting Across the Life Span: Biosocial Dimensions*. New York: Aldine de Gruyter.

Lancaster, J., and C. Lancaster (1987). The watershed: change in parental investment and family formation strategies in the course of human evolution. Pp. 187–206 in J. Lancaster, J. Altmann, A. S. Rossi, and L. R. Sherrod, eds., *Parenting Across the Life Span: Biosocial Dimensions*. New York: Aldine De Gruyter.

Leland, L., T. Struhsaker, and T. H. Butynski (1984). Infanticide by adult males of three primate species of the Kibale Forest, Uganda: a test of hypotheses. Pp. 151–172 in G. Hausfater and S. B. Hrdy, eds., *Infanticide: Comparative and Evolutionary Perspectives*. Hawthorne, NY: Aldine.

Leutenegger, W. (1980). Monogamy in callithricids: a consequence of phyletic dwarfism? *International Journal of Primatology* 1:45–98.

Lindburg, D. G. (1971). The rhesus monkey in North India: an ecological and behavioral study. Pp. 1–106 in L. J. Rosenblum, ed., *Primate Behavior: Developments in Field and Laboratory Research*, Vol. 2. New York: Academic Press.

———— (1977). Feeding behavior and diet of rhesus monkeys *(Macaca mulatta)* in a Siwalik forest in North India. Pp. 223–250 in T. H. Clutton-Brock, ed., *Primate Ecology*. New York: Academic Press.

Lovejoy, C. O. (1981). The origin of man. *Science* 211:341–350.

Maynard Smith, J. (1965). The evolution of alarm calls. *American Naturalist* 100:637–650.

———— (1977). Parental investment: a prospective analysis. *Animal Behaviour* 25:1–9.

McGrew, W. C. (1988). Parental division of infant caretaking varies with family composition in cotton-top tamarins. *Animal Behaviour* 36:285–286.

McGrew, W. C., and E. C. McLuckie (1986). Philopatry and dispersion in the cotton-top tamarin, *Saguinus (o.) oedipus:* an attempted laboratory simulation. *International Journal of Primatology* 7:401–422.

Masataka, N. (1981). A field study of the social behavior of Goeldi's monkeys *(Callimico goeldi)* in North Bolivia, 1: group composition, breeding cycle, and infant development. *Kyoto University Overseas Reports of New World Monkeys 2:*23–41.

Mehlman, P. T. (1986). Population ecology of the Barbary macaque *(Macaca sylvanus)* in the fir forests of the Ghomara, Moroccan Rif mountains. PhD Dissertation, University of Toronto.

―――― (1988). L'evolution des soins paternels chez les primates et les hominides. *Anthropologie et Societes 12*(3):131–149.

Mitani, J. C. (1984). The behavioral regulation of monogamy in gibbons *(Hylobates muelleri). Behavioral Ecology and Sociobiology 15:*225–229.

Mitchell, G. (1969). Paternalistic behavior in primates. *Psychological Bulletin 71:*399–417.

Moehlman, P. D. (1986). Ecology of cooperation in canids. Pp. 64–86 in D. I. Rubenstein and W. R. Wrangham, eds., *Ecological Aspects of Social Evolution: Birds and Mammals.* Princeton, NJ: Princeton University Press.

Mori, U. (1979). Development of sociality and social status. Pp. 125–154 in M. Kawai, ed., Ecological and Sociological Studies of Gelada Baboons, *Contributions to Primatology 16.* Basel: Karger.

Nagel, U., and Kummer, H. (1974). Variation in cercopithecoid aggressive behavior. Pp. 159–184 in R. L. Holloway, ed., *Primate Aggression, Territoriality, and Xenophobia.* New York: Academic Press.

Nicholson, N. (1982). Weaning and the development of independence in olive baboons. PhD Dissertation, Harvard University.

Packer, C. (1980). Male care and exploitation of infants in *Papio anubis. Animal Behaviour 28:*512–520.

Parker, S. (1987). A sexual selection model for hominid evolution. *Human Evolution 2:*235–253.

Parsons, T. (1958). Social structure and the development of personality: Freud's contribution to the integration of psychobiology and sociology. *Psychiatry 21:*321–340.

Pook, A. G. (1978). A comparison between the reproduction and parental behavior of the Goeldi's monkey *(Callimico),* and of the true marmosets (Callitrichidae). Pp. 1–14 in H. Rothe, H. J. Wolters, and J. P. Hearn, eds., *Biology and Behavior of Marmosets.* Gottingen, Federal Republic of Germany: Eignenverlag Hartmut Rothe.

Popp, J. L. (1978). Male baboons and evolutionary principles. PhD Dissertation, Harvard University.

Radcliffe-Brown, A. R. (1935). Patrilineal and matrilineal succession. *Iowa Law Review 20:*286–298.

Ransom, T. W., and B. S. Ranson (1971). Adult male-infant relations among baboons *(Papio anubis). Folia Primatologica 16:*179–195.

Redican, W. K. (1976). Adult male-infant interactions in nonhuman primates. Pp. 345–385 in M. E. Lamb, ed., *The Role of the Father in Child Development.* New York: John Wiley & Sons.

Redican, W. K., and G. Mitchell (1974). Play between adult male and infant rhesus monkeys. *American Zoologist 14:*295–302.

Redican, W. K. and D. M. Taub (1981). Adult male-infant interactions in nonhuman primates. Pp. 203–258 in M. E. Lamb, ed., *The Role of the Father in Child Development,* 2nd edition. New York: John Wiley & Sons.

Rhine, R., and H. Hendy-Neely (1978). Social development of stumptail macaques *(Macaca arctoides):* momentary touching, play and other interactions with aunts and immatures during the infants' first 60 days of life. *Primates 19:*115–123.

Richards, M. P., M. Dunn, and B. Antonis (N.D.). Caretaking in the first year of life: the father's and mother's social solution. Unpublished manuscript.

Robinson, J. G., P. C. Wright, and W. G. Kinzey (1987). Monogamous cebids and their relatives: intergroup calls and spacing. Pp. 44–53 in B. B. Smuts, D. L. Cheney, R. M. Seyfarth, R. W. Wrangham, and T. T. Struhsaker, eds., *Primate Societies.* Chicago: University of Chicago Press.

Rodman, P. S. (1973). Population composition and adaptive organisation among orang-utans of the Kutai Nature Reserve, East Kalimantan. Pp. 384–414 in T. H. Clutton-Brock, ed., *Primate Ecology: Studies of Feeding and Ranging Behavior in Lemurs, Monkeys, and Apes.* London: Academic Press.

Rowell, T. (1974). Contrasting adult male roles in different species of nonhuman primates. *Archives of Sexual Behavior 3:*143–149.

Rylands, A. B. (1981). Preliminary field observations on the marmoset, *Callithrix humeralifer intermedius* (Hershkovitz, 1977) at Dadanelos, Rio Aripuana, Mato Gross. *Primates 22:*46–59.

Schaller, G. B. (1963). *The Mountain Gorilla: Ecology and Behavior.* Chicago: University of Chicago Press.

Shipman, P. (1986). Scavenging or hunting in early hominids: theoretical framework and tests. *American Anthropologist 88:*27–43.

Short, R. (1981). Sexual selection in man and the great apes. Pp. 181–214 in C. Graham, ed., *Reproductive Biology of the Great Apes.* New York: Academic Press.

Smith, E. O., and P. G. Peffer-Smith (1984). Adult male-immature interactions in captive stumptail macaques *(Macaca arctoides).* Pp. 88–112 in D. M. Taub, ed., *Primate Paternalism.* New York: Van Nostrand Reinhold.

Smuts, B. (1982). Special relationships between adult male and female olive baboons *(Papio anubis).* PhD Dissertation, Stanford University.

———— (1985). *Sex and Friendship in Baboons.* Hawthorne, NY: Aldine.

Snowdon, C. T., and S. J. Suomi (1982). Parental behavior in primates. Pp. 63–108 in H. Fitzgerald, J. Mullins, and P. Gage, eds., *Child Nurturance: Studies of Development in Nonhuman Primates,* Vol. 3. New York: Plenum Press.

Soini, P. (1982). Ecology and population dynamics of the pygmy marmoset, *Cebuella pygmaea. Folia Primatologica 39:*1–21.

Suomi, S. (1977). Adult male-infant interactions among monkeys living in nuclear families. *Child Development 48:*1255–1265.

———— (1979). Differential development of various social relationships by rhesus monkey infants. Pp. 219–244 in M. Lewis and L. Rosenblum, eds., *The Child and Its Family.* New York: Plenum Press.

Spencer-Booth, Y. (1968). The behavior of group companions towards rhesus monkey infants. *Animal Behaviour 16:*541–557.

Starin, E. D. (1978). Food transfer by wild titi monkeys *(Callicebus torquatus torquatus). Folia Primatologica 30:*145–151.

Stein, D. M. (1981). The nature and function of social interactions between infant and adult male yellow baboons *(Papio cynocephalus).* PhD Dissertation, University of Chicago.

———— (1984). Ontogeny of infant-adult male relationships during the first year of life for yellow baboons *(Papio cynocephalus).* Pp. 213–243 in D. M. Taub, ed., *Primate Paternalism.* New York: Van Nostrand Reinhold.

Stein, D. M., and P. B. Stacey (1981). A comparison of infant-adult male relations in a one-male group with those in a multi-male group for yellow baboons *(Papio cynocephalus). Folia Primatologica 36:*264–276.

Stevenson, M. F. (1970). Birth and perinatal behavior in family groups of the common marmoset *(Callithrix jacchus jacchus)* as compared to other primates. *Journal of Human Evolution 5:*265–281.

——— (1978). The behavior and ecology of the common marmoset *(Callithrix jacchus jacchus)* in its natural environment. Pp. 298–321 in H. Rothe, H. J. Wolters, and J. P. Hearn, eds., *Biology and Behavior of Marmosets.* Gottingen, Federal Republic of Germany, Eigenverlag Rothe.

Stoltz, L. P., and G. S. Saayman (1970). Ecology and behavior of baboons in the northern Transvaal. *Annals of the Transvaal Museum 26:*99–143.

Strassmann, B. (1980). Sexual selection, paternal care, and concealed ovulation in humans. *Ethology and Sociobiology 2:*21–40.

Struhsaker, T. T., and L. Leland (1985). Infanticide in a patrilineal society of red colobus monkeys. *Zeitschrift für Tierpsychologie 69:*89–132.

——— (1987). Colobines: infanticide by adult males. Pp. 83–97 in B. B. Smuts, D. L. Cheney, R. M. Seyfarth, R. W. Wrangham, and T. T. Struhsaker, eds, *Primate Societies.* Chicago: University of Chicago Press.

Strum, S. C. (1984). Why males use infants. Pp. 146–185 in D. M. Taub, ed., *Primate Paternalism.* New York: Van Nostrand Reinhold.

——— (1985). Use of females by male olive baboons *(Papio anubis). American Journal of Primatology 5:*93–109.

Symons, D. (1979). *The Evolution of Human Sexuality.* New York: Oxford University Press.

Taub, D. M. (1978). Aspects of the biology of the wild Barbary macaque (Primates, Cercopithecinae, *Macaca sylvanus* L. 1758): biogeography, the mating system, and male–infant interactions. PhD Dissertation, University of California, Davis.

——— (1980). Testing the agonistic buffering hypothesis I. The dynamics of participation in the triadic interaction. *Behavioral Ecology and Sociobiology 6:*187–197.

——— (1984). Male caretaking behavior among wild barbary macaques *(Macaca sylvanus).* Pp. 337–406 in D. M. Taub, ed., *Primate Paternalism.* New York: Van Nostrand Reinhold.

——— (1985). Male-infant interactions in baboons and macaques: a critique and evaluation. *American Zoologist 25:*861–871.

——— (1990). The functions of primate paternalism: a cross-species review. Pp. 338–377 in J. Feierman, ed., *Pedophilia: Biosocial Dimensions.* New York: Springer-Verlag.

Taub, D. M., and W. K. Redican (1984). Adult male–infant interactions in Old World monkeys and apes. Pp. 377–406 in D. M. Taub, ed., *Primate Paternalism.* New York: Van Nostrand Rienhold.

Tenaza, R. R. (1975). Territory and monogamy among Kloss' gibbons *(Hylobates klossi)* in Siberut Island, Indonesia. *Zeitschrift für Tierpsychologie 40:*37–52.

Terborgh, J., and A. W. Goldizen (1985). On the mating system of cooperatively breeding saddle backed tamarins *(Saguinus fuscicollis). Behavioral Ecology and Sociobiology 16:*293–296.

Tilson, R. (1981). Family formation strategies of Kloss' gibbons. *Folia Primatologica 35:*259–287.

Trivers, R. L. (1971). The evolution of reciprocal altruism. *Quarterly Review of Biology 46:*35–57.

——— (1972). Parental investment and sexual selection. Pp. 136–179 in B. Campbell, ed., *Sexual selection and the Descent of Man 1871–1971.* Chicago: Aldine.

Vaitl, E. A. (1977). Experimental analysis of the nature of social context in captive groups of squirrel monkeys *(Saimiri sciureus). Primates 18:*849–859.

Van den Berghe, P. (1979). *Human Family Systems.* New York: Elsevier.

Vessey, S., and D. Meikle (1984). Free living rhesus monkeys: adult male interactions with infants and juveniles. Pp. 113–126 in D. M. Taub, ed., *Primate Paternalism.* New York: Van Nostrand Reinhold.

Vogel, C., and H. Loch (1984). Reproductive parameters, adult-male replacements and infanticide among free-ranging langurs *(Presbytis entellus)* at Jodphur (Rajasthan), India. Pp. 237–256 in G. Hausfater and S. B. Hrdy, eds., *Infanticide: Comparative and Evolutionary Perspectives.* Hawthorne, NY: Aldine.

Vogt, J. L. (1984). Interactions between adult males and infants in prosimians and New World Monkeys. Pp. 340–376 in D. M. Taub, ed., *Primate Paternalism.* New York: Van Nostrand Reinhold.

Vogt, J. L., H. Carlson, and E. Menzel (1978). Social behavior of a marmoset *(Saguinus fuscicollis)* group. 1: Parental care and infant development. *Primates 19:*715–726.

Whitten, P. L. (1987). Infants and adult males. Pp. 343–357 in B. B. Smuts, D. L. Cheney, R. M. Seyfarth, R. W. Wrangham, and T. T. Struhsaker, eds., *Primate Societies.* Chicago: University of Chicago Press.

Williams, G. C. (1966). *Adaptation and Natural Selection: A Critique of Some Current Evolutionary Thought.* Princeton, NJ: Princeton University Press.

Wittenberger, J. F., and R. L. Tilson (1980). The evolution of monogamy: hypotheses and evidence. *Annual Review of Ecology and Systematics 11:*197–232.

Wolters, J. (1978). Some aspects of role-taking behavior in captive family groups of the cotton-top tamarin *(Saguinus oedipus oedipus).* Pp. 259–278 in H. Rothe, H. J. Wolters, and J. P. Hearn, eds., *Biology and Behavior of Marmosets.* Gottingen, Federal Republic of Germany: Eigenverlag Rothe.

Wrangham, R. W. (1977). Feeding behavior of chimpanzees in Gombe National Park, Tanzania. Pp. 504–538 in T. H. Clutton-Brock, ed., *Primate Ecology.* New York: Academic Press.

———(1981). An ecological model of female-bonded primate groups. *Behaviour 75:*269–299.

Wright, P. C. (1984). Biparental care in *Aotus trivirgatus* and *Callicebus moloch.* Pp. 59–76 in M. F. Small, ed., *Female Primates: Studies by Women Primatologists.* New York: Alan R. Liss, Inc.

4

Ontogeny of Behavior in Humans and Nonhuman Primates: The Search for Common Ground

JANICE CHISM

Should researchers who study the ontogeny of human behavior be interested in the work done on development in nonhuman primates? Theories of human development are many, diverse, and in some cases conflicting in their predictions or in attribution of causality. Comparisons of the ontogeny of behavior in human and nonhuman primates can contribute to our understanding of human development by providing data that may help us select between competing theories or hypotheses. This chapter discusses some areas in which studies of nonhuman primate development may provide such help:

1. How appropriate are different methodological approaches for investigating particular aspects of behavioral development?
2. How did particular patterns of development evolve and why do humans show the patterns that they do?
3. How do we increase our understanding of the ways in which genetic and environmental factors contribute to the development of behavior?
4. What stages of development in other primate species help us to better understand the role of stages in human development (see also Ambrose [1976] for a discussion of the importance of the concept of stages in development theory)?

In examining each of these areas, examples of how such considerations have been useful in enlarging our understanding of human development will be presented, along with cautions that need to be kept in mind when making such comparisons between human and nonhuman primate studies. In addition, the questions on which developmental studies of nonhuman primates have concentrated are discussed. The approaches to developmental questions taken by researchers with different theoretical viewpoints and backgrounds (e.g., evolutionary biology, behavioral ecology, ethology) provide insights into possible new perspectives on problems in human development. The fruitfulness of this comparative method has been amply demonstrated by work on attachment systems in which human developmentalists drew on the work of ethologists for both methodological and theoretical insight (e.g., Bowlby, 1969, 1973; Main and Weston, 1982).

METHODOLOGICAL CONSIDERATIONS

A fundamental difference between studies of development in humans and in nonhuman primates is the ability of the researcher to obtain retrospective and introspective responses from human subjects. Freudian theories of development were based in large part on introspective accounts of impressions of childhood events by adults, although Freud himself realized that this approach had limitations (see, e.g., Freud, 1920). Piaget's work on periods of cognitive development was based in part on interviews and naturalistic observations of children, but many of his ideas derived from experiments analyzing how children's responses to problem solving differed from those of adults (see, e.g., Piaget, 1952, 1962). Piaget's training as a zoologist undoubtedly influenced his use of the experimental method as well as his belief in the fixity of stages of development (Phillips, 1975). Other researchers, such as Vygotsky (1962) and Kohlberg, (1971), have used similar kinds of data to study the development of sensorimotor competence, reasoning, or problem-solving abilities.

Researchers studying attachment behavior, such as Ainsworth and Main, based many of their conclusions on the reactions of infants and their mothers to a standardized test situation (although these data were supplemented with observations of subjects under more natural conditions; see, e.g., Ainsworth, 1982; Ainsworth et al., 1978; Main and Weston, 1982). Work on the development of temperament in children often utilizes ratings or interviews with mothers (e.g., Schaffer and Emerson, 1964; Maccoby and Masters, 1970; Simpson and Stevenson-Hinde, 1985). When direct observations of behavior are used, they are often viewed as supplementary to other methods, are usually fairly brief, or are conducted under atypical circumstances.

In contrast, methods employed in developmental studies of nonhuman primates are based primarily on observations. Some studies have been experimental (e.g., studies of attachment and separation response in monkeys: Harlow and Harlow, 1965; Hinde et al., 1966; studies based on Piaget's periods of development: Parker, 1977). Most studies, however, have used naturalistic observations of animals freely interacting in at least a close approximation of their normal social environment.

Researchers studying development in nonhuman primates have favored more extensive observations than is typical of observational studies of humans (where, in some cases, 1–2 hours of observation of each subject over one to two sessions may be considered sufficient). Nonhuman primate studies, particularly field studies, usually involve repeated observation sessions for each subject and many hours of observation. A typical observation protocol for the development of behavior in infant monkeys may involve at least 6 hours per month on each subject over the first 6–12 months of life (see, e.g., Hinde and Spencer-Booth, 1967; Rowell et al., 1968; Berman, 1980).

Obviously, researchers studying nonhuman primates cannot use retrospective or interview data since their subjects cannot talk. The rationale for the use of repeated and extensive naturalistic observations of behavior by primatologists emerges from the methodological approaches of ethology, the study of animal behavior. The term ethology comes from the Greek *ethos,* meaning character or disposition; ethologists study the characteristic behavior of a species. As it was developed by Lorenz, Tinbergen, and their colleagues, beginning in the 1930s, ethology emphasized that to understand the meaning of behavior one had to observe it in its normal or natural context. Ethologists emphasize that behavior is as much a part of the biology of a species as any of its physiological or morphological structures (see Lehrman, 1974, for an eloquent discussion of the rationale for this approach).

Primatologists assume that the best data on the behavioral development of nonhuman primates come from field studies, followed by studies of infants living with their mothers in social groups that closely resemble the social structure, or at least the age-sex composition, of wild groups of the species. Least naturalistic are studies conducted on infants living alone with their mothers or in small groups containing one or a few other mother–infant pairs. Nevertheless, for very good reasons, fewer studies of infant development have been conducted on nonhuman primates living in wild groups than in captive ones. Under field conditions, the kind of sustained, detailed observations of infants and their mothers necessary to developmental studies are difficult to obtain. In a field study of patas monkey *(Erythrocebus patas)* infant development, for example, observational problems included trying to observe 45-cm-tall infants in grass that was sometimes 1–1.5 m high, while simultaneously trying to keep track of the focal infant subject in a play group that could include ten infants or more, all born within a few weeks of each other and, thus, effectively the same size and distressingly identical in appearance. Conducting careful, systematic observations on infants in a group that is moving continuously and sometimes rapidly through woodland areas where visibility is often poor can be a frustrating experience. At least patas infants spend most of their time on or near the ground. Observing infants in the crowns of tall trees in tropical forests is even more difficult.

In addition to the difficulties involved in the actual observation of subjects, field studies of nonhuman primates pose other problems. They are expensive. It may be difficult to obtain a sufficient sample size in a wild group. Many species live in relatively small groups, and females have relatively long birth intervals (2–3 years is typical of many monkey species), so few infants may be born into a group in a given year. The expense of doing field work makes it difficult to stay in the field long enough to get enough subjects to make a reasonably balanced and sufficiently large sample or to follow known individuals from birth to, at least, sexual maturity. This minimum period might range from 3 years to a decade or more, depending on the species.

Although field studies present an uncontrollable situation in which observations are difficult, developmental studies of captive infants housed alone with their mothers or in small, carefully designed groups provide controlled observation and sampling conditions. Variability in social or learning experience can be kept to a minimum by ensuring that infants have the same number and age-sex class of social companions, the same opportunities for manipulating objects, etc. Not surprisingly, psychologists used to conducting studies under carefully controlled conditions have favored nonhuman primate developmental studies done under captive conditions, whereas anthropologists and zoologists have favored field studies.

Captive settings can be designed to allow normal group structure and social interactions. The assumption is that this permits observation of development of a full range of normal social interactions. The many studies of development done in such settings have also assumed that cages allowed enough mobility and variety of possibilities for locomotion and manipulation so that researchers were seeing normal development of motor skills. Studies of the development of problem-solving skills have also assumed that the captive setting provided a representative sample of the kinds of problems that primates normally face. Even the most naturalistic captive studies, however, cannot provide the full range of conditions that a species encounters in the wild.

How is life different for wild and captive nonhuman primates? In captivity, food is always plentiful, if monotonous, and requires no finding. Whereas wild primates spend a major portion of the day looking for food and eating, these activities occupy little of a captive primate's time. Because captive animals do not have to search for food, they have more time available for social interaction, both affiliative and aggressive. Captive patas monkeys, for example, spend more time in social activities such as grooming than do wild monkeys (Rowell and Olson, 1983). Another major element missing in captive studies is predator detection and avoidance behavior. Wild primates must constantly be on the lookout for predators while they are moving, foraging, socializing, resting, or playing. Choices of resting sites during the day and sleeping sites at night are constrained by the need to avoid predators.

The captive environment is a very simplified version of that experienced by a wild primate with respect to foraging, locomotion, and vigilance requirements. Captivity provides little opportunity for researchers to observe the sophisticated behavioral systems primates have evolved to cope with these aspects of their environment. Focusing on captive studies overemphasizes the importance of social behavior and underemphasizes the complexity of the problems with which primates have to cope, often ignoring the sophistication and variety of the solutions they have developed.

Because most studies of primate development have been conducted on captive subjects, work has focused on the development of social behavior and little attention has been paid to the development of other behavioral systems. Besides encouraging researchers to discount the usefulness of primate data because of the perceived simplicity of the problem-solving abilities of nonhuman primates, this lack of attention to other behavioral systems has had another effect. Because cage environments are very similar, no matter what the species, they present similar sets of problems. Any differences in response to the environment are viewed, then, as being genetically based (i.e., different species-typical responses) and therefore less complex and variable than equivalent behaviors in humans, which are always viewed as learned. All feeling for the richness, complexity, and changeability of the natural environment is lost along with an appreciation of the necessity for variability in an individual's response to its environment.

When determining what conditions are appropriate for a developmental study, researchers who work on nonhuman primates are really considering what constitutes the appropriate *rearing* conditions for the experiment. Since researchers who study humans cannot vary the rearing conditions of their subjects except by the choice of subjects (home-reared vs. institutionalized subjects, for example), they must choose between observational conditions. The experiences of primate researchers may aid researchers who work on human development in evaluating different methods of data collection and selecting those most likely to reflect normal behavior.

Studies of nonhuman primate development under a wide variety of conditions have shown that, ultimately, understanding of development is dependent on seeing it in the environment to which it is adapted (the "environment of evolutionary adaptation"; Bowlby, 1969). The last 3 decades of work on nonhuman primates suggest that data on human development derived from experimental studies should be compared, whenever possible, with observations of children under natural conditions. Main and Weston (1982), for example, found it useful to compare data derived from observations in home and day care settings with those obtained under experimental laboratory settings in their analysis of avoidance behavior in children.

EVOLUTIONARY THEORY

One of the most important contributions of the ethological approach was the explicit recognition that behavior is a part of the biology of a species and as such has been subject to the same processes of natural selection that have shaped all other aspects of the species' biology. The fact that studies of nonhuman primate development are based on evolutionary theory allows researchers to use these investigations as models for considering the evolutionary basis of patterns of human development.

For primatologists, this background of evolutionary theory has had a major influence on the approaches taken to the question of the development of behavior. A recent example of this is work based on Trivers' (1974) theory of conflicts of interest between parents and their offspring (Altmann, 1980; Nicolson, 1982; Whitten, 1982). Evolutionary theory appears to have had little impact among researchers studying human development, with a few exceptions, such as the work on the role of attachment in development and Kagan's work on the development of personality (e.g., Kagan, 1984; Kagan et al., 1988).

Although a few primatologists and zoologists have made major attempts to bring an evolutionary perspective to bear on human behavior (e.g., Hinde, 1974a; Wilson, 1975; Barash, 1977), human developmentalists seem little affected by these efforts. One reason for this may be the perception that although selection may have operated on human behavior at one time (among early hominids), it is irrelevant now. Another impediment to interest in the evolution of behavior may be that the environment in which humans evolved is viewed as so remote and unlike that in which most humans now live that there can be very little behavioral carryover to modern people. This perception is most common among members of high-technology societies. This view, however, reveals a fundamental misunderstanding of the process of evolution. Human behavior, like all other aspects of human biology, is continuously modified by selection. Furthermore, most humans alive today live in environments that are more similar to the human "environment of evolutionary adaptation" than to the environment of Western technology-based societies.

To say that behavior evolves is not, however, to say that every behavior we see in humans or any other species is adaptive. It is easy and entertaining to make up a story about how any behavior might benefit an individual and so have come to be selected for. Being able to imagine how a behavior might be adaptive is not the same as *demonstrating* that the behavior is adaptive. Some "neutral" behaviors are able to survive, at least for a while, because although they do not contribute to the individual's fitness, neither do they not detract from it. They may be considered a byproduct of selection for some other trait, and they may persist as long as they are not too costly to the organism. Other behaviors may be harmful and may be in the process of being selected against. Selection, even against harmful traits, is not necessarily instantaneous. Harmful behaviors might also be retained as part of a polymorphism in which some aspects of the behavior *are* adaptive, in the same way sickle cell anemia is maintained despite its lethal effects. Behaviors or processes, like development patterns, are more likely to have been selected for if they are phylogenetically old or if they occur in a number of closely related species, since over long periods of time these traits would probably have disappeared if they had been maladapative. They might still be neutral but, the longer they persist, the less likely this is to be true.

Cross-species comparisons, particularly among animals that are relatively closely related, can provide important insights into both the "how" and the "why" of development in humans. Interspecific comparisons among primates can indicate how patterns of behavioral development may have evolved. Human developmentalists who reject such comparisons as not useful usually do so on the grounds that, whereas simpler behaviors such as sucking, grasping, or fear of falling may be readily comparable across species, human thought processes, language, and social organization are so much more complex that there can be no useful comparison with developmental processes in nonhuman species.

A major focus of theories of human development is an attempt to understand the processes by which complex thought and language emerge, and biologists are often puzzled by the apparent lack of interest in what evolutionary theory has to say about these aspects of human development. To zoologists and primatologists, it is clear that human cognitive abilities must have evolved through the normal process of natural selection, just as did the opposable thumb and bipedal locomotion.

The fact that humans have unique adaptations not shared by their closest primate relatives does not negate the usefulness of interspecific comparisons. Even closely related species are unique in some biological aspects, or they would not be separate species. It is often the *differences* that emerge from interspecific comparisons (and an understanding of the reasons for the differences) that are the most instructive aspects of such comparisons (Ambrose, 1976). For example, the ways in which infant attachment behavior and response to separation from mothers differ in pigtailed *(Macaca nemestrina)* and bonnet *(Macaca radiata)* macaques, and differences between these species and rhesus macaques *(Macaca mulatta),* have shown that species-specific patterns of behavioral development could be closely related to the social organization of a species (Kaufman and Rosenblum, 1967a, 1969; Kaufman, 1974).

In fact, most of the comparative work on human and nonhuman primates has focused on infants and such preverbal behaviors as attachment. Part of the reason for the greater acceptance of comparative work on infancy may arise from the view that in the period one is dealing with more "primitive" systems, which are more appropriate for comparison with nonhuman primates. Most of the first 2 years of human development takes place in the absence of language and complex thought, and it is during this period that many of the precursors of behavioral systems present in the adult's repertoire are established.

There are a number of hidden assumptions in theories of human development that interfere with progress that might come from comparative studies. One is the view that human behavior (at least in its adult form) is more "advanced" or more highly evolved than that of other animals. This view has its roots in the concept of a scala natura and in the notion of evolution as progress, and is often expressed in the metaphor of the evolutionary tree with humans at its apex. This leads to such misconceptions as viewing the chimpanzee *(Pan troglodytes,* probably our closest living primate relative) as a more primitive, less evolved form of human, rather than as a species highly adapted to its own way of life. The same insistence on viewing evolution as progress pervaded nineteenth century and later comparisons of Europeans with hunter-gatherers or nomadic pastoralists, or indeed any other human groups who did not share the particular adaptations of northwestern Europeans. Hunter-gatherers were seen as reflecting an earlier form of our species' evolution rather than as having behavior adapted to different ecological conditions. This view has finally begun to fade as we have come to understand the sophistication of lifestyles such as those of hunter-gathers in allowing populations to establish long-term equilibrium with their environments and resources (e.g., Konner, 1972).

RELATIVE INFLUENCE OF GENETIC AND ENVIRONMENTAL FACTORS

A frequent consequence of the view that any trait humans share with other species is "primitive," whereas all uniquely human traits are "advanced," is that traits that we share with other animals are viewed as likely to be under genetic control, whereas our unique traits are seen as subject to limited genetic influence at most. There is special resistance to seeing complex thought as being genetically determined. This resistance is more often based on political beliefs than on scientific paradigms. The controversy raised by the views of psychologist A.R. Jensen (1969) on race and intelligence and the outrage directed at attempts to extend sociobiological theory to human behavior (e.g., Alper et al., 1976; Masters, 1976; Sociobiology Study Group, 1976; Wilson 1976) illustrate the difficulties involved in approaching this issue. The confusion of the political with the scientific in this area seriously hampers any attempt to sort out the contribution of genetics and environmental factors to the process of behavioral development.

Two generations of ethologists have produced a substantial body of work and many careful theoretical discussions of how genetics and environmental factors contribute to the development of behavior (for reviews of this topic, see Lehrman, 1970; Ambrose, 1976). These studies make it clear that all behavior has genetic and environmental components, and that these factors can interact at any stage of the ontogeny of an organism to produce a particular behavior pattern.

Studies of development, which are primarily concerned with how and why behaviors emerge and come to have their adult forms, must be particularly concerned with the relative roles of genetic and environmental factors. Comparative studies have a major role in elucidating the contributions of these factors. If closely related species that inhabit very different ecological or social environments nevertheless show similar patterns of behavioral development, there is a strong likelihood that the environment is not a major determinant of the pattern. If a behavior pattern differs in a number of closely related species, relating the differences to particular features of the environment of each species may provide a better understanding of how developmental processes respond to environmental differences.

Even showing that behavior responds to environmental differences, however, does not mean that the response is *not* genetically determined. An instructive example of how a complex behavior pattern may be genetically determined in one sex and learned in the other sex of the same species is provided by hamadryas *(Papio hamadryas)* and anubis *(Papio anubis)* baboons. Hamadryas social structure consists of a troop composed of a number of one-male, multiple-female units. Males closely herd the females in their own units and any female that fails to follow her male is bitten on the back of the neck (Kummer, 1968). The closely related anubis baboon, with a range that overlaps that of the hamadryas, has a very different social organization, consisting of multimale groups in which individual males do not herd females. When occasional interbreeding occurs between anubis and hamadryas in the zone where the species overlap, all hybrid males show partial, ineffective herding, whether they are raised in an anubis group or a hamadryas group (Nagel, 1973). This indicates that in males this behavior pattern is genetically determined. All females, however, whether anubis, hamadryas, or hybrids, shows the appropriate response to male behavior in whatever social situation they find themselves, indicating that their response is learned.

STAGES OF DEVELOPMENT

The idea that we can clearly identify defined stages of behavioral development has considerable theoretical importance in the field of human development (e.g., Freud's stages of psychosexual development and Piaget's stages of cognitive development). An individual moves to a recognizably new stage when "a new or transformed system becomes dominant and functionally subordinates or incorporates previously existing systems" (Langer, 1969:87). This view suggests that stages represent qualitative changes in behavior; a notion that contrasts with viewing development as the gradual accumulation of change, as suggested by Tanner (1970) for physical development. Viewing development as a series of stages, however, leads to some questions that have proved difficult to answer:

—Is there a sequence of progression through development stages that is absolutely characteristic of a particular species?
—Is it necessary for an individual to go through each stage in succession (without skipping stages, for example)?
—If there are obligate stages, do all individuals of a species (or a population, at least) go through each stage of development at roughly the same age?

As Ambrose (1976) points out, one problem with the concept of stages is that, if an organism is well-adapted to one stage, what pushes it to move to another stage? The possibilities include a genetic program for the emergence of new behavioral systems and the effect of accumulated experience at some point bringing the organism to a new state of equilibrium.

What can the nonhuman primate data on development contribute to the understanding of these questions? The very long period over which development takes place in humans makes it difficult for any one investigator to observe particular individuals progress through all stages of development. Thus studies of humans tend to favor a cross-sectional (rather than a longitudinal) approach. This carries with it the risk that the inability to control the many environmental variables acting over such long periods of time may obscure much of the development process. Nonhuman primates mature much more rapidly, and a researcher can reasonably expect to observe a single set of individuals passing through all stages of development. Shorter periods of development also make it easier to control for environmental variables that may affect development.

If stages of development could be identified reliably in nonhuman primates, this might lend support to the usefulness of the concept in human development, especially if the same stages were widespread throughout the primate order. In addition, identifying stages in nonhuman primates might help us develop better criteria for defining stages of development in humans.

NONHUMAN PRIMATE DEVELOPMENTAL STUDIES

The following sections review studies of nonhuman primate development conducted over the last 4 decades in order to identify questions that primatologists have addressed and to discover how these questions (or areas of interest) can be related to areas of interest to human developmentalists.

Sensorimotor Development

A major focus of primate developmental studies has been the description of "normative" development of sensorimotor behaviors. Such studies provide data on the age at which sensorimotor behavior patterns appear (typically, first occurrence and mean age of occurrence) for a particular species. These studies have concentrated on the appearance and development of behaviors such as the following: clinging to the mother's ventrum unsupported, releasing grip for tactile exploration (e.g., touching another animal or object or stroking the mother's fur), coordinated reaching (looking at and then reaching for an object), eye–hand–mouth coordination (looking at an object, reaching out to grasp it and bringing it to the mouth), eating solid food (obtaining a food item and successfully eating it), independent locomotion (steady walking and competent climbing), mounting another animal, and giving the play face or other marker facial expressions or other species-typical displays or behaviors.

These behaviors are of interest to primatologists because they are crucial to the ability to survive the immediate demands of infancy or because they form the basis of behaviors important later in the life cycle. For example, eye–hand–mouth coordination is of particular interest as a necessary precursor to independent feeding, since most nonhuman primates are not fed solid food by their mothers.

The development of independent walking or climbing is critical; no monkey or ape species has permanent home bases where infants are left during the day while mothers forage and move with their groups. Infants must be carried along on the day range until they are able to keep up with the group on their own. In some species, this means moving several kilometers per day. It may take from several months to several years for infants to become fully independent of mothers for transport. A typical sight is mothers carrying older infants at the end of the day when infants are tired, even though these same infants might move independently during the rest of the day (Altmann, 1980; Chism, personal observations).

The appearance and development of some motor behaviors are particularly important because of their use in social behavior patterns. For example, mounting is an important component of sociosexual behavior in many non-human primates (e.g., Hanby, 1976). The play face is another social signal that occurs early in development. Interest in the play face in nonhuman primate studies is comparable to the interest shown in smiling and its appearance and context in studies of human development. As with smiling, the play face may be given initially by infants outside the appropriate context. In patas infants, the average age for the first appearance of the play face is about 7 weeks, although social play hardly occurs at this age (Chism, 1986; see also Hooff, 1967; Loizos, 1967, Redican, 1975).

Quantitative and descriptive data of this kind are available for many species of nonhuman primates and are derived from studies of captive and wild groups (e.g., captive rhesus macaques: Hinde et al., 1964; Hinde and Spencer-Booth, 1967; wild and captive baboons [*Papio cynocephalus*]: DeVore, 1963; Rowell et al., 1968; Ransom and Rowell, 1972; captive vervets *(Cercopithecus aethiops)*, Sykes' monkeys *(Cercopithecus mitis albogularis)*, De Brazza's monkeys *(Cercopithecus neglectus)*, and gray-cheeked mangabeys *(Cercocebus albigena)*: Chalmers, 1972; wild Hanuman langurs *(Presbytis entellus)*: Jay, 1963; Sugiyama, 1965; wild and captive patas monkeys: Chism, 1986; Chism and Rowell, N.D.; wild and captive talapoin monkeys *(Miopithecus talapoin)*: Hill, 1966; Gautier-Hion, 1971; Chism, 1980; wild chimpanzees: Lawick-Goodall, 1967; Sugiyama, 1972).

Interest in this area among primatologists centers on comparing the rates of sensorimotor development in different species in order to detect relationships between rate of development of these behaviors and other features of the species' ecological or social environment. It was initially assumed that the rate of sensorimotor development was primarily phylogenetically determined, i.e., that closeness of relationship would be the best predictor of species' rates of development (Dunbar, 1988). In a broad sense, this is true. Prosimians do develop more rapidly than most monkeys, which in turn develop more rapidly than the great apes. Nevertheless, when overall rates of sensorimotor development are compared among more closely related groups of species, some interesting differences emerge. For example, among Old World monkeys, talapoins, vervets, and Sykes', De Brazza's and patas monkeys all belong to the supergenus *(Cercopithecus)* (Gautier-Hion et al., 1988), and are all more closely related to each other than any of them is to the macaques, baboons, and mangabeys, which belong to another group of Old World monkeys. When the developmental rates of *(Cercopithecus)* species were compared with those of rhesus macaques, savannah baboons, and gray-cheeked mangabeys, it was found that closeness of evolutionary relationships was not the main factor determining similarity of development of all behaviors (Chism, 1980). Regardless of classification and phylogenetic relations, competent climbing appeared earlier in infants of arboreal species than in those of more terrestrial species (Hinde, 1971; Chalmers, 1972) suggesting that at least some differences in developmental rates are related most closely to the ecological adaptations of particular species.

On the other hand, talapoin monkey infants are precocious in their motor development, unlike their *Cercopithecus* relatives (Chism, 1980). Their pattern is more like that of the more distantly related baboons and rhesus macaques. This result does not correlate with any obvious ecological variables; talapoins are highly arboreal residents of swamp forests, whereas baboons are a highly terrestrial, open-country species. Rhesus are found in a variety of habitats that are generally dry and wooded, in which they tend to move and forage on or near the ground (Lindburg, 1977).

Such complex and subtle differences suggest that the particular pattern of sensorimotor development shown by a species is the result of a whole series of selection pressures, including the need to move competently in a difficult habitat (such as in the crowns of tall trees), the need to learn how to obtain food, and the need to evade predators.

Development of Independence

The process by which a completely dependent infant gradually becomes independent in feeding, locomotion, and social interaction has been another major interest among those researching primate development. The major theoretical focus here has been the relative roles of mothers and infants in the development of independence. The most carefully studied parts of this process involve the development of the infant's ability to transport itself (rather than being carried) and the young primate's integration into the social network of its group.

The development of independent feeding involves two distinct processes. One of these processes is the gradual decrease and eventual cessation of nursing by the infant (commonly referred to as weaning), and the second is learning to recognize, locate, and obtain solid food (independent foraging). The first process has been carefully documented in many primate species under captive and wild conditions, but the second has received little attention until recently (see, e.g., Boinski and Fragaszy, 1989).

Virtually all infant monkeys and apes show a high rate of nursing in the first few months of life (this period of strong infant dependency is longer in the great apes than in monkeys), followed by a gradual decline to a lower level and then maintenance of this low plateau for some additional months (see, e.g., Hinde et al., 1964, for captive rhesus; Berman, 1980, for free-ranging rhesus; Hiraiwa, 1981, for wild Japanese macaques *(Macaca fuscata):* Nash, 1978; Altmann, 1980; Nicolson, 1982, for wild baboons *(Papio anubis);* Struhsaker, 1971; Whitten, 1982, for wild vervets; Chism, 1986; Chism and Rowell, N.D., for captive and wild patas monkeys). In many of these species mothers continue to nurse infants at a low level until the next infant is born. Thus for most primate infants, weaning is a long, gradual process, often completed only at the birth of the following sibling.

Theoretical discussions of weaning, however, especially those based on parent–offspring conflict (Trivers, 1974), have tended to view this process very differently. The relatively brief period during which nursing scores decline to a low level is treated as being synonymous with "weaning," and little attention is paid to the fact that the infant keeps on nursing for months or years after this. In species in which the decline in nursing is accompanied by a peak in maternal rejection of nursing attempts, the responsibility for the decline in nursing is assumed to be mostly the mother's; she is viewed as forcing the infant to shift to independent feeding by rejecting its attempts to nurse.

In fact, observations of infant monkeys, especially in wild groups, indicate that infants begin to feed themselves at a very early age without any prodding from their mothers and at a time when maternal rejection is infrequent (Altmann, 1980; Nicolson, 1982; Whitten, 1982; Chism, 1986). Then, as their motor skills improve, infants increase both their feeding time and the range of foods they eat. At some point in the process infants acquire both motor skills and knowledge of the location of resources sufficient to feed themselves completely independently.

One way to assess when the point of full independence is reached is to observe whether an orphaned infant of a particular age is able to survive in the wild without begin nursed by a foster mother. (An infant capable of foraging independently that is orphaned but not adopted may still not survive for other reasons such as intragroup agression or predation, however.) Patas monkey infants can survive the death of their mothers without adoption after about 7 months of age, even though they would normally continue to nurse until age 12 months. For a species such as patas with a highly seasonal pattern of reproduction, survival at 7 months is probably reasonably predictable. Infants are almost always born at the same time of the year and so always reach this age at a time when food is most abudant (Chism et al., 1984). For infants of species such as baboons, which are not so strictly seasonal in their reproduction, the age at which an infant reaches feeding self-sufficiency may be more variable. Altmann (1980) has pointed out the importance of the availability of "weaning foods," which are easy for infants to obtain and process, at the time when infants are decreasing their dependence on milk. If an infant baboon is orphaned at a time of year when these foods are not available, it may not survive even if it has the skills and knowledge to feed itself.

In wild patas groups, maternal rejection, although never frequent, peaks when infants are about 4 months old, at least 3 months before infants are known to be capable of self-sufficiency (Chism et al., 1984). Similar observations have been made on wild baboons (Altmann, 1980) and free-ranging rhesus macaques (Berman, 1980). What, then, is the reason for this apparent push from mothers at a time when infants are almost certainly not capable of wholly independent feeding? Altmann (1980) has suggested that increased maternal rejections at this point function to shift the infant's *pattern* of nursing so that it occurs during rest periods or at night rather than while the mother is foraging or socializing. The point then of the rejection peak is to alter the timing of nursing rather than to limit it or prohibit it altogether.

This suggestion fits well with some observations indicating that, by the time infants of some monkey species are about 4–6 months old, the increased energetic costs of feeding and transporting them are approaching the limits of what mothers can provide (Dunbar, 1988; Fig. 4–1). If infants can be induced, via rejection, not to interrupt their mothers' foraging with demands to nurse,

Fig. 4–1 Female patas monkey *(Erythrocebus patas)* forages in savannah woodland habitat in Kenya with her young infant on her back. (Photo: Dana Olson.)

mothers may be able to forage more efficiently and thus continue to provide infants with the nutritional cushion they need while developing the skills necessary to become independent foragers.

The situation described above regarding maternal rejection and the shift to independent feeding may represent the kind of compromise between an infant's demand for maximal maternal investment and a mother's need to maintain her own fitness for future reproductive efforts that would be predicted by Trivers' (1974) model of parent–offspring conflict. Observations of infant monkeys strongly suggest that mothers do not have to force independence on infants through rejection of contact or nursing. From the earliest weeks of life, infants are struggling to climb off their mothers, to move out of reach, and to move out of sight to play with other animals. Data from captive and field studies of several species indicate that, in the early weeks, infants leave their mothers and mothers maintain or reestablish proximity by restraining, retrieving, or approaching infants. Over the next few months, infants assume a greater and greater role in regulating proximity as they increasingly become responsible for returning to their mothers or staying near them (Hinde and Atkinson, 1970; Altmann, 1980; Berman, 1980; Chism, 1986; Dunbar, 1988).

In most cage situations, it is virtually impossible for the infant to get out of sight of the mother, so that the critical importance of an efficiently functioning proximity-maintenance mechanism may not be obvious. Under field conditions the risk of a poorly functioning proximity-regulation system becomes immediately obvious (as it is to any human who has tried to shop in a crowded store with an unrestrained toddler). Thus, whereas mothers have clearly been selected to be vigilant about their infants' whereabouts, especially in the early weeks when infants have limited locomotor abilities, the burden seems to shift to infants after that, freeing mothers to concentrate more on foraging and less on keeping track of their babies. In a wild patas group, when a predator approaches the group, mothers stand up and look around for their infants, but it is the infants that approach their mothers. If an infant does not return to its mother rapidly, the mother flees with the group, returning later to retrieve the infant after the predator has left the area (Chism, personal observations).

In nonhuman primates, independent social relations develop over several years. In most species of monkeys and apes, other group members, particularly adult and immature females, show interest in and give attention to young infants. Infants reciprocate this interest by attempting to approach other animals as soon as they can crawl off their mothers. In some species, certain macaques and baboons, for example, mothers restrict access to their infants for at least the first few weeks of life. In other species (e.g., Hanuman langurs, vervet monkeys, and patas), infants are frequently taken, carried off, and handled by other group members in the first few days of life (for reviews of primate caretaking patterns, see Hrdy, 1976; McKenna, 1979; Quiatt, 1979).

For most species, the mother's social relations are primary in determining the rate at which her infant develops independent relations with other group members and the form these relations take (Fig. 4–2). Infants of species with extensive early allomaternal care (i.e., caretaking by group members other than the mother) begin to interact with others independently at an early age, since their mothers are not always nearby. In species in which mothers are more restrictive of contact, it may be several months before infants have much independent contact with others.

For most of the first year, monkey mothers actively intervene in encounters involving their infants, threatening other animals that are aggressive to their infants and even chasing off lower ranking animals that are attempting friendly interactions (Fig. 4–3). As infants spend increasing amounts of time out of their mothers' immediate proximity, they must assume more and more responsibility for their own social interactions. Infants and juveniles should not be perceived as interacting randomly with other group members, however. Their interaction patterns are to a greater or lesser extent determined by their family and rank (e.g., Berman, 1982). In many species the rank of females is determined primarily by that of their mothers (Sade, 1967; Missakian, 1972; Che-

Fig. 4–2 Adult female patas monkey with her infant sits in close proximity to another female. Patas females form close relationships with one another, and these associations are passed on to offspring. (Photo: Dana Olson.)

ney, 1977; Dunbar, 1980; Horrocks and Hunte, 1983). In species with allomaternal caretaking, infants have an opportunity to develop relations with a number of other group members, which may affect later social interactions (see, e.g., Chism, 1986).

Development of Attachment

Studies of nonhuman primates (as well as developmental studies of other animals) have made an important contribution to the understanding of attachment behavior in humans. Bowlby (1969, 1973) based many of his ideas about the function of attachment behavior in humans on the work of ethologists on imprinting behavior in birds and mammals and on the work of Harlow and his colleagues (e.g., Harlow, 1961; Harlow et al., 1963) on the development of attachment in infant monkeys. Harlow studied infant rhesus monkeys raised on artifical mothers consisting of wire figures with or without soft covers or attachments for feeding bottles. Harlow then tested infants in mildly fear-provoking situations analogous to those that evoke attachment responses in human children to see if the presence of the artifical mother alleviated the infant's distress. Harlow's work indicated that an infant's attachment to a

Fig. 4–3 Adult female patas monkey threatens a groupmate (out of view) away from her newly mobile infant. At the same time, she tolerates the close proximity of another adult female with a young infant. (Photo: Dana Olson.)

mother figure was based mainly on the contact comfort it provided. He further showed that proximity to the mother-figure alleviated the infant's distress in fear-provoking situations. Demonstrating that contact comfort was the basis for attachment in infancy cast serious doubt on Freud's theory that the bond between mother and infant was a result of a secondary drive, (the idea that as the mother provides the infant with food and warmth, satisfying its physiological needs, she becomes associated with satisfaction in the infant's mind) (Bowlby, 1969; Maccoby and Masters, 1970).

Bowlby theorized that attachment in infancy has the primary function of keeping the vulnerable infant in close proximity to its mother. He emphasized that the development of attachment behavior could be understood only in the context of humans' "environment of evolutionary adaptation" (Bowlby, 1969). An important feature of this environment was that it contained predators capable of taking young humans who strayed too far from the protection of their mothers or other caretakers. Thus, in Bowlby's view, attachment behavior in humans had been selected far over at least the last several million years. Since virtually all primates, except for a few species of prosimians that cache their young, exhibit very similar patterns of attachment behavior, it is likely that this is a very old pattern, on the order of 20–30 million years at least.

Bowlby's theory that attachment behavior was related to proximity mainte-nance, rather than resulting from gratification of primary physiological needs or reinforcement via feeding, contradicted both Freudian theory and behavior-ist interpretations of dependency (Maccoby and Masters, 1970) and so was not readily accepted into human child development theory. Part of the difficulty lay in experimentally testing the basis for attachment behavior in humans. For ethical reasons, some experiments simply cannot be performed on humans, and it was the extensive experimental work on attachment and separation in nonhuman primates that provided the most compelling evidence that attach-ment to the mother was based on need for proximity rather than reinforcement from feeding. In addition, studies of nonhuman primates indicated to Bowlby that the common fears of childhood (of the dark, of being left alone, of strang-ers, etc.) were not necessarily either irrational or due to some suppressed exper-ience but were highly adaptive when viewed in their original evolutionary con-text. In fact, it is still maladaptive for a small child to wander out of its parent's proximity, especially at night, and the problem of how to balance a child's war-iness of strangers with the need to accept caretaking from a number of initially strange adults is one of considerable discussion regarding the effects of wide-spread use of day care.

Studies of attachment and separation response in monkeys highlight the way in which uncovering species differences in behavioral developmental patterns can lead to a better understanding of the mechanisms of development. Studies of mother–infant separation in rhesus macaques by Harlow and his colleagues (Seay et al., 1962; Seay and Harlow, 1965) and by Hinde et al. (1966) showed that rhesus infants responded to separation in ways that closely resembled the responses of humans (Spitz, 1946). Furthermore, these studies suggested that the effects of separation in monkeys could be long-lasting and that there were individual differences in both the strength of the initial response and the dura-tion of long-term effects (Hinde, 1974a, 1986).

Studies of two other closely related species of monkeys, bonnet macaques and pigtailed macaques, demonstrated that species-specific interaction patterns of adults also influenced infants' responses to separation from their mothers (Kaufman and Rosenblum, 1967a,b). Bonnet macaques have a social organi-zation characterized by close physical contact among adults and immatures, and bonnet mothers allow a large amount of physical contact between their infants and other group members. Pigtailed macaques, on the other hand, do not spend much time in close contact, and mothers are more restrictive of con-tact with their infants than are bonnets. Details of interaction between mothers and infants also differ between the two species; pigtailed mothers are more restrictive of their infants in general, more often reject contact, and are more punitive during weaning than are bonnet mothers.

All these differences in species interaction patterns are reflected in the responses of infants of these two species to separation from their mothers. Whereas infants of both species show initial agitation to separation, pigtailed infants, but not bonnet infants, go into a state of depression. Bonnet infants seek and are given contact comfort and are even nursed by other group mem-bers, whereas pigtailed infants are not. Bonnet infants' behavior quickly returns to normal whereas pigtailed infants go through a period of several days of deep depression before their activity levels begin to return to normal.

Kaufman (1974) argued that the responses of infant monkeys to separation are adaptive in the wild. The initial agitated activity and vocalization will serve to reunite the infant with its mother if she is nearby. If the mother does not return within a few hours, the infant would quickly exhaust itself if it kept up this level of activity. Thus, Kaufman suggested, depression may serve to conserve energy while the infant adjusts to its loss. Bonnet infants, who can elicit caretaking, feeding, and carrying from other adults, do not need to conserve energy and so do not go into a state of depression.

Why do infants of any of the species go into a prolonged period of depression? Is it not more likely that infants would have been selected to recover quickly from loss of the mother after the period of initial agitation, since an infant in a depressed state for up to a week is in great danger of succumbing to predators or losing contact with its group in the wild. A further refinement of these separation studies (Hinde, 1974b) using rhesus monkeys provides a possible answer. In these experiments, one set of infants stayed in the home cages while their mothers were removed and later returned. The other set of infants was removed from their home cages while the mothers remained. Surprisingly, rhesus infants that remained in their home cages were more distressed than those who were removed. The "mother-removed" infants changed more quickly from agitation to depression and had more difficulty in reestablishing their relationships with their mothers on reunion. This contrasts with human children, who respond with more distress to separation coupled with removal from their familiar environment (e.g., hospitalization) than to separation while remaining in the familiar environment (Bowlby, 1973). Hinde explained the difficulty in reestablishment of the mother–infant relationship as being due to the fact that, when rhesus mothers were returned to the groups, they had to reestablish social relations with all the other gorup members as well as their own infants, a difficult and stressful task for most monkeys. Hinde was puzzled, however, over the more rapid movement into depression shown by the "mother-removed" group (Hinde, 1986).

An understanding of the rhesus infants' response to the removal of their mothers, as well as a better understanding of the general phenomenon of depression in primate infants separated from their mothers can be gained from field studies of mother–infant relations in monkeys. In the wild, an infant will experience separation from its mother only under two conditions. The first is if the infant somehow becomes separated from its mother *and* her group, a condition corresponding to the "infant-removed" situation of Hinde's study. Under such conditions, the initial agitated searching and vocalizing seen in separated infants is clearly adaptive. If the mother or the group is nearby, these are the behaviors most likely to reunite the infant. If the infant is not reunited with its mother (or group) fairly quickly, such agitation should cease; a continuation would serve only to exhaust the infant and attract predators. At that point an infant would be most likely to survive for a few days (and possibly regain contact with its mother) if it reduced its activity somewhat and remained quiet and inconspicuous. If it becomes so depressed that it does not eat or drink, it is unlikely to survive long enough to be reunited in any case (Rhine et al., 1980; Hamilton et al., 1982).

The second circumstance under which an infant may be separated from its mother occurs when the mother is no longer in the group. Since it is extremely unlikely that a mother would emigrate from her group leaving a dependent infant behind, this would probably occur only if the mother died. An infant that has not yet achieved independent foraging must be adopted almost at once or it will not survive its mother's death, yet adoption has rarely been reported in wild primate groups (Quiatt, 1979; Nicolson, 1986). Thus depression may not have been selected for and probably does not promote survival in and of itself. It may simply be the best an infant can manage in the absence of adoption. Hinde's rhesus infants that were left in the group while their mothers were removed were experiencing conditions that normally would only occur if their mothers died. Their more rapid movement into depression is thus understandable.

The area of attachment theory and research is thus exemplary for showing the usefulness and importance of cross-species comparisons in development: Observation of a particular behavior in humans (anaclitic depression) led to a series of studies in nonhuman primates, and these studies helped to stimulate a new approach to the question of mother–infant attachment in humans. Attempts to refine the initial studies on monkeys to make them more comparable to the conditions experienced by human children led to discoveries of species and individual differences, some of which were explainable from the results of other captive studies and some of which were not. Finally, field work on nonhuman primates was able to provide additional information about the normal context of the behavior that helped to explain some of the puzzling results of the captive studies.

Ontogeny of Social Behavior

One of the striking characteristics of most monkey and ape species is their sociality. As Rowell (1975) points out, it is not necessary to make a young monkey social; infants energetically seek out social interaction with other group members as soon as they are capable of moving away from their mothers. The development of social behavior can be viewed as the gradual incorporation of behavior patterns characteristic of adults into the infant's behavioral repertoire.

One of the striking characteristics of higher primates is the length of the period of immaturity. This means that primates spend a relatively long time (from 2 years in some rapidly developing monkeys to 10 years or more in the great apes) acquiring adult behavior patterns. It has been suggested that long periods of immaturity allow for more learned behavior and that a greater reliance on learned behavior allows for a more flexible response to a changing environment (Baldwin and Baldwin, 1979) Although this may well be the case for some nonhuman primates (e.g., savannah monkeys such as baboons, patas, and vervets, which live in seasonal or unpredictable environments), many monkey species inhabit tropical forests, which are often characterized as being among the most stable of environments. Alternatively, a long period of immaturity might be necessary for learning more complex behavior, but there is no a priori reason why an animal has to learn everything it will ever need to know during the period of immaturity. Mason (1979) points out that some studies of learning have shown that learning efficiency may continue to increase in adulthood. In fact, another characteristic of primates (especially humans) is that they reach reproductive maturity and reproduce successfully long before they show fully adult patterns of social behavior.

Mason (1979) also suggests that the real advantage of the lengthy period of immaturity in primates has to do with the openness of behavioral systems to change during this period; it is easier to incorporate new information and to change an open system than a closed one. In primates, openness increases as animals move from infancy to the juvenile period, then decreases (but does not completely disappear even in adulthood; see below) as sexual and social maturity are reached.

There are excellent reasons why infants' behavioral systems should be less open than those of juveniles. Because of their small size, less developed motor abilities, and inexperience, infants are more vulnerable to predators, to becoming separated from their groups, to eating food items that are poisonous, and to intragroup aggression. Their behavior may respond less readily to changes in the environment than that of juveniles or adults. Comparisons of development patterns in captive and wild patas monkeys suggest that it is mainly through changes in maternal behavior that the mother–infant pair adjusts its responses to differences in the enviornment (Chism and Rowell, N.D.)

To eat or not to eat

In primates, the juvenile period appears to be the least conservative in terms of responsiveness to new stimuli, and during this period behavioral systems appear most open to change. The classic example of juvenile innovation is that of the juvenile female Japanese macaque that invented two new methods of processing food items, sweet potato washing and wheat-washing (throwing wheat mixed with sand into the sea to separate the two), in response to experimentally induced changes in the environment (Kawai, 1965).

What benefit does the curiosity and innovativeness of the juvenile confer on either the individual itself or its close relatives? In a highly stable, predictable environment where successive generations of animals experience the same living conditions, willingness to innovate and explore may entail more risk than benefit to the young animal. In environments that are unpredictable or are experiencing rapid change, however, the benefits of discovering new food sources or ways to process known food items, for example, might be great enough to overcome the risk of eating an unpalatable item. Similar benefits might accrue to close relatives of the juvenile innovator. The food innovations of the young Japanese macaque described above were passed throughout the social group as mothers observed and learned the new behaviors from their offspring, as juveniles learned from other juveniles, and as infants learned from their mothers. The last to pick up the new behavior were the adult males, possibly because they had the least contact with the juveniles (Kawai, 1965).

The above example indicates how adaptive it can be for a species to retain at least some openness in its behavioral systems, even in adulthood. Still, there are excellent reasons for behavioral conservatism, especially with regard to food preferences. Over the long periods during which most species have occupied a particular habitat, it is likely that they would have discovered most of the edible food items in the area. New items are more likely to be dangerous than edible, on average. The risks of trying new items can clearly be seen in the way young primate infants learn which food items are to be eaten. Infants of many species begin eating solid food by picking up bits of food dropped by their mothers or by trying to take food from their mothers' hand or mouths (Rowell, 1975). Infants and juveniles also carefully sniff the mouths of their mothers and other group members while they are eating (Gartlan, 1969; Fig. 4–4). Thus, learning to feed oneself is very much a social activity. It is doubtful that a young monkey or ape, left to its own devices, would be able to survive long enough to determine what food items were safe to eat and to acquire a balanced diet. When a group of patas monkeys housed in an outdoor field cage at the University of California, Berkeley, encountered poison hemlock for the first time in their enclosure, about half the group tried it, with fatal results for some of these less conservative animals. Observations of captive or hand-raised monkeys and

Fig. 4–4 Juvenile patas monkey sniffs the mouth of an adult female eating sunflower seeds from a roadside plant in Kenya. This may have been the first time the juvenile had encountered this food item. (Photo: Dana Olson.)

apes that have been released back into the wild indicate that foraging and feeding is an activity that must be learned from experienced social companions.

Play

Akin to the exploratory and innovative behaviors of immature primates is their propensity to play. Play is an important component of the behavioral repertoire of young monkeys and apes, as it is for human children. Extensive amounts of play appear to be correlated with long periods of immaturity (Fagen, 1978). In most nonhuman primates, the earliest form of play to appear is solitary or exploratory play (Wolfheim, 1977; Chism, 1986) in which the infant mouths, handles, or manipulates objects or engages in solitary motor or locomotor activity. The next type of play to appear is parallel play, in which infants engage in synchronous exploratory or locomotor activities with other animals. Observations of patas and talapoin infants indicate that, once social play began, parallel play was infrequent (Chism, 1980), and it may represent an early, fragmentary form of social play in nonhuman primates occurring when infants are beginning to be interested in playing with other animals but not yet having sufficient coordination or social skills to do so. Once social play began in infant talapoins and patas (in the second month), it quickly eclipsed solitary and parallel play and was at least three times as frequent as the other two types combined throughout the rest of the first year (Chism, 1980, 1986). Parallel play may be more important in humans than in nonhuman primates.

Many functions have been suggested for play, including fulfilling a need for exercise; providing practice of skills such as running, climbing, jumping and fighting that are important during daily activities (e.g., evading predators, negotiating the home range, or engaging in aggressive encounters); learning features of the home range; establishing dominance relations; and establishing bonds with other group members (for reviews, see Loizos, 1967; Bekoff, 1978; Martin and Caro, 1985). Young primates appear to be strongly motivated to engage in play. The extent to which their play activity may be costly has been debated. Although field observatons indicate that, when under nutritional stress, young monkeys may stop playing, presumably to conserve energy for foraging or keeping up with the group (Lee, 1984, for vervets; Chism, personal observations, for patas monkeys), recent work by Martin (1982) and Martin and Caro (1985) questioned whether play is energetically costly. If there are costs, either energetic or in heightened risks of injury or predation due to lack of vigilance or attracting predators (Biben et al., 1989), the opportunities play provides to learn socially useful skills in a protected setting could well outweigh them.

It has been suggested that during play young primates have an opportunity to engage in activities that are typical of aggressive interactions (grappling, chasing, hitting, etc.) and, in so doing, to test their skills against other group members in preparation for adult dominance encounters. In fact, it has been suggested that play encounters may actually help establish adult dominance relationships since animals will learn which group members they can readily outrun or intimidate and which they cannot. Although young animals may indeed practice aggressive behaviors during play, there are several observations suggesting that establishing adult dominance relations is not the main function of social play. For example, in many species female dominance relationships are rather strictly determined by mother's rank and are therefore unlikely to be altered by testing skills that occur in play fighting. Furthermore, in species in which most, if not all, males leave their natal groups before reaching adulthood and eventually take up residence in other groups, long-lasting adult male dominance relationships also are unlikely to be determined during the play of youngsters.

Aggression

Aggressive behaviors appear during infancy in most primates. Infants may begin to take an active part in aggressive encounters after about 6 months of age. In these early aggressive encounters, infants are more likely to join in with mothers, frequent caretakers, or siblings in an attack than they are to act alone. Patas infants for example, incite their relatives or caretakers to attack other group members that threaten or interfere with them and attempt to defend their mothers when the latter are under attack (Chism, 1986).

Affiliative behavior

Discussion of affiliative behavior among nonhuman primates usually concentrates on grooming relations, although other behaviors can also be considered affiliative, such as sitting with another animal or aiding another animal in infant caretaking or during agonistic encounters. Discussion of aiding behavior brings up the questions of altruism and kin selection, but both are well outside the scope of this chapter (for a recent discussion, see Dunbar, 1988).

Primate infants are groomed by their mothers, caretakers, and other group members. In most species, mothers are the most frequent groomers of infants. Although grooming of others begins during infancy (in monkeys, during the second half of the first year), infants groom rarely and briefly (Pereira and Altmann, 1985; Chism, 1986). Furthermore, to a large extent, infant grooming relations are mediated through mothers or other caretakers. It is during the juvenile period that independent grooming relations of young primates really begin to develop.

Grooming in primates serves at least two functions. One function is to rid the skin and pelage of ectoparasites, dirt, and other foreign objects. The second function of grooming is affiliative in nature. How much grooming an animal receives and from whom is intimately connected to the place of that animal in the social organization of its group or family. For most group-living monkeys and apes, establishment of adult affiliative patterns can be understood by looking at the development of grooming relations. Data from the most extensively studied species, the baboons and macaques, indicate that initially mothers are the most frequent grooming partners of their juvenile offspring, that mothers groom juvenile daughters and sons roughly equally, but that daughters groom mothers more than sons do (Missakian, 1974; Cheney, 1978; Pereira and Altmann, 1985). The same pattern has been reported for patas monkeys (Rowell and Chism, 1986).

As juveniles age, they begin to direct more of their grooming to other group members (particularly adult females) and less of it to their mothers, although for patas monkeys and baboons, at least, mothers continue to receive a disproportionate share of the grooming. This shift appears to signal the beginning of the process by which juveniles establish their own relationships with other group members, as opposed to interacting with others as social extensions of their mothers. The process of establishing independent social relations is a relative one, however, since in many primate species an individual's rank and social relationships are determined largely by those of its mother. This in more true in some species (macaques, for exmple) than in others, and more so for one sex than the other (it is most likely to be true for the nondispersing sex).

Learning maternal skills

Rudimentary forms of maternal behavior (here taken as synonymous with infant caretaking behavior) appear as early as the second half of the first year in some monkey species (vervets: Gartlan, 1969; Lancaster, 1971; patas; Chism, 1986). One-year-old patas infants hold and attempt to carry younger infants and may be left as caretakers with a crèche of younger infants while the mothers forage nearby (Chism and Rowell, N.D.). Patas yearlings have even been observed trying to assist younger infants in distress (temporarily left behind by their mothers or stuck in a tree, for example).

Juveniles of most species show strong interest in infants and act as caretakers when mothers will permit it. Female juveniles show more interest and are more frequent caretakers than males in most Old World monkey and ape species (e.g., Spencer-Booth, 1968; Lancaster, 1971; Hrdy, 1976; Rowell and Chism, 1986).

One function suggested for allomaternal caretaking systems, such as those of patas, vervets, and many colobine monkey species, is that they provide immature females with an opportunity to practice their mothering skills before giving birth to an infant of their own (Lancaster, 1971). Although it is unlikely that allomaternal behavior evolved to provide such practice (see Hrdy, 1976, for a critique of this view), some limited data suggest that species with such caretaking systems may experience lower rates of infant mortality among firstborn infants than do species in which juvenile females do not get an opportunity to practice handling newborns. Among rhesus monkeys, which do not permit handling of newborns, primiparous females have higher rates of infant mortality than do experienced mothers (Drickamer, 1974). In contrast, among wild patas monkeys (a species with extensive allomaternal care by immature females), primiparous females do not differ from multiparous females in the rate of infant mortality (Chism et al., 1984). These observations indicate, at least, that the opportunity to practice mothering skills on young infants might enhance a female's chance of having her first infant survive.

CONCLUSIONS

This review has focused on two issues in considering the contribution that studies of the ontogeny of behavior among nonhuman primates might make to our understanding of human development. First, it has examined the considerations (both methodological and theoretical) an evolutionary biological perspective might bring to the understanding of human development; second it has reviewed the kinds of questions on which researchers have concentrated in their studies on nonhuman primate development. These two aspects are closely related, since many of the approaches of primatologists to the question of behavioral ontogeny derive from an interest in the evolution of behavior and patterns of behavioral development.

Benefits to be derived from comparing the results of studies on human and nonhuman primates flow both ways. Human developmental research has been significantly influenced by the work on attachment in nonhuman primates, for example. In this and other areas of developmental research, methods and theory derived from ethology have been incorporated into studies of humans. Primatologists, however, find that their own work is stimulated by questions posed by researchers working on humans. The desire to apply what is learned from nonhuman primate studies in a meaningful way to humans is not just lip service to granting agencies for most primatologists. There is an active effort to bring the two areas closer together, as witnessed by the large number of papers reviewing this subject. From the viewpoint of biologists in general, the more students of human development incorporate evolutionary theory into their thinking about behavior, the greater will be the contribution that each area can make to the other.

ACKNOWLEDGMENTS

The author wishes to thank William Rogers for his thoughtful comments on this chapter and for easing the burdens of parent–offspring conflict during its preparation. She is grateful to Michael Rodriguez for his asistance in the preparation of the manuscript and to Roy Henrickson for his patience and tolerance of the writing process.

REFERENCES

Ainsworth, M.D.S. (1982). Attachment: retrospect and prospect. Pp. 3–30 in C. M. Parkes and J. Stevenson-Hinde, eds., *The Place of Attachment in Human Behavior.* New York: Basic Books.

Ainsworth, M.D.S., M. C. Blehary, E. Waters, and S. Wall (1978). *Patterns of Attachment: A Psychological Study of the Strange Situation.* Hillsdale, NJ: Lawrence Erlbaum Associates.

Altmann, J. (1980). *Baboon Mothers and Infants.* Cambridge, Mass.: Harvard University Press.

Alper, S., J. Beckwith, S. I. Chorover, J. Hunt, H. Inouye, T. Judd, R. V. Lange, and P. Sternberg (1976). The implications of sociobiology. *Science 192:*424–427.

Ambrose, A. (1976). Methodological and conceptual problems in the comparison of developmental findings across species. Pp. 269–317 in M. Von Cranach, ed., *Methods of Inference from Animal to Human Behavior.* Chicago: Aldine.

Baldwin, J. D. and J. I. Baldwin (1979). The phylogenetic and ontogenetic variables that shape behavior and social organization. Pp. 89–116 in I. S. Berstein and E. O. Smith, eds., *Primate Ecology and Human Origins.* New York: Garland.

Barash, D. P. (1977). *Sociobiology and Behavior.* New York. Elsevier.

Bekoff, M. (1978). Social play: structure, function, and the evolution of a cooperative social behavior. Pp. 367–383 in G. M. Burghardt and M. Bekoff, eds., *The Development of Behavior: Comparative and Evolutionary Aspects.* New York: Garland.

Berman, C. M. (1980). Mother-infant relationships among free-ranging rhesus monkeys on Cayo Santiago: a comparison with captive pairs. *Animal Behaviour 28:*860–873.

———— (1982). The ontogeny of social relationships with group companions among free-ranging infant rhesus monkeys I. Social networks and differentiation. *Animal Behaviour 30:*149–162.

Biben, M., D. Symmes, and D. Bernhards (1989). Vigilance during play in squirrel monkeys. *American Journal of Primatology 17:*41–49.

Boinski, S., and D. M. Fragaszy (1989). The ontogeny of foraging in squirrel monkeys, *Saimiri oerstedi. Animal Behaviour 37:*415–428.

Bowlby, J. (1969). *Attachment and Loss, Vol. I: Attachment.* London: The Hogarth Press.

———— (1973). *Attachment and Loss, Vol. II: Separation, Anxiety and Anger.* New York: Basic Books.

Chalmers, N. R. (1972). Comparative aspects of early infant development in some captive cercopithecines. Pp. 63–82 in F. Poirier, ed., *Primate Socialization.* New York: Random House.

Cheney, D. L. (1977). The acquisition of rank and the development of reciprocal alliances among free-ranging immature baboons. *Behavioral Ecology and Sociobiology 2:*303–318.

———— (1978). Interactions of immature male and female baboons with adult females. *Animal Behaviour 26:*389–408.

Chism, J. (1980). A comparison of development and mother-infant relations in patas and talapoin monkeys. PhD dissertation, University of California, Berkeley.

———— (1986). Development and mother–infant relations among captive patas monkeys. *International Journal of Primatology 7:*49–81.

Chism, J., and T. E. Rowell (N.D.). Mothers, infants and environments: development of patas infants in the wild and in captivity. Unpublished manuscript.

Chism, J., T. E. Rowell, and D. K. Olson (1984). Life history patterns of female patas monkeys. Pp. 175–190 in M. Small, ed., *Female Primates: Studies by Women Primatologists.* New York: Alan R. Liss, Inc.

DeVore, I. (1963). Mother-infant relations in free-ranging baboons. Pp. 305–335 in H. Rheingold, ed., *Maternal Behavior in Mammals.* New York: John Wiley & Sons.

Drickamer, L. C. (1974). A ten-year summary of reproductive data for free-ranging *Macaca mulatta. Folia Primatologica 21:*61–80.

Dunbar, R.I.M. (1980). Determinants and evolutionary consequences of dominance among female gelada baboons. *Behavioral Ecology and Sociobiology 7:*253–265.

———— (1988). *Primate Social Systems.* Ithaca, NY: Comstock/Cornell University Press.

Fagen, R. M. (1978). Evolutionary biological models of animal play behavior. Pp. 385–404 in G. M. Burghardt and M. Bekoff, eds., *The Development of Behavior: Comparative and Evolutionary Aspects.* New York: Garland.

Freud, S. (1920). *Beyond the Pleasure Principle. The Complete Psychological Works of Sigmund Freud,* Vol. 18. London: The Hogarth Press.

Gartlan, J. S. (1969). Sexual and maternal behavior of the vervet monkey. *Cercopithecus aethiops. Journal of Reproduction and Fertility, Supplement 6*:137–150.

Gautier-Hion, A. (1971). Repertoire comportemental du Talapoin *(Miopithecus talapoin). Biologica Gabonica. 7*:295–391.

Gautier-Hion, A., F. Bourliere, J. P. Gautier, and J. Kingdon (eds.) (1988). *A Primate Radiation.* London: Cambridge University Press.

Hamilton, W. J., and C. Busse, and K. S. Smith (1982). Adoption of infant orphan chacma baboons. *Animal Behaviour 30*:29–34.

Hanby, J. (1976). Sociosexual development in primates. Pp. 1–67 in P.P.G. Bateson and P. Klopfer, eds., *Perspectives in Ethology,* Vol. 2. New York: Plenum Press.

Harlow H. F. (1961). The development of affectional patterns in infant monkeys. Pp. 75–97 in B. M. Foss, ed., *Determinants of Infant Behaviour,* Vol. I. London: Methuen.

Harlow, H. F., and M. K. Harlow (1965). The affectional systems. Pp. 287–334 in A. M. Schrier, H. F. Harlow, and F. Stollnitz, eds., *Behavior of Nonhuman Primates,* Vol. II. New York: Academic Press.

Harlow, H. F., M. K. Harlow, and E. Hansen (1963). The maternal affectional system of rhesus monkeys. Pp. 254–281 in H. Rheingold, ed., *Maternal Behavior in Mammals.* New York: John Wiley & Sons.

Hill, W. C. O. (1966). Laboratory breeding, behavioural development and relations of the Talapoin *(Miopithecus talapoin). Mammalia 30*:353–370.

Hinde, R. A. (1971). Some problems in the development of social behavior. Pp. 411–432 in E. Tobach, L. R. Aronson, and E. Shaw, eds., *The Biopsychology of Development.* New York: Academic Press.

———— (1974a). Mother–infant relations in rhesus monkeys. Pp. 29–46 in N. White, ed., *Ethology and Psychiatry.* Toronto: University of Toronto Press.

———— (1974b). *Biological Bases of Human Social Behaviour.* New York: McGraw-Hill.

———— (1986). Can nonhuman primates help us understand human behavior? Pp. 413–420 in B. B. Smuts, D. L. Cheney, R. M. Seyfarth, R. W. Wrangham, and T. T. Struhsaker, eds., *Primate Societies.* Chicago: University of Chicago Press.

Hinde, R. A. and S. Atkinson (1970). Assessing the roles of social partners in maintaining mutual proximity, as exemplified in mother-infant relations in monkeys. *Animal Behaviour 18*:169–176.

Hinde, R. A., T. E. Rowell, and Y. Spencer-Booth (1964). Behaviour of socially living rhesus monkeys in their first six months. *Proceedings of the Zoological Society of London 143*:609–649.

Hinde, R. A., and Y. Spencer-Booth (1967). The behaviour of socially-living rhesus monkeys in their first two-and-a-half years. *Animal Behaviour 15*:169–196.

Hinde, R. A., Y. Spencer-Booth, and M. Bruce (1966). Effects of six-day maternal deprivation on rhesus monkey infants. *Nature 210*:1021–1023.

Hiraiwa, M. (1981). Maternal and alloparental care in a troop of free-ranging Japanese monkeys. *Primates 22*:309–329.

Hooff, J.A.R.A.M. van (1967). The facial displays of the catarrhine monkeys and apes. Pp. 9–88 in D. Morris, ed., *Primate Ethology.* New York: Doubleday.

Horrocks, J., and W. Hunte (1983). Maternal rank and offspring rank in vervet monkeys: an appraisal of the mechanisms of rank acquisition. *Animal Behaviour 31*:772–781.

Hrdy, S. (1976). Care and exploitation of nonhuman primate infants by conspecifics other than the mother. *Advances in the Study of Behavior 6*:101–158.

Jay, P. C. (1963). Mother-infant relations in langurs. Pp. 282–304 in H. Rheingold, ed., *Maternal Behavior in Mammals.* New York: John Wiley and Sons.

Jensen, A. R. (1969). How much can we boost IQ and scholastic achievement? *Harvard Educational Review 39:*1–123.

Kagan, J. (1984). *The Nature of the Child.* New York: Basic Books.

Kagan, J., J. S. Reznick, and N. Snidman (1988). Biological bases of childhood shyness. *Science 240:*167–171.

Kaufman, I. C. (1974). Mother-infant relations in monkeys and humans: a reply to Professor Hinde. Pp. 47–68 in N. F. White, ed., *Ethology and Psychiatry.* Toronto: University of Toronto Press.

Kaufman, I. C., and L. A. Rosenblum (1967a). Depression in infant monkeys separated from their mothers. *Science 155:*1030–1031.

——— (1967b). The reaction to separation in infant monkeys: anaclitic depression and conservation-withdrawal. *Psychosomatic Medicine 29:*648–675.

——— (1969). The waning of the mother-infant bond in two species of macaque. Pp. 41–59 in B. M. Foss, ed., *Determinants of Infant Behaviour,* Vol. IV. London: Methuen.

Kawai, M. (1965). Newly acquired pre-cultural behavior of the natural troop of Japanese monkeys on Koshima island. *Primates 6:*1–30.

Kohlberg, L. (1971). From is to ought. Pp. 150–200 in T. Mischel, ed., *Cognitive Development and Epistomology.* New York: Academic Press.

Konner, M. J. (1972). Aspects of the developmental ethology of a foraging people. Pp. 285–304 in N. Blurton-Jones, ed., *Ethological Studies of Child Behavior.* London: Cambridge University Press.

Kummer, H. (1968). *Social Organization of Hamadryas Baboons.* Basel: S. Karger.

Lancaster, J. (1971). Play-mothering: the relations between juvenile females and young infants among free-ranging vervet monkeys. *(Cercopithecus aethiops). Folia Primatologica 15:*161–182.

Langer, J. (1969). *Theories of Development.* New York: Holt, Rinehart and Winston.

Lawick-Goodall, J. Van (1967). Mother-offspring relations in free-ranging chimpanzees. Pp. 365–436 in D. Morris, ed., *Primate Ethology.* New York: Doubleday.

Lee, P. C. (1984). Ecological constraints on the social development of vervet monkeys. *Behaviour 90:*245–261.

Lehrman, D. S. (1970). Semantic and conceptual issues in the nature-nurture problem. Pp. 17–52 in L. R. Aronson, E. Tobach, D. S. Lehrman, and J. S. Rosenblatt, eds., *Development and Evolution of Behavior.* San Francisco: Freeman.

——— (1974). Can psychiatrists use ethology? Pp. 187–196 in N. F. White, ed., *Ethology and Psychiatry.* Toronto: University of Toronto Press.

Lindburg, D. G. (1977). Feeding behavior and diet of rhesus monkeys *(Macaca mulatta)* in a Siwalik forest in North India. Pp. 223–249 in T. H. Clutton-Brock, ed., *Primate Ecology.* New York: Academic Press.

Loizos, C. (1967). Play behaviour in higher primates: a review. Pp. 226–282 in D. Morris, ed., *Primate Ethology.* New York: Anchor Books.

Maccoby, E. E., and J. C. Masters (1970). Attachment and dependency. Pp. 73–157 in P. Mussen, ed., *Carmichael's Manual of Child Psychology,* 3rd edition. New York: John Wiley & Sons.

Main, M., and D. Weston (1982). Avoidance of the attachment figure in infancy: descriptions and interpretations. Pp. 31–59 in C. M. Parkes and J. Stevenson-Hinde, eds., *The Place of Attachment in Human Behavior.* New York: Basic Books.

Martin, P. (1982). The energy cost of play: definition and estimation. *Animal Behaviour* 30:294–295.

Martin, P., and T. M. Caro (1985). On the function of play and its role in behavioral development. *Advances in the Study of Behavior 15*:59–103.

Mason, W. (1979). Ontogeny of social behavior. Pp. 1–28 in P. Marler and J. G. Vandenberg, eds., *Handbook of Behavioral Neurobiology, Vol. 3: Social Behavior and Communication.* New York: Plenum.

Masters, R. D. (1976). The implications of sociobiology. *Science 192*:427–428.

McKenna, J. J. (1979). Aspects of infant socialization, attachment, and maternal caregiving patterns among primates: a cross-disciplinary review. *Yearbook of Physical Anthropology 22*:250–286.

Missakian, E. A. (1972). Genealogical and cross-genealogical dominance relations in a group of free-ranging rhesus monkeys *(Macaca mulatta)* on Cayo Santiago. *Primates 13*:169–180.

——— (1974). Mother-offspring grooming relations in rhesus monkeys. *Archives of Sexual Behavior 3*:135–141.

Nagel, U. (1973). A comparison of anubis baboons, hamadryas baboons and their hybrids at a species border in Ethiopia. *Folia Primatologica 19*:104–165.

Nash, L. T. (1978). Development of mother-infant relationships in wild baboons *(Papio anubis). Animal Behavior 26*:747–759.

Nicolson, N. A. (1982). *Weaning and the development of independence in olive baboons.* PhD dissertation, Harvard University.

——— (1986). Infants, mothers and other females. Pp. 330–342 in B. B. Smuts, et al., eds., *Primate Societies.* Chicago: University of Chicago Press.

Parker, S. T. (1977). Piaget's sensorimotor series in an infant macaque: a model for comparing unstereotyped behavior and intelligence in human and nonhuman primates. Pp. 43–112 in S. Chevalier-Skolnikoff and F. E. Poirier, eds., *Primate Biosocial Development.* New York: Garland.

Pereira, M. E., and J. Altmann (1985). Development of social behavior in free-living non-human primates. Pp. 217–309 in E. S. Watts, ed., *Nonhuman Primate Models for Human Growth and Development.* New York: Alan R. Liss, Inc.

Phillips, J. L. (1975). *The Origins of Intellect: Piaget's Theory,* 2nd edition. San Francisco: W. H. Freeman.

Piaget, J. (1952). *The Origins of Intelligence in Children.* New York: International Universities Press.

——— (1962). *Play, Dreams and Imitation in Childhood.* New York: W. W. Norton and Co.

Quiatt, D. (1979). Aunts and mothers: adaptive implications of allomaternal behavior of nonhuman primates. *American Anthropologist 81*:311–319.

Ranson, T. W., and T. E. Rowell (1972). Early social development of feral baboons. Pp. 105–144 in F. E. Poirier, ed., *Primate Socialization.* New York: Random House.

Redican, W. K. (1975). Facial expressions in nonhuman primates. Pp. 103–194 in L. A. Rosenblum, ed., *Primate Behavior.* New York: Academic Press.

Rhine, R., G. W. Norton, W. J. Roertgen, and H. D. Klein (1980). The brief survival of free-ranging baboons *(Papio cynocephalus)* after separation from their mothers. *International Journal of Primatology 1*:401–409.

Rowell, T. E. (1975). Growing up in a monkey group. *Ethos 3*:113–128.

Rowell, T. E., and J. Chism (1986). The ontogeny of sex differences in the behavior of patas monkeys. *International Journal of Primatology 7*:83–106.

Rowell, T. E., N. A. Din, and A. Omar (1968). The social development of baboons in their first three months. *Journal of Zoology, London 155*:461–483.

Rowell, T. E., and D. K. Olson (1983). Alternative mechanisms of social organization in monkeys. *Behaviour 86*:31–54.

Sade, D. S. (1967). Determinants of dominance in a group of free-ranging rhesus monkeys. Pp. 99–114 in S. A. Altmann, ed., *Social Communication Among Primates.* Chicago: University of Chicago Press.

Schaffer, H. R., and P. E. Emerson (1964). The development of social attachments in infancy. *Monographs of the Society for Research in Child Development 29*:1–77.

Seay, B., E. W. Hansen, and H. F. Harlow (1962). Mother-infant separation in monkeys. *Journal of Child Psychology and Psychiatry 3*:123–132.

Seay, B., and H. F. Harlow (1965). Maternal separation in the rhesus monkey. *Journal of Nervous and Mental Disorders 140*:434–441.

Simpson, A. E., and J. Stevenson-Hinde (1985). Temperamental characteristics of three-to-four-year-old boys and girls and child-family interactions. *Journal of Child Psychology and Psychiatry 26*:43–53.

Spencer-Booth, Y. (1968). The behaviour of group companions towards rhesus monkey infants. *Animal Behaviour 16*:541–557.

Sociobiology Study Group of Science for the People (1976). Sociobiology—Another biological determinism. *Bioscience 26*:182, 184–86.

Spitz, R. A. (1946). Anaclitic depression. *Psychoanalytical Study of the Child 2*:313–342.

Struhsaker, T. T. (1971). Social behaviour of mother and infant vervet monkeys *(Cercopithecus aethiops). Animal Behaviour 19*:233–250.

Sugiyama, Y. (1965). Behavioral development and social structure in two troops of Hanuman langurs *(Presbytis entellus). Primates 6*:213–247.

————— (1972). Social characteristics and socialization of wild chimpanzees. Pp. 145–165 in F. E. Poirier, ed., *Primate Socialization.* New York: Random House.

Tanner, J. M. (1970). Physical growth. Pp. 77–155 in P. Mussen, ed., *Carmichael's Manual of Child Psychology,* Vol. 1. New York: John Wiley & Sons.

Trivers, R. L. (1974). Parent-offspring conflict. *American Zoologist 14*:249–264.

Vygotsky, L. S. (1962). *Thought and Language.* Cambridge, MA: MIT Press.

Whitten, P. L. (1982). Female reproductive strategies among vervet monkeys. PhD dissertation, Harvard University.

Wilson, E. O. (1975). *Sociobiology: The New Synthesis.* Cambridge MA: The Belknap Press.

————— (1976). Academic vigilantism and the political significance of sociobiology. *Bioscience 26*:183, 187–190.

Wolfheim, J. (1977). Sex differences in behaviour in a group of captive juvenile Talapoin monkeys *(Miopithecus talapoin). Behaviour 63*:110–128.

Human Evolution and the Sexual Behavior of Female Primates

LINDA D. WOLFE

This chapter explores how the study of the sexual behavior of female alloprimates (primates other than ourselves, Count, 1973; Wolfe and Gray, 1989) contributes to understanding the sexual behavior of women. Among primates, the Old World monkeys and apes are our closest evolutionary relatives, and therefore the discussion in this chapter will, for the most part, be limited to the behavior of the allocatarrhines (the catarrhine primates other than ourselves).

It is my belief that, if we want to know about the specifics of the behaviors (e.g., courtship, female proceptivity, mating practices, communication patterns, social systems, etc.) of any given primate species, we should study the species in question. That is, primatologists do not study gorillas *(Gorilla gorilla)* to understand the details of the mating patterns or social systems of baboons *(Papio* sp.). In the same sense, the study of the behavior of the alloprimates cannot be expected to reveal much about the *specifics* of human behavior. On the other hand, the comparative study of alloprimate behavior can illuminate those features of behavior common to all primates, including humans, because of a shared biological inheritance. (For example, the formation of strong emotional attachments between mothers and their infants is a ubiquitous primate trait that undoubtedly has a common psychophysiological genetic basis.) Furthermore, a knowledge of the behaviors of the alloprimates may foster speculations about the behaviors of the early members of the human family and about the evolution of human behavior.

As we engage in such speculations, however, we should always remain skeptical of musings on our own evolutionary history, because cultural biases have often influenced the way the behavior of the alloprimates is viewed (Strum, 1987), and, all too frequently, our own culturally derived patterns of behavior have been mistaken for that elusive quality called "human nature." For example, in his article entitled "The Origin of Man," Lovejoy (1981) made the assumption that a nuclear family consisting of a pair-bonded male and female (i.e., mates with a strong and long-lasting closeness [Money, 1980a]; note that this is the sort of family idealized in middle- and upper-class U.S. society) is a universal human trait. An examination of the ethnographic literature, however, suggests that neither male–female pair bonding nor the nuclear family is universal among humans (Leibowitz, 1978). When reconstructing our own evolution, we need to take into account the cross-cultural range of human behaviors and not just our own society's patterns of behavior.

The topics I have chosen to discuss in this chapter are estrus and concealed ovulation, the female orgasm, female homosexual interactions, and the social and sexual behavioral patterns of females. These topics, I believe, reflect those behavioral features of human and nonhuman primates that seem to be held in common because of a shared evolutionary history and/or those that have been used by other authorities to reconstruct the behaviors of our early ancestors and speculate on human evolution.

ESTRUS AND CONCEALED OVULATION

The estrus concept as applied to primates has created a quagmire of misunderstanding about their mating habits (see Loy, 1987; Hrdy, 1981). Estrus is usually defined as the "periodical state of receptivity in female animals accompanied by physiological changes in the reproductive organs [such that the female is fertile]" (Wolman, 1973:128). Although this definition may suffice for nonprimates, it is not sufficient for the anthropoid primates, because they may mate at times when the female is not ovulating. To escape the morass of misunderstandings, we must distinguish among three intricately intertwined components of estrus: sexual attractivity, proceptivity, and receptivity (Beach, 1976). These three components have different implications for the likelihood of a copulation occurring and are influenced differently by the sex hormones.

1. Attractivity: This concept refers to a female's effectiveness at sexually stimulating a male (Beach, 1976). Primate males may be stimulated by factors such as female appearance, vaginal odors and textures, and female behavior. Using ovariectomized female rhesus monkeys *(Macaca mulatta)*, Johnson and Phoenix (1976) demonstrated that estrogens (against a background of a normal level of adrenal androgens) are the hormones that most affect female attractiveness (Johnson and Phoenix, 1976). Beach (1976:107) noted that since "estrogen production is closely associated with the timing of ovulation, sexual attractivity is essential to the survival of the species because it maximizes the probability of copulation when the female is fertile and susceptible to impregnation."

2. Proceptivity: This concept comprises the sexual appetitive behaviors of a primate female. Although Johnson and Phoenix's (1976:481) experiments on rhesus monkeys indicated that "estrogen and androgens in appropriate balance" are important in controlling proceptivity, other researchers have not found a clear role for androgens (Wallen and Goy, 1977). The proceptive

behavior of primate females is, however, more conspicuous than that of most nonprimate mammals (Beach, 1976). These proceptive behaviors bring the female into contact with the male so that copulation can occur and are associated with females initiating sexual intercourse. For example, among gorillas, almost all copulations are initiated by females (Harcourt et al., 1980). When field primatologists define estrus as that period when the female actively seeks sexual interaction, that definition is framed in terms of proceptivity, not receptivity.

An important influence on female proceptivity in some species is the attractiveness of males. Some males elicit more proceptive behaviors than other males, and this results in "situation-dependent" estrus (a term coined by Hrdy, 1981). Situation-dependent estrus refers to proceptive behaviors that occur when the female is probably not ovulating; such behaviors are typically found in species whose females lack sex swellings (Hrdy and Whitten, 1987). In the wild rhesus and Japanese macaques *(Macaca fuscata)* that I have observed, novel males often elicit more proceptive behaviors (i.e., they have more stimulus value) than familiar males (Wolfe, 1986; see also Hrdy and Whitten, 1987, on this point). Recently, Boinski (1987) found that among squirrel monkeys *(Samiri oerstedi)* the largest, most "fattened" male attracts the most female mating partners. There are undoubtedly species-specific patterns in the ingredients of proceptivity and in what constitutes male and female attractiveness.

3. Receptivity: As defined by Beach (1976:125), receptivity "comprises those feminine reactions which are necessary and sufficient for fertile copulation with a potent male ... [including] adoption of a posture facilitating the male's achievement of insertion plus maintenance of appropriately oriented contact long enough so that intravaginal ejaculation can occur." In some prosimians and in many mammals, the vagina can be penetrated only when the female is in estrus. In anthropoid primates, however, the vagina can be penetrated at all times, a fact meaning that copulation is always possible (Loy, 1987). It has been reported that ovariectomizing rhesus females does not necessarily reduce their acceptance of male mounting attempts. Moreover, the administration of estrogen to ovariectomized rhesus females does not always raise the already high level of female acceptance over that shown without hormone treatment, nor does removing adrenal androgens interfere with receptivity (Johnson and Phoenix, 1976). Thus, for higher primates, sexual receptivity depends more on social stimuli and individual preference than on hormones. Johnson and Phoenix (1976:482) concluded that "though rhesus female receptivity shows evidence of emancipation from hormonal dependence, the other aspects of female sexuality, attractiveness and proceptivity, remain strongly influenced by the female's hormonal state." In other words, although higher primate females continuously maintain the ability to engage in receptive behavior (unlike species in which the vagina closes when the female is not in estrus), attractivity and proceptivity of higher primate females change over the course of their menstrual cycle, with sexual intercourse most likely to occur—but not occurring exclusively—at midcycle, when estrogen levels are relatively high and females are ovulating.

Copulations tend to be most frequent during the periovulatory period, but matings are also observed during nonovulatory phases of the menstrual cycle. For example, rhesus and Japanese monkeys have been observed copulating during menstruation (Loy, 1970; Wolfe, 1979), and Goodall (1986) reported that 4% of the copulations she observed among the common chimpanzees *(Pan troglodytes)* of the Gombe National Park in Tanzania occurred outside the periovulatory period. Thomspson-Handler et al. (1984) found a tendency for bonobos *(Pan paniscus)* to mate throughout the menstrual cycle. Moreover, postconception estrus periods have been reported for several species of monkeys and apes. (See Loy [1987] for a broad discussion of the occurrence of sexual intercourse among primates at times other than ovulation.)

Is there evidence that human females show a typical catarrhine pattern of continuous sexual receptivity throughout the menstrual cycle with fluctuation in proceptive behaviors and attractiveness? The literature on variations in female sexual behavior during the menstrual cycle (reviewed by Gray and Wolfe, 1983; Hill, 1988) provides such evidence. For example, an experiment conducted by Hill and Wenzel (1981) suggests that estrogen fluctuations may influence a woman's proceptive behaviors and/or attractiveness. They found that men at a nightclub were more likely to touch women who were in the middle of their menstrual cycles than they were to touch women in other phases. Adams et al. (1978) found that married women (not on birth control pills) exhibited elevated frequencies of both autosexual and female-initiated heterosexual behavior at midcycle.

Campbell (1988) recently reported a midcycle increase in female sexual behavior correlated with changes in salivary androgens. Bancroft et al. (1983) found that women tend to initiate sexual behavior during the midfollicular stage, whereas men are the primary initiators during the midluteal phase. A research project involving lesbians living in stable pairs also found that "women, like other female mammals, show midcycle increases in sexual behavior [i.e., orgasms and self-initiated sexual encounters] and decreases in behavior associated with the luteal phase.... The data further suggest that women's interest in sexual thoughts and behaviors is not influenced by their partners' cycles but by their own cycles" (Matteo and Rissman, 1984:254). Although human females may show more proceptive behaviors or may be more attractive at midcycle, Schreiner-Engle (1981:199) found that "[s]ubjective reports of sexual arousal [by the 32 women in their study] did not differ among menstrual cycle phases...."

A possible conclusion from this still preliminary evidence is that humans and their early ancestors did not require regularly and blindly repeated copulation to ensure pregnancy. In other words, it appears that human females have not completely concealed ovulation, since the accompanying changes in hormones tend to produce changes in both attractiveness and proceptivity that increase the probability of periovulatory copulation. The differences between the sexual behavior of women and female alloprimates are differences of degree, not kind, and may be related to the increase in the size of the human brain and a concomitant increase in the role of learning and individual preference (see Spuhler, 1979).

Concealed Ovulation

Estrogens are present at high levels during that phase of the menstrual cycle when the female is ovulating, and these are responsible for the periovulatory anogenital swellings or "sex swellings" observed in about half of the allocatarrhine species (Dixson, 1983). Human females experience high periovulatory estrogen levels but do not develop sex swellings. Explaining humans' lack of swelling is challenging. Deciding whether early human females exhibited sex swellings is difficult because of the problems in determining whether traits shared by closely related species are homologous features of common origin or are convergent traits that evolved independently. Among the hominoids, only common chimpanzee females have a large, conspicuous swelling of the external genitalia associated with ovulation (Goodall, 1986). Bonobos (pygmy chimpanzees) have a sex swelling, but it is present for about 75% of the monthly menstrual cycle (de Waal, 1989). Gorillas, orang-utans *(Pongo pygmaeus)*, and humans lack a visible swelling. If the lack of a visible sex swelling among the three hominoid lineages just mentioned came about independently after divergence from a common ancestor who had a sex swelling, then this shared trait would, of course, be an example of independent convergence. Sex swellings may, however, be a recently acquired trait among chimpanzees and bonobos that was lacking in the last common ancestor of the living hominoids. If that were the case, then the lack of swellings among gorillas, orang-utans, and humans would be a homologous trait, and the swelling of the chimpanzee and bonobo a uniquely derived trait acquired after divergence from the common hominoid ancestor.

Parsimony suggests that the living, nonswelling hominoids resemble the last common hominoid ancestor in this regard. In other words, it cannot be assumed that human ancestors showed swellings, and the real possibility that sex swellings were never a feature of our lineage must be considered (see Dixson, 1983, for a further discussion of this point.)

THE FEMALE ORGASM

So female invention went on. And the latest—though I should never consider the last—was the female orgasm. The capacity varies so widely among individuals that one must suspect a very recent evolutionary heritage. The male's capacity, two hundred million years old, is automatic. The female's, varying so remarkably, one must suspect is fairly new in evolutionary terms and so is subject to the inhibitions of culture. But this we may suspect: The female orgasm through enhancement of female desire provided one further guarantee that the males would return from the hunt. The male might be tired; female desire would refresh him. The male's orgasm, perfected through the ages, is a reflex; the female's demands a certain discipline, a concentration on the part of the central nervous system. I should doubt very much that female reward preceded by long the enlargement of the great human brain.

The female orgasm was the last—at least until now—of that long line of female inventions contributing to human sexual uniqueness, and at the same time—quite in accord with Darwinian thought—advancing the probability that her young would survive (Ardrey, 1976:88–89).

Ardrey's remarks typify many of the arguments in the literature on the evolution of female sexual behavior. These arguments share some common assumptions: that the female's orgasm is different from—and perhaps even inferior to—the male's; that female orgasm enhances the woman's sexual desire; and that female orgasm is unique to humans and of recent origin. A close examination of the evidence suggests that we must maintain an open mind to the possibility that one or more of these propositions is false.

Recent work raises questions about the contention that male and female orgasms are essentially different. For example, Vance and Wagner (1976) presented descriptions of orgasms written by 24 men and 24 women to a panel of professional judges whose assignment it was to detect which descriptions were written by men and which by women. The authors reported that the "judges could not correctly identify the sex of the person describing an orgasm" and suggested, contrary to popular belief, "that the experience of orgasm for males and females is essentially the same" (Vance and Wagner, 1976:87). Likewise, Masters, Johnson, and Kolodny (1988) reported that the sexual fantasies of men and women are more similar than they are different.

Despite the similarity between men and women in the sensations of orgasm, women believe that it takes them 40–80% more time than their husbands to attain orgasm during heterosexual intercourse (Fisher, 1973). When masturbating, however, women reach orgasm in less than 4 minutes (mean time), a fact that has given rise to the belief that it takes women longer than men to orgasm during heterosexual intercourse due to the "relative inefficiency with which stimulation [is] applied during intercourse" (Fisher, 1973:191; see also Hite, 1976). In other words, men and women are similar in the length of time it takes to orgasm when the intensity of the stimulation is similar. Thus it would seem that orgasm for men and women is more alike than different and that arguments to the contrary—such as Ardrey's—should be rejected.

The other issue raised by Ardrey, that female orgasm enhances sexual desire, is not supported by evidence on two counts. It is known that women—about 60% of the U.S. women (Fisher, 1973)—do not consistently experience orgasm during heterosexual intercourse, yet these same women usually indicate that they enjoy intercourse because of the associated affection and closeness (Hite, 1976). Unfortunately, information obtained directly from women in other societies on how they experience orgasm and heterosexual intercourse, which might inform the question of the relationship between female orgasm and sexual desire, is scarce (Gray and Wolfe, 1988). Mead (1961:1466) indicated that there is "great variability in the female sex in regard to this [orgasmic] capacity, so it may be that societies will disallow the possibility of orgasmic response for all women, stylize vigorous response as appropriate for the prostitute but not for the respectable woman, or insist upon response from all women as a sign of affection or assent to male sex activity."

Evidence that allocatarrhine females exhibit the capacity for orgasm (see Table 5–1, Fig. 5–1) suggests that human female orgasm is an expression of an ancient catarrhine trait, not a recently evolved adaption enhancing women's sexual desire to create exchanges of sex for food or protection from males. Symons (1979:82) examined much of the evidence listed in Table 5–1 and stated that "there is no compelling evidence" that allocatarrhine females experience orgasm during heterosexual intercourse because in those cases when orgasm had been reported it involved "direct and prolonged stimulation of [the] clitoris or clitoral area, either by experimental design or by rubbing against another animal" (Symons, 1979:83). Symons assumed that during intercourse allocatarrhine females experience only indirect stimulation of the clitoris, as is true of human females. He therefore suggested that when mating with males (as opposed to receiving mechanical stimulation) allocatarrhine females would not receive enough stimulation during intercourse to produce orgasm.

A comparison of the genital anatomy of allocatarrhine and human females

Table 5–1 A sample of the evidence for orgasm in alloprimate females.

Species	Evidence	Reference
Macaca arctoides (stumptail macaques)	Facial expressions during homosexual encounters	Chevalier-Skolnikoff (1974)
	Uterine contractions and increases in heart rate during heterosexual intercourse and homosexual mounting	Goldfoot et al. (1980)
	Clutching reaction declines following ovariectomy	Mitchell (1979)
	Once in every six matings on average, the females show the orgasm facial expression	de Waal (1989)
Macaca fuscata (Japanese macaques)	Contraction of perineal muscles of the base of the vagina and anus following self-masturbation	Wolfe (1984a)
Macaca mulatta (rhesus macaques)	Female clutching reaction during heterosexual intercourse	Zumpe and Michael (1968)
	Vaginal contractions and increased heart rate in response to stimulation by a mechanical penis	Burton (1971)
Pan troglodytes (chimpanzees)	Contraction of perineal muscles of the base of the vagina following digital masturbation by a researcher	Allen and Lemmon (1981)
Papio ursinus (chacma baboons)	Vocalizations and postcoital reactions	Saayman (1970)

Fig. 5–1 Japanese macaque female masturbating by rubbing her clitoris between her index finger and her ischial callosity. Contractions of the perianal area indicative of orgasm were observed. Note the female's hand under her tail. (Photo: Linda Wolfe.)

leads me to reject Symons' suggestion. In the allocatarrhine female, the urinary meatus is found either inside the vagina or near the base of the vagina (Eckstein, 1958). The allocatarrhine clitoris, in turn, is found near the base of the vagina, and in some allocatarrhines, such as the chimpanzee, it may be relatively larger than the human clitoris (Graham and Bradley, 1972). Dahl noted that in orang-utan females the clitoris is "relatively large and intimately associated with the relatively small vaginal aperture so that stimulation during intromission is highly likely" (Dahl 1988:141). Given that her clitoris is near the base of the vagina, it is likely that the allocatarrhine female receives some degree of direct clitoral stimulation during heterosexual intercourse and that this stimulation is similar to that received under experimental conditions listed in Table 5–1.

For human females, on the other hand, the urinary meatus is located in the vulva between the vagina and the clitoris. The location of the clitoris away from the base of the vagina means that human females experience only indirect stimulation during heterosexual intercourse regardless of coital position (Masters and Johnson, 1966) and often need some form of direct manual stimulation to reach orgasm (Fisher, 1973). The difference in genital anatomy between humans and allocatarrhine females suggests that the latter may find it easier to achieve orgasm through heterosexual intercourse than do women.

The most likely explanation for the genital anatomy of women is that as the size of the neonatal head increased there was selection to move the urinary meatus away from the vagina to protect it from trauma during child birth. The result of the present placement of the urinary meatus in the vulva is that the clitoris is located away from the base of the vagina in human females. During hominid evolution, the nature of clitoral stimulation during heterosexual intercourse changed from some degree of direct stimulation, as it likely still is in the allocatarrhines, to the indirect (and often inadequate) stimulation experienced by women.[1]

Symons (1979) also argued that copulation in most allocatarrhine species does not last long enough to produce female orgasm, but his argument is not convincing. Copulation durations for hominoid species vary greatly. Hunt (1975) reported that the median duration of human coitus (excluding foreplay) was 10 minutes. This compares with medians of 7 seconds for chimpanzees, 96 seconds for gorillas (Harcourt et al., 1980) and 13 seconds for bonobos (de Waal, 1989). Symons' argument ignores the fact that the amount (including the duration) of sexual stimulation necessary for male and female orgasms is species-specific. For example, chimpanzee, bonobo, and gorilla males achieve orgasm with copulations that do not last long enough to produce ejaculation in human males. One cannot judge the possibility of orgasm in one species by comparing it with the stimulation needs of another.

The relationship between sexual stimulation and the neurological pathways or orgasmic experiences remains a mystery for both humans and allocatarrhines. There is evidence suggesting that such relationships are very complex. For example, in Japanese and rhesus monkeys, the stimulation pattern leading to ejaculation in heterosexual intercourse is very different from the patterns creating ejaculation during masturbation. These monkeys are known as multimount ejaculators because heterosexual intercourse involves a series of mounts, each of which lasts about 5 seconds. Thirty to fifty seconds elapse between each mount, and the sequence from first mount to ejaculation may take from 10 to 15 minutes. In contrast, when these same males masturbate, they stroke their penises continuously and ejaculate in less than 1 minute.

Until the relationship between genital stimulation and the neurology of orgasm in different species of primates is better known, we cannot settle definitely the issue of female orgasm. Nonetheless, behavioral observations do provide evidence for female orgasms in many alloprimates (Allen and Lemmon, 1981; Mitchell, 1979; Hrdy, 1981, Wolfe, 1984a). Such evidence argues that human female orgasm is most likely a trait inherited from an ancient primate ancestor, not a human invention that evolved because it played a significant role in the development of the relationship between men and women.

HOMOSEXUAL INTERACTIONS

Alloprimates

Anthropology has failed to devise its own theory of homosexuality; as a result, anthropologists have been heavy borrowers of theories and ideas from other disciplines on this topic. Since such borrowed ideas are inevitably based upon studies of European or United States homosexual activities and expectations of what is "normal" homosexual activity, they may not be relevant to cross-cultural analysis. Contemporary evidence demonstrates that there is a much greater range of homosexual behavior (and local attitudes about it) elsewhere in the world than is typically found in Euro-American cultures. . . . Indeed, it has been argued that extreme homophobia is a distinctly Anglo-American trait (Davis and Whitten, 1987:80).

Davis and Whitten (1987) suggest that Euro-American anthropologists do not understand homosexuality in other societies because they have applied a Western folk model of "normal" sexuality to the study of other societies' sexual behavior. The unconscious application of this folk model means that we have either ignored or misinterpreted the range of sexual behavior in other societies. For example, as Davis and Whitten point out, Frayser's (1985) "recent HRAF style study of human sexuality does not discuss homosexuality at all" (Davis and Whitten, 1987:82).

Similarly, recent texts on primate behavior that explore sexual behavior (e.g., Smuts et al., 1986; Dunbar, 1988) also fail to discuss homosexuality among the alloprimates. Primatology has failed to create a framework for observing, reporting, or analyzing data on homosexual behavior among the alloprimates. For example, I have talked with several (anonymous at their request) primatologists who have told me that they have observed both male and female homosexual behavior during field studies. They seemed reluctant to publish their data, however, either because they feared homophobic reactions ("my colleagues might think that I am gay") or because they lack a framework for analysis ("I don't know what it means"). If anthropologists and primatologists are to gain a complete understanding of primate sexuality, they must cease allowing the folk model (with its accompanying homophobia) to guide what they see and report. In place of the folk model, a holistic analytic model needs to be developed that encompasses the totality of primate sexual experiences.

Primatologists can develop analytical models to explain the structure, function, and evolution of primate sexuality only after they have described the full range of sexual behavior. Field primatologists must record and report all aspects of the sexual behavior of the species they are observing, including male-female mounting, female-male mounting, same-sex mounting, all forms of oral-genital contact, male courtship and female approach patterns, and masturbation. We should expect that each species will have its own species-specific pattern of sexual behavior and that it will not necessarily fit the Euro-American folk model of sexuality. For example, among the macaques, it appears that stumptail macaques *(Macaca arctoides)* are the most varied in their sexual behavior, displaying both male-male and female-female homosexual interactions and oral-genital contact (Chevalier-Skolnikoff, 1976), and that bonnet macaques (*Macaca radiata:* Simonds, 1965) are the least varied. Japanese, rhesus, and pigtail *(Macaca nemestrina)* macaques seem to be somewhere between stumptail and bonnet macaques regarding variety of sexual behavior.

Although there is a dearth of information on the total sexual repertoire of the alloprimates, some data on female homosexual behavior do exist. Before I discuss these data, I must point out that the term "female homosexual behavior" refers to two females engaging in one or more mounts resembling a heterosexual copulation (Fig. 5–2). For primatologists, the label "homosexual" denotes nothing about the "sexual object orientation" of individual animals. Because the alloprimates do not possess language, it is impossible to inquire into their sexual eroticism. In other words, homosexual and heterosexual behaviors can be observed, recorded, and analyzed, but we cannot infer either homoeroticism or heteroeroticism from such behaviors.

Female homosexual behavior has been described among a wide variety of free-ranging and captive alloprimates, and I believe it is likely that female homosexual behavior occurs in other species but that it has not been reported (Table 5–2). Of the 11 species showing female homosexual behavior that are listed in Table 5–2, five were observed in the wild and six in captive situations. In two cases (chimpanzees and squirrel monkeys), there were no males available in the study group, and it seemed that female homosexual behavior was a substitution for heterosexual activities. In seven cases (gorillas, stumptail macaques, Japanese macaques, rhesus macaques, pigtail macaques, bonobos, and langurs [*Presbytis entellus*]), it appeared that the homosexual activity was the result of female preference. In the two remaining cases (talapoins [*Miopitheaus talapoin*] and baboons), the published information does not permit us to judge the cause of the behaviors.

My own field work among two troops of Japanese macaques has produced some information on the patterning of female homosexual behavior in this species (Wolfe, 1984a, 1986). For example, my data indicate that the frequency of

Fig. 5–2 A mount between two young adult female rhesus macaques. (Photo: James Loy.)

Table 5–2   Female homosexual behavior

Species	Comments	Reference
Gorilla gorilla	Ten cases of female homosexual interactions are reported among wild gorillas; generally associated with estrus; males present; behavior attributed to female preference	Harcourt et al. (1981)
Macaca arctoides	Female-female mounting among captive stumptail macaques; associated with estrus; males present; behavior attributed to female preference	Chevalier-Skolnikoff (1974, 1976) Goldfoot et al. (1980)
Macaca fuscata	Homosexual interactions observed among free-ranging Japanese macaques; loosely associated with estrus; males present; attributed mostly to female preference and (secondarily) to a skewed sex-ratio in some cases	Hanby and Brown (1974) Wolfe (1979, 1984a, 1986)
Macaca mulatta	Homosexual interaction observed among free-ranging rhesus macaques; loosely associated with estrus; males present; attributed to a complex set of individual female characteristics	Akers and Conaway (1979)
Macaca nemestrina	Female-female mounting sometimes associated with estrus; captive situation, males present; attributed to female preference	Tokuda et al. (1968)
Miopithecus talapoin	Female-female mounting mentioned briefly; captive situation, males present	Wolfheim and Rowell (1972)
Saimiri sciureus	Female-female mounting; captive situation, no males present	Talmage-Riggs and Anschel (1973)
Pan paniscus	Female-female genital rubbing observed among wild animals; perhaps associated with a greeting ritual	Kano (1980)
Pan troglodytes	Female-female mounting associated with sex swellings; captive situation, no males present Female-female genital rubbing associated with sex swellings; captive situation, males present	Yerkes (1939) Kohler (1959)
Papio hamadryas	Female-female mounting is briefly mentioned; captive situation, males present	Zuckerman (1932)
Presbytis entellus	Female-female mounting reported among wild animals; loosely but not entirely associated with estrus	Hrdy (1977)

female-female homosexual consortships in a breeding season may be affected by at least three variables: female preference; troop traditions; and social and demographic factors, especially a lack of novel males as mating partners. A homosexual consortship between two female Japanese macaques may result in the development of a lengthy nonsexual friendship between them. Moreover, Japanese monkey females seem to avoid homosexual interactions with close relatives so that homosexual dyads involving mothers and daughters, sisters, first cousins, and grandmothers and granddaughters have not been observed (Wolfe, 1979). Finally, no female raised in a group composed of adults and peers of both sexes has exhibited exclusive homosexual behavior.

Akers and Conaway (1979) reported on female homosexual behavior in semifree-ranging rhesus monkeys. They argued that female homosexual behavior in rhesus monkeys is not preparation or substitution for sexual intercourse with males. They concluded that such behavior can be explained by analyzing the same factors one would examine when explaining heterosexual behaviors: the social relationship between the partners, individual preference, past sexual experience, and reproductive status.

In summary, homosexual behavior is part of the repertoire of sexual behavior of both free-ranging and captive alloprimates. Until homosexual behaviors are thoroughly observed and consistently reported, however, a complete understanding of the sexual repertoire of the alloprimates will not be possible.

Homosexual Behavior of Women

What similarities are there between the homosexual behavior of female alloprimates and those of women? First, homosexual behavior in both groups is not terribly unusual. Of the 4,500 U.S. women in Hite's 1987 sample, 11% reported that they have "love relationships only with other women. An additional 7% sometimes have relationships with women" (Hite 1987:542). Hunt (1975:312) found that "7 percent of single women and 2 percent of married women [out of a sample 1044 females] had homosexual contact beyond [age 19] . . . and if we project these figures upward to make them approximately equivalent to life-time accumulative incidences we get something on the order of 10 to 12 percent for single females and 3 percent for married females." Hunt (1975) argued that his figures were similar to those reported by Kinsey et al. (1953). Little is known concerning the homosexual behavior of women in other cultures. Ford and Beach (1951) found only 17 societies (of approximately 200 surveyed) in which female homosexual behavior was reported. Second, humans share with the alloprimates the practice of engaging in homosexual behavior under conditions in which heterosexual behavior is not possible (e.g., prison for women and captivity for the alloprimates). Third, homosexual behavior for both human and alloprimate females can be a matter of individual preference. Hite (1987) noted that 46% of the women engaging in homosexual behaviors in her sample said that being lesbian was a choice or a matter of preference. Of the women in Hite's sample who engaged in homosexual activities, 24% were women over 40 (most of whom had been married before), who had chosen a homosexual relationship for the first time in their lives. In comparing 407 lesbians and 370 heterosexual women, Coleman et al. (1983) found that women who identified themselves as lesbians reported experiencing more sexual satisfaction than the women who identified themselves as heterosexuals.

On the other hand, there are women (e.g., 54% of the lesbians in the Hite [1987] sample) who maintain that they are lesbians because of "biological" factors. For them, their sexual fantasies involve having sex with other women, and being a lesbian is not a choice but a "biological calling."

Interestingly, no consistent hormonal differences between self-identified heterosexual women and self-identified lesbians have been found. As Money (1980b:70) noted, " . . . available evidence supports a nongenetic hypothesis for the origin [i.e., cause] not only of homosexuality, but of psychosexual differences and variations of all types. Prenatal hormonal determinants probably do no more than create a predisposition on which the postnatal superstructure of psychosexual status differentiates, primarily, like native language, under the programming of social interaction." The capacity to engage in sexual behavior with members of the same sex appears to be part of the primate phylogenetic heritage, so there is no need for special ultimate explanations for its occurrence among human females.

SOCIAL AND SEXUAL BEHAVIORAL PATTERNS OF FEMALES

Research has demonstrated that in humans and allocatarrhines sexual behavior and the process of sexual arousal have become at least partially decoupled from hormonal control (Loy, 1987). Socialization in a stable social unit, sexual rehearsal with both same- and opposite-sex partners by infants and juveniles, and sexual experience while growing to full adulthood have all come to play important roles in the development and maintenance of sexual behavior among humans, Old World monkeys, and apes.

The study of allocatarrhine sexual behavior also suggests that sexual intercourse has become linked to the development and/or existence of social relationships between individuals. For example, the development of emotional ties between two individuals early in life may prohibit later mating between these individuals during adulthood. Goodall (1986) reported that among common chimpanzees in the wild, mothers and sons as well as brothers and sisters rarely mate. Wolfe (1979) reported that mating avoidance was observed between an adult Japanese monkey female orphaned as an infant and the adult male who raised her. In contrast, the development of positive social relationships *between adults* is often prerequisite to mating. Smuts (1985) reported that among savannah baboons *(Papio cynocephalus)*, adult males cultivate adult female "friends" with whom they eventually mate.[2] In fact, she argued that the formation of "friendships" with adult females is a regular aspect of the reproductive strategy of adult male baboons.

Among orang-utans, the most solitary of hominoids, mating involves a complex set of interaction patterns and depends on relationships developed during the period of sexual quiescence (Galdikas, 1981; Mitani, 1985). For orangutans, sexual behavior is, in other words, more than just copulation; it also involves males and females choosing among possible partners, establishing a positive social relationship, displaying the proper sexually stimulating behaviors, and mutually cooperating in mating. Schurmann (1981:131–133) described the development of a heterosexual relationship among orang-utans that eventually included mating:

> From adolescence onwards Yet [the female] was already keenly interested in the adult male male Jon, but it took a long time, namely at least 5 years, for her to build up a relationship which involved . . . mating with him. . . . Yet often tried to attract the attention of Jon. . . . At first Jon avoided all . . . advances but gradually he began to show more tolerance towards her. . . . Yet would often feed very close to Jon and sometimes she was allowed to pick some pieces of fruit directly out of his hand. . . . She often presented in front of Jon, but initially Jon merely looked at her. She touched Jon . . . often, investigating his genitals manually and sometimes orally. . . . Jon started to react to Yet's behavior with "male presenting." . . . In response to this male presenting Yet would touch Jon again, look at him and position herself in a hanging posture above or in front of him and lower herself on his penis, often aiding intromission with her fingers. Female Yet . . . was the active partner and did all the pelvic thrusting. Most copulations took some 15 minutes or more. Positions varied greatly. Most frequently seen was the ventro-ventral position, but often latero-ventral and dorso-ventral positions occurred during the same copulation. . . . Many copulations occurred, up to 24 in 10 days.

Social Systems and Reproductive Success

A feature of the large social units of many alloprimates is a dominance hierarchy in which rank is often measured by priority of access to food sources or favorite sitting places. Dominance hierarchies may be present among both the males and females. The assertion is often made that female rank and reproductive success are correlated. For example, Dunbar (1980) reported that, among gelada baboons *(Theropithecus gelada),* high-ranking females produce more offspring than low-ranking females and he suggested that low-ranking females experience frequent anovulatory cycles due to the stress of harassment. When a variety of alloprimate species are considered, however, the data are equivocal, and it is doubtful that there is a correlation between rank and reproductive success when the entire life spans of all female members of a group are considered (Gray, 1985). For example, Wolfe (1984b), and Wolfe and Noyes (1981) found no relationship between female rank and the number of offspring for Japanese macaques, and, similarly, Cheney, Lee, and Seyfarth (1981) reported no correlation between female rank and reproductive success for vervet monkeys *(Cercopithecus aethiops).* Altmann (1980) stated that a correlation between female rank and reproductive success had not been demonstrated for savannah baboons, and Goodall (1986) failed to find such a relationship among the common chimpanzees at Gombe.

There are differences in the social systems of gelada baboons, Japanese macaques, vervets, savannah baboons, and chimpanzees. Gelada baboons live in harems in which female mobility is restricted, and low-ranking females cannot avoid higher-ranking females. Japanese macaques, vervets, savannah baboons, and chimpanzees, on the other hand, live in social units in which there are minimal restrictions on the activities and movements of females. Low-ranking females in the less restrictive social units have more options available to them than their gelada counterparts and can avoid the stress of competition with higher ranking females. Whether rank influenced the reproduce success of our ancestral female hominids is unknown. Indeed, we do not know whether they lived in social units characterized by a dominance hierarchy, nor is it clear to what degree the activities of early hominid females were restricted.

Social Systems and Mating Choice

Although we cannot assume that social systems are the same as mating patterns (Rowell, 1988), social systems do affect the availability of mating partners and, thus, the frequency of female choice. Some females, for example, the monogamous gibbons (*Hylobates* sp.), make relatively few decisions with regard to the choice of a sexual partner during their entire lives. On the other hand, primates such as female rhesus monkeys may make several mate choices over the course of 1 day. Social systems are also important because it is primarily through these systems that primates adapt to their environment and socialize their young. Because they are important to survival, primate social systems and mating patterns have been used to guide speculations on ancient human social systems and mating patterns. I have listed below examples of the social systems that exist among the allocatarrhines and the ways each influences mating patterns.

1. Solitary: Orang-utans, the Asian great apes, are the most solitary of the allocatarrhines. Adult males and adult females forage alone, although females may be accompanied by offspring. The foraging areas of females are small compared to those of males. Adult males defend their territories against one another, and each male's territory overlaps extensively with the foraging areas of several females. Females tend to mate with the male with whom they share a foraging area and to avoid undesirable males (Galdikas, 1985). It has been reported that adolescent males forcibly copulate with adult females. These matings rarely result in pregnancy, however (Schurmann and van Hoof, 1986).

2. Territorial pairs: Among the catarrhine primates, monogamy is rare, although it is more common among the New World primates (Kinzey, 1987). The nine species of gibbons and siamangs, all small Asian apes, are an exception. Gibbons and siamangs live in stable, lifelong, monogamous families whose members forage and sleep together as a unit. Territories are primarily defended by male-female dueting. Upon sexual maturity, offspring are excluded from the territories of their parents, although parents may assist them in finding their own territories (Tilson, 1981). Both males and females choose their lifelong mates, and, once a monogamous unit is formed, mating occurs only within the context of the monogamous pair bond (Preuschoft et al., 1984).

3. One-male social groups: Gorillas live in one-male or unimale social units in which there are several adult females and their offspring, one dominating silverback male who is physically and socially mature, and one or two younger

blackback males. The unimale group forages and sleeps together as a unit. Females leave their natal group at puberty and eventually join a new one. Once a female joins a new group, the resident silverback will most likely father all her subsequent offspring. In other words, female gorillas exercise mate choice by choosing which unimale group to join. Gorilla females tend to be monandrous (i.e., they mate with only one male), whereas male gorillas are polygynous (i.e., they mate with several females). Gorillas are less sexually active than chimpanzees, and most copulations are initiated by females (Harcourt et al.,1980). Gelada and hamadryas baboons also live in one-male social groups in which females are monandrous and males polygynous (Kummer, 1971).

Not all of the females who live in one-male social groups are monandrous, however. For example, patas monkeys *(Erythrocebus patas)* and blue monkeys *(Cercopithecus mitis)* live in social units in which there are multiple females and their young but only one resident male. During the breeding season, however, females exercise mate choice and copulate with both resident and nonresident males (Harding and Olson, 1986; Chism, et al., 1984; Cords, et al., 1986). Thus, in this situation, both males and females are polygamous (i.e., have multiple mating partners), but heterosexual social units are generally composed of multiple females and one resident male. It is unclear what factors influence female choice of a mating partner in these species.

4. Multimale/multifemale troops: The multimale/multifemale troop is probably the most common form of primate social unit. Japanese and rhesus macaques are examples of monkeys who live in stable troops composed of several adult males, several adult females, and their offspring. The troop stays together throughout the day and night and forages and rests as a unit. Females remain in their natal troop for life and form tightly knit kin units within the troop known as matrifocal units. Males, on the other hand, emigrate from their natal troop at puberty and either become solitary or join a new troop. Savannah baboons also live in multimale/multifemale troops similar to those of Japanese and rhesus macaques.

In multimale/multifemale troops, both males and females are polygamous, exercise partner choice, and initiate sexual intercourse. Adult Japanese macaque females lack a visible sex swelling, and they mate with both nonnatal, immigrant troop males and with the solitary males who appear on the periphery of a troop during the breeding season. This pattern of mating with resident and nonresident males is similar to that noted above for patas and blue monkeys, which live in one-male social units. Wolfe (1984a, 1986) has argued that the sexual physiology and psychology of adult Japanese macaque females evolved because they created and maintain genetic diversity.

There are two forms of Japanese and rhesus macaque matings: brief, opportunistic matings and matings within consortships (exclusive mating units). Consortships may last a few hours to several days. During a consortship, the male and female mate exclusively with each other and coordinate their movements so that other troop members do not disrupt their interactions.

5. Fission-fusion social system: Chimpanzees (Goodall, 1986) live in loosely knit communities composed of several males and females and their

dependent offspring. They have been characterized as having a fission-fusion social system, within which members are scattered when food resources are scarce and larger groups are formed when food resources are more abundant. Female chimpanzees and their dependent offspring spend most of their time foraging alone in small core areas, and males spend their time with other males traveling over larger areas (Wrangham and Smuts, 1980). Both males and females may emigrate from their natal communities. Bonobos (White, 1988) have also been described as having a fission-fusion social system similar to that of the common chimpanzee.

Three mating patterns exist among chimpanzees: promiscuous, noncompetitive, opportunistic matings; possessive matings; and consortships. Opportunistic matings are frequent, brief copulations that may be initiated by the male or the female. Possessive matings occur in the context of short-term relationships between a male and female during which the male prevents lower-ranking rivals from mating with the female. Consortships occur when the male and female leave the group and avoid other chimpanzees. Conceptions are most likely to occur during consortships. Females exercise mate choice first through their selection of a community in which to live and second by cooperating in the formation of consortships (Wrangham and Smuts, 1980; Tutin, 1980).

RECONSTRUCTING EARLY HUMAN BEHAVIOR

Each allocatarrhine social system has different implications for both how a female selects mating partners and how many partners she may have. The ancestral form(s) of the human social system will never be known with certainty, because it is nearly impossible to reconstruct social behavior from fossils. If alloprimate social systems and mating patterns were optimal adaptations, it might be a relatively simple matter to identify the living alloprimate type whose habitat most closely matches that of early hominids and then to assume that the social system and mating patterns of our ancestors were the same as those shown by that species. Primates with different social systems and mating patterns have, however, been found in the same habitat, and, furthermore, primatologists have not demonstrated that primate social systems are optimal adaptations. Rowell (1979:20) has noted that the "null hypothesis which must be accepted, until an alternative can be demonstrated to fit the observations better, is that the differences in social organization are the result of random drift, by which I understand that each species could use the same food sources and avoid predation as successfully if it were organized like one of the others."

We cannot, therefore, easily reconstruct the social systems or mating patterns of our early hominid ancestors, because we cannot at this time match a primate social system or mating pattern with any particular habitat. That being the case, we are left to speculate. The problem with speculation is that "we are cursed with our own mental agility . . ." (Rowell, 1979:20). That is, our mental agility allows us to imagine adaptive explanations for almost any set of circumstances.

Nonetheless, we may be able to reach some conclusions about ancestral humans' social and mating systems.

Because a lack of sexual dimorphism in body size is associated with monogamy among the alloprimates, the high degree of such sexual dimorphism in the Plio-Pleistocene hominids suggests that our early ancestors had a social system that included nonmonogamous mating patterns. Ruling out a monogamous "gibbon-type" social system, however, leaves several other possibilities: polygynous male/monandrous female social units (e.g., like gorillas and geladas); one-male social units in which mating is polygamous for both males and females (e.g., as in patas and blue monkeys); multi-male/multi-female troops in which mating is promiscuous (e.g., similar to rhesus and Japanese macaques, and savannah baboons); a fission-fusion social system in which both males and females have multiple partners (e.g., like chimpanzees and bonobos); a solitary social system in which the home ranges of males and females overlap, mating is opportunistic, and adults forage alone and only rarely congregate for even short periods (e.g., as in orang-utans).

Primate social systems in which there is more than one adult male and in which both males and females have multiple mating partners are more common than the one-male systems with restricted options for females. Plio-Pleistocene hominids, therefore, probably lived in social units containing more than one resident male member, and both females and males had multiple sex partners. They likely exhibited both opportunistic matings and consortships. Group type was probably either fission-fusion like that of chimpanzees or multimale/multifemale like that of savannah baboons. These speculations are supported by the finding that humans and chimpanzees have large specialized penises and large testes compared to gorillas and orang-utans (Smith, 1984). This suggests that, unlike gorillas and orang-utans, human males have evolved within a promiscuous mating structure similar to—but not nearly as extreme as—that of chimpanzees. According to Harris (1989:196), it "is certainly human nature to have a powerful sex drive and appetite, and it is certainly human nature to be able to find diverse ways of satisfying these species-given needs and appetites. But is not human nature to be exclusively promiscuous or polyandrous or monogamous or polygynous."

Most of the extant allocatarrhine females maintain contact with female friends (perhaps agemates) and with their female kin. Female kin groups take the form of matrifocal units. Plio-Pleistocene hominid females also likely maintained contact with female friends, agemates, and female kin. Depending on the local traditions and/or environmental conditions, social systems may have varied over the areas inhabited by our early ancestors.

In general, when male scholars discuss human evolution they tend to emphasize an economic role for males and a sexual and reproductive role for females. Females are often portrayed as passive and economically dependent on males (e.g., see Lovejoy, 1981). Because Old World monkey and ape females are neither passive nor economically dependent on males, I reject as unfounded any speculations based on the view that early hominid females' roles were limited to their sexual services and reproductive capacities.

In contrast, female scholars tend to downplay the role of female sexuality in human evolution. Following a model based on the lifeways of chimpanzees (generally thought by anthropologists to be our closest evolutionary relatives), female scholars often view early human females as relatively independent of males and emphasize an economic and socialization role for females in early human societies (Tanner, 1988). Occasionally, however, it has been argued that female mating choice of sociable males played an important role in human evolution because it reduced the physical and behavioral differences between males and females (Tanner, 1981).

One of the most resilient interpretations of human evolution involves the idea that our ancestors evolved psychological structures that inclined them (via pair-bonding or "falling in love") toward the formation of long-term monog-amous relations for mating and raising young. According to this interpretation, the results of our past adaptation are reflected in today's nuclear family, a gen-der-based division of labor, and the institution of marriage. The basis for this interpretation seems to be a belief among students of human evolution that the formation of a pair-bonded nuclear family is a universal aspect of human behavior that evolved due to its beneficial effect on offspring survival (Hara-way, 1988). Part of this interpretation is the idea that a male–female pair bond is necessary to ensure female fidelity and male parental certainty. The assump-tion is that a human male will heavily invest in his wife's offspring only if he is reasonably certain that he is the father (Tooby and DeVore, 1987). Further-more, early hominid archeological sites (such as those at Olduvai Gorge in East Africa) have traditionally been though to be "home bases." From these find-ings, scholars have inferred that gender-based divisions of labor and food shar-ing were established early in the evolution of our species (Potts, 1988).

I suggest, in light of recent discoveries, that few of these propositions survive careful scrutiny. For example, Potts (1988) questioned the assumption that the Plio-Pleistocene archeological sites at Olduvai Gorge were home bases and sug-gested instead that they were places where bones accumulated and stone tools were cached, somewhat analogous to the stone tool caches used by nut-crack-ing chimpanzees. In other words, evidence is lacking that the lives of our early ancestors revolved around a home base similar to those of modern hunter-gath-erers. That being the case, the arguments that our early ancestors had a division of labor based on gender or that they shared food lose most of their force.

Furthermore, although all extant human cultures have marriage practices, neither the nuclear family nor love/pair bonding is a universal human trait or a necessary aspect of marriage. For example. Leibowitz (1978) reported that 42% of hunter-gatherers do not favor nuclear families (e.g., see Goodale, 1971, on the Tiwi), and in many societies marriages are arranged (to a greater or lesser degree) and love between the husband and wife is not a concern (Frayser, 1985). An examination of the role of women and polygyny in Islamic societies led Mernissi (1987) to argue that love between a husband and wife has been viewed as detrimental to society. Moreover, I am not convinced that paternal confidence is an important variable in the equation of male–female relation-ships and child care. In a holocultural study, Gray and Wolfe (1984) found no correlation between the amount of interaction between fathers and offspring and the degree to which a society tolerated female extramarital sexual activity.

I am not arguing that men and women do not pair bond, only that pair bonding is not a necessary or universal feature of marriage. Western culture's insistence that love, marriage, female fidelity and dependence, and a nuclear family should occur together leads us to assume all too quickly, and without convincing evidence, that our early ancestors must have exhibited these traits. Fox (1980:140) has clearly stated the relationship between marriage, pair bonding, and mating systems.

> In particular, it seems to be downright unimaginative to push dreary monogamy onto these enterprising creatures as so many commentators seem to want to do—reflecting their own culture-bound obsession with the nuclear family more than anything in the record of nature. It also does not help, and is positively misleading, to pose the problem as the origin of the pair bond in hominids. This is simply a bad analogy with an instinctive process in lower animals. Humans do not pair bond in this sense at all. Marriage is a legal, rule-governed institution, not a direct expression of instinctive drives.

Data from some modern societies suggest that men and women sometimes have difficulty cooperating in child care and subsistence activities, a fact that calls into question the assumption that humans are a species in which males have evolved through natural selection a tendency to provision women and children. For example, analyses of development schemes have shown that in many cases women and their children are worse off nutritionally under conditions of cash cropping than under the traditional system of women's subsistence farming. Often land traditionally used by women for growing food for household consumption is diverted to cash crops. Furthermore, a woman forced to labor on cash crops frequently neglects subsistence crops destined for the household. Money earned from the cash crops goes to the husband, who expects the woman to feed the household from her own farming projects on lands not devoted to cash crops. Unfortunately, women are often unable to do both cash cropping and traditional subsistence farming, and, consequently,the nutrition of both women and children suffers (e.g., see Engberg, et al., 1988).

The subsistence relationships between men, women, and children may take many forms, however. For example, many societies have living arrangements such that several women cooperate in the securing and processing food. The women may be neighbors (related and nonrelated), cowives, and/or sisters- and mothers-in-law in extended families. (See Turnbull [1981] on the Mbuti; Brown [1975] on Iroquois women; and Draper [1975] on the !Kung for examples of such societies.) O'Connell et al. (1988) reported that among the Hadza of Tanzania women forage in groups for plant foods, tortoises, and small mammals. They reported, moreover, that a group of adult Hadza women drove a leopard from a kill with their digging sticks.

Unlike our society, which views child care as primarily the responsibility of the child's mother, many societies have alternative child-care arrangements.

Weisner and Gallimore (1977) researched child care cross-culturally using a holocultural sample of 186 societies. They reported that for infancy the mother was the primary caretaker in 46.2% of the societies. In 40% of the societies, however, infants were cared for by persons other than the mother (12.9% of the societies could not be coded). Older children and other female family members accounted for most of the caretaking not provided by the mother. For early childhood (i.e., toddlers), the mother was the principal caretaker in fewer than 20% of the societies, and "about a third of the societies were rated as settings where children spent half or less of their time with the mother" (Weisner and Gallimore, 1977:170). Peer groups, older children, and adults were found to be the principal caretakers of children in early childhood (also see Minturn and Lambert, 1964). Weisner and Gallimore (1977:170) concluded that "in the majority of societies mothers are not the principal caretakers or companions of young children." Subsistence activities most often remove the mother from the home and away from her children. Oakley (1985:141) reported that "studies of women's role in agriculture undertaken between 1940 and 1962 show that . . . women did between 60% and 80% of the total agricultural work in a sample of African peoples living in Senegal, Gambia, Nigeria, Uganda and Kenya," Weisner and Gallimore (1977) described child caretakers as having the responsibility of bathing, toilet training, and in peasant societies carrying a hungry infant to the mother who is laboring in her fields. In addition, "[p]assing babies from hand to hand and breast to breast is commonly done in non-Western societies. . . ." (Leibowitz, 1978:201; see also Oakley, 1985). From these studies, it seems clear that many societies include alternative child care arrangements so that women can carry on subsistence activities such as foraging, farming, and marketing.

In addition to women assisting each other in child care and subsistence activities, there is the suggestion that women need other women friends for their psychological well-being. Hite (1987) found that a majority of U.S. women—married or unmarried—count other women as their best friends and feel more comfortable talking to other women than to men, even their husbands. She concludes that it "is vital to have women's friendships, groups, and even communities, or just time for women to spend together, be friends, because in this way women can be themselves, think their own thoughts, say them—and in sharing come up with new insights, ideas, and possibilities" (Hite, 1987:720).

From the data on nonhuman primates and living humans, I conclude that the social system of our Plio-Pleistocene ancestors was one in which females provided for themselves and their offspring by foraging for plants and insects, much as chimpanzee females do today (Tanner, 1988). There is also the possibility that they scavenged animal carcasses. Mating was based on short-term relationships, and long-term pair bonds between males and females did not form. As brain size began to increase and infants were born more helpless (perhaps at the early *Homo erectus* stage), the ability to form stable cohorts of cooperating females based on matrifocal units and female–female friendships was exploited and became vital to the survival of females and their offspring. As brain size increased further and social systems and culture became more complex (perhaps beginning with archaic *H. sapiens*), several changes occurred: clans and lineages arose, as well as methods by which members were inducted; intrasocietal differences in wealth, power, and/or status developed; trade networks were established; populations expanded, weapons became more lethal, and warfare increased; and marriage came into existence. Marriage, in other words, is not an ancient adaptation for harvesting food resources for women and children; rather, it is a relatively recent institution forged to cope with the political and economic problems confronting archaic and modern *Homo sapiens*. Those political and economic problems were inducting offspring into families, lineages, clans, etc.; the regulation of warfare and the distribution of wealth and trade networks through the control of the number and acquisition of mating partners; and the facilitation of alliance formation mainly through the exchange of women (Fox, 1980).

Modern societies exhibit much variation in their political and economic systems, and it should not surprise us that both the presence and the intensity of pair bonding within marriages and the extent to which the sexual activities of spouses are controlled vary cross culturally. Our Western image of marriage and the importance of a sexual division of labor affects our interpretation of how women gather resources by emphasizing the role of husband–wife transactions. I believe that a more careful examination of the domestic economies of nonstate societies would demonstrate that many societies still exhibit the ancient practice of women working in groups and provisioning their offspring more or less independently of men (e.g., see Turnbull [1981] on the Mbuti; Brown [1975] on Iroquois Women; Draper [1975] on the !Kung; O'Connell et al. [1988] on the Hadza; and the articles in Poats et al. [1988]).

CONCLUDING REMARKS

The evidence presented in this chapter indicates that human females' sexual behavior does not set them dichotomously apart from allocatarrhine primate females. Human and other primate females experience orgasm when adequately stimulated and engage in heterosexual and homosexual behaviors depending on species-specific mating patterns and social systems, their particular social situation, their hormonal status, and individual preferences. Important social variables influencing female sexuality include whether the females are captive/imprisoned or wild, the presence or absence of novel males or other appropriate partners, and the degree to which males and/or society restrict their behavior.

Ovulation in human females and many other catarrhine primates is not marked by visible sexual swellings of the external genitalia, changes in the color of the face, etc. Nonetheless, ovulation in women and in other nonswelling allocatarrhines may be marked by changes in attractiveness and proceptivity (but not necessarily by changes in receptivity, which is more or less constant over the monthly menstrual cycle).

Sexual interactions among humans and other primates involve more than merely mating for reproduction. Mating also provides pleasure, functions to reconciliate dyads after a dispute, forges social bonds, and can be used to manipulate social and feeding situations (see de Waal [1989] on the Kama Sutra primates). The trend in catarrhine evolution is in the direction of decoupling the control of sexual behavior from hormones and increasing the role of learning and individual preference. This trend parallels increases in the size, complexity, and learning capacities of the brain. I leave it to paleoneurologists to clarify the details of the relationship between the evolution of the primate brain and the development of sexual behavior in primates, including humans.

It is significant that the same theorists who create a gulf between the sexual behavior of humans and the alloprimates are often quick to equate human marriage with the pair bonding observed in some alloprimates. It is an oversimplification to equate marriage (even monogamous marriages) and pair bonding. Marriage is a legal contract usually made between two kin groups, specifying the disposition of offspring, distribution of wealth, roles and obligations of spouses, etc. Pair bonding is an emotional connection or closeness between two individuals that is long-lasting (Money, 1980a). The Euro-American folk model assumes that marriage will be based on love (or pair bonding), but this is not true in the folk models of other societies. In many societies, marriages are arranged and a pair bond is never formed between husband and wife. Marriages are stable in the absence of pair bonds, however, because of legal contracts that spell out the roles and obligations of the husband, the wife, and their kin groups. I have not claimed, however, that humans are incapable of pair bonding. Clearly, pair bonds are formed between mothers (and occasionally fathers) and their offspring and between friends (both opposite gender and same gender).

The notion that male–female pair bonds evolved in humans to attach a male to the mother–infant(s) unit needs to be critically evaluated by questioning the assumption that males play(ed) a significant tole in the actual care of infants and children. Within certain social strata in industrialized societies (such as our own), males do play a significant role in providing women and their offspring with food and other needs because women were (are) denied an economic role. We cannot, however, generalize from our folk models of human behavior to the rest of humanity. Instead, we need to examine how women in other societies mate (as well as marry), feed themselves and their offspring, and form and maintain female work groups. In addition, we need to explain the reason(s) why the social role of the "father" was created by humans, because this social role, and the cultural and mythical importance attached to it, is one of the unusual features of human society.

In our search for the evolutionary origins of human sexual behavior, we frequently forget that there is a unique influence on human sexuality: culture (which I understand to mean a set of shared symbols based on language that forms human reality and is used to generate behavior). As was pointed out by Gray and Wolfe (1988:668),

> [I]n every society, sexuality has a social and cultural meaning beyond the biological role of reproduction. The classification of sexual relationships as correct or incorrect

and of specific sexual acts as normal or abnormal always carries over into nonsexual areas such as gender roles and images of the world. Thus, there is no such thing as meaningless sex. Even the briefest sexual encounter between strangers forces them to reflect on cultural themes such as the distinction between sex within marriage and sex outside marriage, the distinction between masculine and feminine gender roles (even if the sexual behavior is homosexual) and so on. Even if a couple is not aware of these themes, they influence the couple's interaction by affecting everything from the manner in which they use their bodies to give and to receive pleasure to how they will think about the sex act and about each other when the physical act is over.

Genital mutilation rituals are a clear example of how culture creates a manifestation of sexuality that has no analogues among the alloprimates. In general, genital mutilation and other painful rituals are associated with rites of initiation in those horticultural societies in which gender is an important principle of social organization and strong fraternal interest groups are present (see discussion and references in Gray and Wolfe, 1988). Because the other primates lack culture, they do not mutilate their genitals as a means of expressing power, sexuality, and other concerns.

As was noted by Haraway (1988), the 1980s have been a time of dismantling the 1950s and 1960s "man, the hunter" models of human evolution. I hope this chapter's exploration of female sexual behavior will aid in dismantling culturally biased models of human evolution. I believe a greater familiarity with the cross-species and cross-cultural literature on female sexuality, combined with more attention to avoiding our folk models, will help scholars in the 1990s produce more complete and more accurate discourses on both primate behavior and human evolution.

ACKNOWLEDGMENTS

I thank Jim Loy and Breck Peters for their suggestions and editorial comments. I also thank my friends and colleagues, especially Robert Lawless, Maxine Margolis, Jean Gearing, and Pat Gray, who read the manuscript at various stages and offered their support, comments, disagreements, and editorial suggestions. Any errors or misinterpretations of the works of other researchers, however, are my own.

NOTES

1. Two reviewers of this chapter suggested that the position of the human clitoris might be related to ventral-ventral copulation in humans. I do not, however, believe that there is such a relation, because it appears that ventral-ventral copulations in humans and bonobos (de Waal, 1989) are due mainly to the forward position of the vagina. My understanding of Masters's and Johnson's (1966) research on humans is that no position results in direct stimulation of the clitoris during heterosexual intercourse. Orgasms do, however, occur in human females in response to manual stimulation of the vagina (Alzate, 1985) and during sexual intercourse presumably from the indirect stimulation of the clitoris and vaginal stimulation. On the other hand, if the clitoris of alloprimate females is located near the vaginal opening (as suggested in this chapter), then it would

seem that any copulatory position would result in some degree of direct stimulation to the clitoris and the facilitation of orgasm.

2. Smuts (1985:52) defined "friends" as adult male-female dyads that "showed high frequencies of grooming and spatial proximity throughout most or all of the study period."

REFERENCES

Adams, D. B., A. R. Gold, and A. D. Burt (1978). Rise in female-initiated sexual activity at ovulation and its suppression by oral contraceptives. *New England Journal of Medicine 299:*1145–1150.

Akers, J. S., and C. H. Conaway (1979). Female homosexual behavior in Macaca mulatta. *Archives of Sexual Behavior 8:*63–80.

Allen, M. L., and W. B. Lemmon (1981). Orgasm in female primates. *American Journal of Primatology 1:*1–15.

Altmann, J. (1980). *Baboon Mothers and Infants.* Cambridge, MA: Harvard University Press.

Alzate, H. (1985). Vaginal eroticism: a replication study. *Archives of Sexual Behavior 14:*529–537.

Ardrey, R. (1976). *The Hunting Hypothesis.* New York: Atheneum.

Bancroft, J., D. Sanders, D. Davidson, and P. Warner (1983). Mood, sexuality, hormones, and the menstrual cycle. III. *Psychosomatic Medicine 45:*509–516.

Beach, F. A. (1976). Sexual attractivity, proceptivity, and receptivity in female mammals. *Hormones and Behavior 7:*105–138.

Boinski, S. (1987). Mating patterns in squirrel monkeys *(Saimiri oerstedi). Behavioral Ecology and Sociobiology 21:*13–21.

Brown, J. K. (1975). Iroquois women: an ethnohistoric note. Pp. 235–251 in R. R. Reiter, ed., *Toward an Anthropology of Women.* New York: Monthly Review Press.

Burton, F. D. (1971). Sexual climax in female *Macaca mulatta.* Pp. 180–191 in *Proceedings of the Third International Congress of Primatology.* Basel: Karger.

Campbell, B. (1988). Testosterone, loss of estrus, and female proceptivity. Paper presented at the American Anthropological Association meetings.

Cheney, D. L., P. C. Lee, and R. M. Seyfarth (1981). Behavioral correlates of non-random mortality among free-ranging female vervet monkeys. *Behavioral Ecology and Sociobiology 9:*153–161.

Chevalier-Skolnikoff, S. (1974). Male-female, female-female, and male-male sexual behavior in the stumptail monkeys, with special attention to the female orgasm. *Archives of Sexual Behavior 3:*95–116.

——— (1976). Homosexual behavior in a laboratory group of stumptail monkeys *(Macaca arctoides):* forms, contexts, and possible social functions. *Archives of Sexual Behavior 5:*511–527.

Chism, J., T. E. Rowell, and D. Olson (1984). Life history patterns of female patas monkeys. Pp. 175–190 in M. F. Small, ed., *Female Primates: Studies by Women Primatologists.* New York: Alan R. Liss, Inc.

Coleman, E. M., P. W. Hoon, and E. F. Hoon (1983). Arousability and sexual satisfaction in lesbian and heterosexual women. *Journal of Sex Research 19*(1):58–73.

Cords, M., B. J. Mitches, H. M. Tsingalia, and T. E. Rowell (1986). Promiscuous mating among blue monkeys in the Kakamega Forest, Kenya. *Ethology 72*(3):214–226.

Count, E. W. (1973). On the idea of protoculture. Pp. 1–25 in E. W. Menzel, Jr., ed., *Precultural Primate Behavior*. Basel: Karger.

Dahl, J. F. (1988). External genitalia. Pp. 133–144 in J. H. Schwartz, ed., *Orang-utan Biology*. New York: Oxford University Press.

Davis, D. L., and R. G. Whitten (1987). The cross-cultural study of human sexuality. *Annual Review of Anthropology 16*:69–98.

Dixson, A. F. (1983). Observations on the evolution and behavioral significance of "sexual skin" in female primates. *Advances in the Study of Behavior 17*:63–106.

Draper, R. (1975). !Kung women: contrasts in sexual egalitarianism in foraging and sedentary contexts. Pp. 77–109 in R. R. Reiter, ed., *Toward an Anthropology of Women*. New York: Monthly Review Press.

Dunbar, R.I.M. (1980). Determinants and evolutionary consequences of dominance among female gelada baboons. *Behavioral Ecology and Sociobiology 7*:253–265.

——— (1988). *Primate Social Systems*. Ithaca, NY: Cornell University Press.

Eckstein, P. (1958). Internal reproductive organs. Pp. 542–629 in H. Hofer, A. H. Schultz, and D. Starck, eds., *Primatologia*. Basel: Karger.

Engberg, L. E., J. H. Sabry, and S. A. Beckerson (1988). A comparison of rural women's time use and nutritional consequences in two villages in Malawi. Pp. 99–110 in S. V. Poats, M. Schmink, and A. Spring, eds., *Gender Issues in Farming Systems Research and Extension*. Boulder, CO: Westview Press.

Fisher, S. (1973). *The Female Orgasm*. New York: Basic Books.

Ford, C. S., and F. A. Beach (1951). *Patterns of Sexual Behavior*. New York: Harper and Brothers.

Fox, R. (1980). *The Red Lamp of Incest*. New York: E. P. Dutton.

Frayser, S. G. (1985). *Varieties of Sexual Experience*. New Haven, CT: HRAF Press.

Galdikas, B.M.F. (1981). Orangutan reproduction in the wild. Pp. 281–300 in C. E. Graham, ed., *Reproductive Biology of the Great Apes: Comparative and Biomedical Perspectives*. New York: Academic Press.

——— (1985). Adult male sociality and reproductive tactics among orangutans at Tanjung Puting. *Folia Primatologica 45*:9–24.

Goldfoot, D. A., H. Westerborg-Van Loon, W. Groeneveld, and A. K. Slob (1980). Behavioral and physiological evidence of sexual climax in the female stumptailed macaque *(Macaca arctoides)*. *Science 208*:1477–1478.

Goodale, J. C. (1971). *Tiwi Wives*. Seattle: University of Washington Press.

Goodall, J. (1986). *The Chimpanzees of Gombe*. Cambridge, MA: Harvard University Press.

Graham, C. E., and C. F. Bradley (1972). Microanatomy of the chimpanzee genital system. Pp. 77–126 in G. H. Bourne, ed., *The Chimpanzee: Histology, Reproduction, and Restraint*. Baltimore: University Park Press.

Gray, J. P. (1985). *Primate Sociobiology*. New Haven, CT: HRAF Press.

Gray, J. P., and L. D. Wolfe (1983). Human female sexual cycles and the concealment of ovulation problem. *Journal of Social and Biological Structures 6*:345–352.

——— (1984). Correlates of monogamy in human groups: tests of some sociobiological hypotheses. *Behavior Science Research 18*:123–140.

——— (1988). An anthropological look at human sexuality. Pp. 650–678 in W. H. Masters, V. E. Johnson, and R. C. Kolodny, *Human Sexuality*, 3rd. Boston: Little, Brown and Co.

Hanby, J. P., and C. E. Brown (1974). The development of sociosexual behaviours in Japanese macaques *(Macaca fuscata)*. *Behaviour 49*:152–196.

Haraway, D. (1988). Remodeling of human way of life. Pp. 206–259 in G. W. Stocking,

Jr., ed., *Bone, Bodies, Behavior: Essays on Biological Anthropology*. Madison: The University of Wisconsin Press.

Harcourt, A. H., D. Fossey, K. J. Steward, and D. P. Watts (1980). Reproduction in wild gorillas and some comparisons with chimpanzees. *Journal of Reproduction and Fertility, Supplement 28:*59–70.

Harcourt, A. H., K. J. Steward, and D. Fossey (1981). Gorilla reproduction in the wild. Pp. 265–279 in C. E. Graham, ed., *Reproductive Biology of the Great Apes: Comparative and Biomedical Perspectives*. New York: Academic Press.

Harding, R. S., and D. K. Olson (1986). Patterns of mating among male patas monkeys *(Erythrocebus patas)* in Kenya. *American Journal of Primatology 11:*343–358.

Harris, M. (1989). *Our Kind*. New York: Harper and Row.

Hill, E. M. (1988). The menstrual cycle and components of human female sexual behavior. *Journal of Social and Biological Structures 11:*443–455.

Hill, E. M., and P. A. Wenzel (1981). Variation in ornamentation and behavior in a discoteque for females observed at differing menstrual phases. Paper presented at the Animal Behavior Society meetings.

Hite, S. (1976). *The Hite Report*. New York: MacMillan Publishing Co.

———— (1987). *Women and Love: A Cultural Revolution in Progress*. New York: Alfred A. Knopf.

Hrdy, S. B. (1977). *The Langurs of Abu*. Cambridge, MA: Harvard University Press.

———— (1981). *The Woman that Never Evolved*. Cambridge, MA: Harvard University Press.

Hrdy, S. B., and P. L. Whitten (1987). Patterning of sexual activity. Pp. 370–384 in B. B. Smuts, et al., eds., *Primate Societies*. Chicago: University of Chicago Press.

Hunt, M. (1975). *Sexual Behavior in the 1970s*. New York: Dell Books.

Johnson, D. F., and C. H. Phoenix (1976). Hormonal control of female sexual attractiveness, proceptivity, and receptivity in rhesus monkeys. *Journal of Comparative and Physiological Psychology 90:*473–483.

Kano, T. (1980). Social behavior of wild pygmy chimpanzees *(Pan paniscus)* of Wambe: A preliminary report. *Journal of Human Evolution 9:*243–260.

Kinsey, A. C., W. B. Pomeroy, D. E. Martin, and P. H. Gebhard (1953). *SexualBehavior in the Human Female*. Philadelphia: W. B. Saunders Co.

Kinzey, W. G. (1987). Monogamous primates: a primate model of human mating systems. Pp. 105–114 in W. G. Kinzey, ed., *The Evolution of Human Behavior: Primate Models*. Albany: State University of New York Press.

Kohler, W. (1959). *The Mentality of Apes*. New York: Vintage Books.

Kummer, H. (1971). *Primate Societies: Group Techniques of Ecological Adaptation*. Chicago: Aldine.

Leibowitz, L. (1978). *Females, Males, Families: A Biosocial Approach*. North Scituate, MA: Duxbury Press.

Lovejoy, C. O. (1981). The origin of man. *Science 211:*341–350.

Loy, J. (1970). Peri-menstrual sexual behavior among rhesus monkeys. *Folia Primatologica 13:*286–297.

———— (1987). The sexual behavior of African monkeys and the question of estrus. Pp. 175–195 in E. L. Zucker, ed., *Comparative Behavior of African Monkeys*. New York: Alan R. Liss, Inc.

Masters, W. H., and V. E. Johnson (1966). *Human Sexual Response*. Boston: Little, Brown and Co.

Masters, W. H., V. E. Johnson, and R. C. Kolodny (1988). *Sex and Human Loving*. Boston: Little, Brown and Co.

Matteo, S., and E. F. Rissman (1984). Increased sexual activity during the midcycle portion of the human menstrual cycle. *Hormones and Behavior 18*:249–255.

Mead, M. (1961). Cultural determinants of sexual behavior. Pp. 1433–1479 in W. C. Young, ed., *Sex and Internal Secretions.* Baltimore: Williams and Wilkins.

Mernissi, F. (1987). *Beyond the Veil: Male-female Dynamics in Modern Muslim Society.* Bloomington, IN: Indiana University Press.

Minturn, L., and W. W. Lambert (1964). *Mothers of Six Cultures.* New York: John Wiley and Sons.

Mitani, J. C. (1985). Mating behavior of male orangutans in the Kutai Game Reserve, Indonesia. *Animal Behavior 33*:392–402.

Mitchell, G. (1979). *Behavioral Sex Differences in Nonhuman Primates.* New York: Van Nostrand Reinhold Co.

Money, J. (1980a). *Love and Love Sickness.* Baltimore: Johns Hopkins University Press.
——— (1980b). Genetic and chromosomal aspects of homosexual etiology. Pp. 59–72 in J. Marmor, ed., *Homosexual Behavior.* New York: Basic Books.

Oakley, A. (1985). *Sex, Gender and Society.* London: Gower Publishing Co.

O'Connell, J. F., K. Hawkes, and N. Blurton Jones (1988). Hadza scavenging: implications for plio/pleistocene hominid subsistence. *Current Anthropology 29*:356–363.

Poats, S. V., M. Schmink, and A. Spring, eds. (1988). *Gender Issues In Farming Systems Research and Extension.* Boulder: Westview Press.

Potts, R. (1988). *Early Hominid Activities at Olduvai.* New York: Aldine De Gruyter.

Preuschoft, H., D. J. Chivers, W. Y. Brockleman, and N. Creel, eds. (1984). *The Lesser Apes: Evolutionary and Behavioural Biology.* Oxford: The Alden Press.

Rowell, T. E. (1979). How would we know if social organization were not adaptive? Pp. 1–22 in I. S. Bernstein and E. O. Smith, eds., *Primate Ecology and Human Origins.* New York: Garland STPM Press.
——— (1988). Beyond the one-male group. *Behavior 104*:189–210.

Saayman, G. S. (1970). The menstrual cycle and sexual behavior in a troop of free ranging chacma baboons *(Papio ursinus). Folia Primatologica 12*:81–110.

Schreiner-Engle, P., R. C. Schiavi, H. Smith, and D. White (1981). Sexual arousability and the menstrual cycle. *Psychosomatic Medicine 43*:199–214.

Schurmann, C. L. (1981). Courtship and mating behavior of wild orangutans in Sumatra. Pp. 130–135 in A. B. Chiarelli and R. S. Corruccini, eds., *Primate Behavior and Sociobiology.* Berlin: Springer-Verlag.

Schurmann, C. L., and J.A.R.A.M. van Hoof (1986). Reproductive strategies of the orang-utan: new data and a reconsideration of existing sociosexual models. *International Journal of Primatology 7*:265–287.

Simonds, P. E. (1965). The bonnet macaque in south India. Pp. 175–196 in I. DeVore, ed., *Primate Behavior: Field Studies of Monkeys and Apes.* New York: Holt, Reinhart and Winston.

Smith, R. L. (1984). Human sperm competition. Pp. 601–659 in R. L. Smith, ed., *Sperm Competition and the Evolution of Animal Mating Systems.* New York: Academic Press.

Smuts, B. B. (1985). *Sex and Friendship in Baboons.* New York: Aldine.

Smuts, B. B., D. L. Cheney, R. M. Seyfarth, R. W. Wrangham, and T. T. Struhsaker, eds. (1986). *Primates Societies.* Chicago: University of Chicago Press.

Spuhler, J. N. (1979). Continuities and discontinuities in anthropoid-hominid behavioral evolution: bipedal locomotion and sexual receptivity. Pp. 454–461 in N. A.

Chagnon and W. G. Irons, eds., *Evolutionary Biology and Human Social Behavior: An Anthropological Perspective.* North Scituate, MA: Duxbury Press.

Strum, S. C. (1987). *Almost Human: A Journey into the World of Baboons.* New York: Random House.

Symons, D. (1979). *The Evolution of Human Sexuality.* New York: Oxford University Press.

Talmage-Riggs, G., and S. Anschel (1973). Homosexual behavior and dominance hierarchy in a group of captive female squirrel monkeys *(Saimiri sciureus). Folia Primatologica 19:*61–72.

Tanner, N. M. (1981). *On Becoming Human.* Cambridge, England: Cambridge University Press.

——— (1988). Becoming human: our links with our past. Pp. 127–140 in T. Ingold, ed., *What Is an Animal?* Boston: Unwin Hyman.

Thompson-Handler, N., R. K. Malenky, and N. Badrian (1984). Sexual behavior of *Pan paniscus* under natural conditions in the Lomako Forest, Equateur, Zaire. Pp. 347–368 in R. L. Susman, ed., *The Pygmy Chimpanzee: Evolutionary Biology and Behavior.* New York: Plenum Press.

Tilson, R. L. (1981). Family formation strategies of kloss' gibbons. *Folia Primatologica 35:*259–287.

Tokuda, K., R. C. Simons, and G. D. Jensen (1968). Sexual behavior in a captive group of pigtailed monkeys *(Macaca nemestrina). Primates 9:*283–294.

Tooby, J., and I. DeVore (1987). The reconstruction of hominid behavioral evolution through strategic modeling. Pp. 183–237 in W. G. Kinzey, ed., *The Evolution of Human Behavior: Primate Models.* Albany: State University of New York Press.

Turnbull, C. M. (1981). Mbuti womenhood. Pp. 205–219 in F. Dahlberg, ed., *Women the Gatherer.* New Haven, CT: Yale University Press.

Tutin, C.E.G. (1980). Reproductive behaviour of wild chimpanzees in the Gombe National Park, Tanzania. *Journal of Reproduction and Fertility, Supplement 28:*43–57.

Vance, E. B., and N. W. Wagner (1976). Written description of orgasm: a study of sex differences. *Archives of Sexual Behavior 5:*87–98.

Waal, F.B.M. de (1989). *Peacemaking among Primates.* Cambridge, MA: Harvard University Press.

Wallen, K., and R. W. Goy (1977). Effects of estradiol benzoate, estrone, and propionates of testosterone or dihydrotestosterone on sexual and related behaviours of ovariectomized rhesus monkeys. *Hormones and Behavior 9:*228–248.

Weisner, T. S., and R. Gallimore (1977). My brother's keeper: child and sibling caretaking. *Current Anthropology 18:*169–190.

White, F. J. (1988). Party composition and dynamics in *Pan paniscus. International Journal of Primatology 9:*179–193.

Wolfe, L. D. (1979). Behavioral patterns of estrous females of the Arashiyama West troop of Japanese macaques *(Macaca fuscata). Primates 20:*525–534.

——— (1984a). Japanese macaque female sexual behavior: a comparison of Arashiyama East and West. Pp. 141–157 in M. F. Small, ed., *Female Primates: Studies by Women Primatologists.* New York: Alan R. Liss, Inc.

——— (1984b). Female rank and reproductive success among Arashiyama B Japanese macaques *(Macaca fuscata). International Journal of Primatology 5:*133–143.

——— (1986). Sexual strategies of female Japanese macaques *(Macaca fuscata). Human Evolution 1:*267–275.

Wolfe, L. D. and J. P. Gray (1989). Alloprimate: a useful alternative. *Pan 8:*15–16.

Wolfe, L. D., and M.J.S. Noyes (1981). Reproductive senescence among female Japanese macaques *(Macaca fuscata fuscata). Journal of Mammology 62*:698–705.

Wolfheim, J. H., and T. E. Rowell (1972). Communication among captive Talapoin monkeys *(Miopithecus talapoin). Folia Primatologica 18*:224–255.

Wolman, B. B. (1973). *Dictionary of Behavioral Science.* New York: Van Nostrand Reinhold Co.

Wrangham, R. W., and B. B. Smuts (1980). Sex differences in the behavioral ecology of chimpanzees in the Gombe National Park, Tanzania. *Journal of Reproduction and Fertility, Supplement 28*:13–31.

Yerkes, R. (1939). Social dominance and sexual status in the chimpanzee. *Quarterly Review of Biology 14*:115–136.

Zuckerman, S. (1932). *The Social Life of Monkeys and Apes.* London: Routledge Kegan Paul.

Zumpe, D., and R. P. Michael (1968). The clutching reaction and orgasm in the female rhesus monkeys *(Macaca mulatta). Journal of Endocrinology 40*:117–123.

6

Male Sexual Behavior: Monkeys, Men, and Apes

RONALD D. NADLER
AND CHARLES H. PHOENIX

As was stated in the introductory chapter, the objective of the remaining individual chapters is to describe results of research on nonhuman primates from which we have learned something of significance about human behavior. If we hope to map the specifics of behavior, anatomy, and physiology, it is of course true that we cannot learn about one species by studying another. Nonetheless, if we take a wider, and perhaps more meaningful, view, it is possible to learn something about one species that is generally applicable to others. If, for example, we understand certain aspects of one biological system in a particular species, we can construct a hypothesis to test the applicability of our understanding of that system to another species. The extent to which results or hypotheses, in fact, have generality depends on a demonstration of consonance between data derived from the two species. This approach is especially productive when employed among species of close phyletic relationship. It is in this sense that the topic of male sexual behavior is addressed in this chapter.

Our approach to the study of behavior is essentially that of comparative psychology. Although we examine herein the research on nonhuman primates to derive hypotheses relevant to human sexual behavior, the process, in practice, does not function in one direction; the comparative approach is an interactive process among investigators using two or more different species, who cooperate by design or circumstance, in the study of similar phenomena. With respect to the study of male sexual behavior in humans and nonhuman primates, there is clearly no grand plan agreed upon by researchers. Furthermore, as suggested by the editors, there may even be serious reservations among some investigators of human behavior regarding the extrapolation of data on nonhuman primates to humans. Our purpose, in part, is to present the case for the usefulness of the nonhuman primate as a model for the human in the area of our interest.

It is noteworthy that several of the issues of current interest in the study of male sexual behavior, among both humans and nonhuman primates, have a history extending back more than 70 years. Two major issues reflected in the earliest studies of sexual behavior in nonhuman primates were derived, in fact, from a primary interest in human sexual behavior: (1) the significance of the male's role to the detection of cyclicity in female sexual behavior during the menstrual cycle (Hamilton, 1914) and (2) the biological basis of male homosexual behavior (Kempf, 1917). Both Hamilton and Kempf were psychiatrists who were interested in obtaining comparative data relevant to humans. Hamilton's position is stated quite clearly: ". . . we still lack that knowledge of infrahuman sexual life without which we may scarcely hope to arrive at adequately comprehensive conceptions of abnormal human sexual behavior" (1914:295). Others, such as Sokolowsky (1923), saw considerable value in the close relationship between humans and apes for the study of sexual behavior from the zoological perspective, "not merely with the description of the morphological construction of the sexual organs, but also with the investigation into the secondary sex characters, the physiological processes which make themselves evident in the sex life of these apes as well as the biological, psychological phenomena which are to be observed in the association of the sexes with each other" (p. 612).

With respect to the question of cyclicity in sexual behavior of nonhuman primates, research has shown that much of the early controversy and the disparate results derived in large part from problems of experimental design, methodology, and interpretation (e.g., inadequate knowledge of behavioral repertoires, imprecise definitions of behavior, and inadequate consideration of the differences in conditions under which observations were made). Other aspects of the controversy can be traced to the special characteristics of these species (e.g., their considerable behavioral adaptability, in particular the relative independence of their sexual behavior from hormonal regulation, and the relatively high and continuous sexual motivation of males, *under certain conditions*). Although this issue has been investigated extensively, one question that remains unanswered is the specific component of male sexual motivation implicated by the research, i.e., sexual arousal, the momentary level of sexual excitation, or sexual arousability, the rate of approach to orgasm (Whalen, 1968).

Sometimes, when research is progressing simultaneously in different species, it is not always clear who is learning from whom. In general, however, scholars working among human and nonhuman primates have obtained analogous and, perhaps, homologous data that have facilitated the progress of research by both. Comparable research on different species contributes to elucidation of their common characteristics as well as areas of difference. The common characteristics ideally reflect broad trends in biology, whereas the differences are important for describing diverse species-typical adaptations. When we have evidence suggesting the functional or adaptive significance of behavior in one species, we can apply that perspective to the comparable investigation of another species. The finding, for example, that different patterns of male sexual initiation among the great apes are related to differences in the number of males simultaneously competing for access to estrous females (Harcourt, 1981) not only contributes to clarification of the data on apes but also suggests that consideration of such an influence may be useful in characterizing the regulation of human sexual behavior.

In this chapter, we present data and analysis on several dimensions of sex research related to male primates.

1. The early organizational influences of hormones on sexual behavior of the adult
2. The influence of early social experience on sexual behavior of the adult
3. The contributions of adrenal and gonadal hormones to sexual behavior at the time of adrenarche and puberty, respectively
4. The phenomenon of estrus (or the hormonal regulation of female sexual behavior during the menstrual cycle) as it influences and is influenced by the male under various environmental conditions
5. The effect on sexual behavior of gonadectomy and hormone replacement
6. The influence of aging on sex hormone levels and sexual behavior
7. The impact of sexual selection on interspecies differences in the regulation of sexual behavior

We examine, therefore, both proximate and ultimate mechanisms but, in the tradition of psychology, focus primarily on the proximate ones. We also focus on those areas of research with which we are most familiar and in which we have conducted a significant amount of research. Therefore, primarily research on the rhesus monkey *(Macaca mulatta)* is described to address issues 1, 5, and 6 cited above, whereas research on the great apes is applied to the others.

ORGANIZATIONAL INFLUENCES OF HORMONES

Among mammals, including primates, the sex of an individual is dependent on the contribution of either an X or a Y chromosome by the male. The chromosomal sex of the individual, established at conception, determines whether the indifferent gonad develops into an ovary or a testis. If testes develop, their hormonal secretions result in the subsequent development of a phenotypic male, whereas, if an ovary develops, or in the absence of a gonad, the individual develops as a phenotypic female (Jost, 1953). This broad outline of sexual differentiation in mammals was established primarily from research on rodents, and, although many details remain to be elucidated, the general outline would seem to apply to sexual differentiation in primates as well, including humans (Wilson et al., 1984; Page et al., 1987).

The original hypothesis that androgenic substances present during prenatal development have an organizing action on the tissues mediating mating behavior was also based on research with rodents. In those experiments, genetic female guinea pigs treated prenatally with testosterone propionate (TP) (1) displayed mating behavior as adults that was similar to that of castrated males, when both were treated with TP, and (2) showed a reduced behavioral responsiveness to exogenous female hormones (Phoenix et al., 1959). The organizational hypothesis led to the prediction that genetic males deprived of endogenous testosterone before the period of sexual differentiation was complete would display a lower capacity for the display of male mating behavior and a greater capacity to display female receptive behavior when appropriately treated with hormones in adulthood (Young et al., 1964). Support for the hypothesis was obtained in a series of experiments in which male rats were castrated during the perinatal period and later (Grady et al., 1965).

What proved to be an important aspect of the research on sexual differentiation was the hypothesis that "the masculinity or femininity of an animal's behavior beyond that which is purely sexual . . . developed in response to certain hormonal substances within the embryo and fetus" (Phoenix et al., 1959). It was largely this aspect of the organizational hypothesis that led to the subsequent research on sexual differentiation in nonhuman primates, since it was reasoned that they would be better suited than rodents to test the hypothesis. Whether or not that reasoning was correct, research was initiated using rhesus macaques.

Previous research had shown that treatment with TP of female rhesus macaque fetuses masculinized the genital morphology of the animals (variously referred to as an intersex, female pseudohermaphrodite, and androgenized female). Whether there were any changes in the morphology of the brain or in behavior had not been investigated (van Wagenen and Hamilton, 1943; Wells and van Wagenen, 1954). To assess these latter issues, a series of behavioral studies was undertaken on young genetic rhesus females that were treated prenatally with TP by injecting their mothers during pregnancy. All these females were born with a well-formed penis, a scrotum, and no vagina. These androgenized females displayed more social threats, initiated play more frequently, and engaged in rough-and-tumble play more frequently than did control females. Mounting behavior also was displayed at an earlier age, with frequencies comparable to those of same-age males rather than to those of females (Phoenix et al., 1968; Goy and Resko, 1972).

The young androgenized females were studied initially with their ovaries intact and without exogenous hormone treatment. Genetic males that had been similarly treated prenatally with testosterone did not differ from untreated males in play behavior or in the display of sexual behavior. Attempts to castrate male rhesus macaques prior to the time that sexual differentiation was complete, as we had done in the rat, were not successful. We did succeed, however, in studying the behavior of one male that was castrated in utero on day 100 of the normal 168 day gestation period. The play and sexual behavior patterns of that male were comparable to those of the control group (Phoenix, 1974b). Quite obviously, castration earlier in gestation would have increased the probability of achieving the same effect in rhesus as had been obtained with perinatal castration in the rat (Grady et al., 1965).

When the androgenized female rhesus were between five and seven years old and ovariectomized, they were paired with ovariectomized, estrogen-treated control females. Under these conditions, the androgenized females displayed higher levels of aggression than the control females (Eaton et al., 1973). When the two groups were treated with TP, the prenatally treated females again displayed higher levels of aggression, but the rate of mounting did not differ significantly between the groups. One prenatally treated female, however, did achieve intromission and even ejaculated in two separate tests when paired with a sexually receptive female. Two other androgenized females, moreover, masturbated to ejaculation, but they did not achieve intromission or ejaculate in the tests of sexual behavior. The female that had ejaculated in copula later had a vagina surgically constructed and, after being treated with estradiol, was paired with adult males. Under these conditions, the female displayed the "present" response and one of the males mounted and achieved intromission (Phoenix et al., 1984).

As indicated earlier, when the androgenized rhesus females were infants and juveniles, they displayed social and sexual behavior characteristic of males. When they were between the ages of 15 and 17 years, however, their behavioral responsiveness to testosterone was much less striking (Phoenix and Chambers, 1982), and their mounting rate did not differ from that of control females. The basis of their diminished responsiveness to testosterone is not known, but it may have simply reflected the same kind of decreased responsiveness to this hormone that is observed in normal aging rhesus males. The year following the tests for male sexual behavior, the same prenatally androgenized females were treated with estradiol and paired with males in tests of female sexual behavior. They proved to be as receptive to male invitations to copulate as the similarly treated control females (Phoenix et al., 1983). The capacity of these females to respond to estradiol by displaying female receptive and proceptive behaviors when paired with adult males had not been assessed when they were younger, but it is unlikely that responsiveness to estradiol would increase in old age. Indeed, a later study with younger androgenized female rhesus showed that receptivity did not differ between them and untreated females (Pomerantz et al., 1985).

Elsewhere, it was reported that, when paired with adult males, androgenized and control females differed significantly only in mean frequency of solicits (Thornton and Goy, 1986). In this study, solicits represented a combination of eight different responses and, thus, a combined category of behaviors that had not been examined in other related research (Phoenix and Chambers, 1982; Phoenix et al., 1983). It is of some interest to note that no reliable differences were found between female rhesus monkeys androgenized by prenatal treatment with the nonaromatizable androgen dihydrotestosterone propionate (DHTP) and those treated with TP. This is in contrast to rodents, which do depend on the actions of an aromatizable androgen for masculinization and defemininization (Pomerantz et al., 1985, 1986, 1988).

Whether female rats ovulate in adulthood when treated perinatally with TP is dose dependent (Gorski and Barraclough, 1963). Female rhesus macaques, however, treated prenatally with various amounts of TP sufficient to produce morphological and behavioral masculinization are not prevented from ovulating, but the age at menarche is greater (Goy and Resko, 1972; Phoenix, 1974b;

Phoenix et al., 1968). Also relevant to differences between rodents and rhesus macaques in the organizing action of testosterone is that the neuroendocrine system of the male rhesus is capable of supporting normal endocrine cyclicity of transplanted ovarian tissue (Norman and Spies, 1986). This capacity is not present in the male rat unless it is castrated shortly after birth (Harris, 1964). This information on differences between rodents and primates was not available at the time when research was initiated on the masculinizing action of testosterone in rhesus macaques.

We were interested, however, in the view that the human being is psychosexually neutral at birth, a concept expounded by Hampson and Hampson (1961) and Money (1961). We reasoned that, if what had been shown to be true for the guinea pig and rat was also true for the rhesus macaque, it might also apply to humans (Young et al., 1964). If this were the case, gender role and sexual orientation of the human might not be determined solely by the sex of assignment at birth and life experiences as a child. This question has been reviewed critically by Diamond (1965). Indeed, a later report of genetic human males born with 5-alpha reductase deficiency and reared as girls indicated that these individuals routinely changed gender identity when they reached puberty. This finding suggests rather clearly that the development of gender role in humans involves something more than the sex of rearing (Imperato-McGinley et al., 1974). Most investigators agree that no one factor determines gender role, including hormones, at any stage of development.

The original "organizational hypothesis" stimulated considerable research, and modifications and reanalysis of the hypothesis have been offered (Arnold and Breedlove, 1985). What may be more appropriate at this time, however, is a redirection of efforts. A nod to genetic factors is no longer enough. Increased research into the specific roles played by genes may give us new insights into the process of psychosexual differentiation.

EARLY SOCIAL EXPERIENCE

Sustained interest in the role of early social experience on adult sexual behavior of nonhuman primates was initiated during H. C. Bingham's (1928) visit to the common chimpanzee *(Pan troglodytes)* colony maintained at the Abreu estate in Cuba. He was particularly curious about the sexual indifference of one captive-born adult male and speculated that its lack of sexual responsiveness might have resulted from a failure during development to observe copulation and intermale competition for females. Bingham (1928) subsequently conducted a study of four wild-born chimpanzees, 2 to 3 years of age, in an effort to document early experiential contributions to the development of sexual competency. He found that there was considerable variability in the behavior of his animals, reinforcing his view that the development of sexual behavior in this species was strongly influenced by early learning. His major finding was that elements of behavior that were originally observed in nonsexual interactions were observed at a later age in novel combinations and integrated into relations that were clearly sexual. He proposed that the capacity to perform adult sexual behavior competently developed through an association between certain aspects of early social interaction, especially those involving bodily contact, and the increase in sexual responsiveness that occurs with physical maturation. The main hypothesis derived from this research is that early social interaction and the opportunity to practice initially primitive patterns of behavior are essential to the development of mature patterns of sexual interaction in chimpanzees.

As a result of Bingham's inspiration and research, a comprehensive program of research was initiated with chimpanzees (Nissen, 1954) "to provide information about the physical, physiological, and behavioral development of chimpanzees, reared from birth under controlled and uniform conditions" (Riesen and Kinder, 1952:x). All the chimpanzees in this initial study (Nissen, 1954) were separated from their mothers within 1–2 days of birth and reared alone in separate cages such that they experienced no physical contact with conspecifics for the first 2 years of life. When these animals were 3 years of age, they were housed as pairs and then as small groups. They remained in these conditions until tests of sexual behavior were begun when they were 6–7 years of age. The unexpected finding was that, despite extensive pair tests of oppositely sexed partners within the group and between the socially deprived animals and wild-born and sexually experienced adults, none of the deprived animals showed any sexual behavior. Even after 2 years of cohabiting with oppositely sexed, sexually experienced partners, only one pair copulated among the 29 that were tested; the male of this pair was the only animal that had lived with older animals prior to its second year.

There followed a concerted effort to improve the sexual performance of the socially deprived animals by housing them individually with sexually experienced adults. Although all but one of the females were eventually "induced" to mate by experienced adult males, only one additional male was found to copulate. This animal was the only other male with some social experience prior to 2 years of age. These results supported Bingham's (1928) hypothesis that "a social learning process is critical for the full development of chimpanzee reproductive behavior" (Riesen, 1971:16). That captive rearing per se was not the cause of the inadequate sexual behavior was indicated by the finding that wild-born chimpanzees reared in captivity from infancy, such as those studied initially by Bingham (1928), generally exhibited apparently normal sexual behavior. The comparison of these wild-born, captive-reared chimpanzees with the early deprived ones suggests that it is the first 2 years of life during which social experience is most significant.

A final study of early social deprivation in chimpanzees investigated the effects of several different types of social restriction, ranging from a cage completely devoid of visual and manipulable stimuli to another that permitted the animals to interact through the bars. The animals were removed from these cages at age 3 years and housed in pairs and small groups until they were tested for sexual behavior at age 10 years. There appeared to be no difference among animals reared in the different conditions of social deprivation and all showed frequent and varied behavioral stereotypies (Davenport and Rogers, 1970). As in the previous study, most of the deprived females eventually "mated" with experienced males, and three of five deprived males eventually achieved intromission and two ejaculated (Rogers and Davenport, 1969). Initially, the females neither approached nor solicited males, and they responded sexually only to the advances of experienced males. With time, however, their frequency of presenting reached the level of wild-born adult females. The males that were subsequently successful sexually were tutored by experienced and persistent females that essentially effected intromission for the males.

The results of these studies (Nissen, 1954; Riesen, 1971; Rogers and Davenport, 1969) and data on other captive-born chimpanzees (Kollar et al., 1968; Lemmon, 1971) are consistent in showing that (1) social deprivation during the first 2 years of life has a profound and deleterious effect on the sexual performance of chimpanzees and (2) the deleterious effects appear to be greater and more enduring in the male than in the female. Although it was originally suggested that rearing by humans was even more detrimental to behavioral development than rearing in total social isolation, the later data did not support this conclusion. There is, in fact, some reason to propose the contrary, i.e., that providing infant male chimpanzees with "good human relationships" (Lemmon, 1971:436) may be, in some respects, an adequate substitute for early conspecific socialization. Moreover, in apparent contrast to the conclusion of Bingham (1928), Lemmon (1971) reported that sexual experience per se was not required during infancy and the juvenile period for the later performance of adequate sexual behavior.

A considerable number of comparable studies of early social deprivation and sexual behavior have been conducted on rhesus monkeys, with basically similar results (e.g., Mason et al., 1968; Missakian, 1969). In the rhesus macaques, moreover, the effects of being reared alone with the mother, in isosexual groups, and in heterosexual groups were also assessed (Goy et al., 1974; Wallen et al., 1981; Berkovitch et al., 1988). The results of this research indicate that males reared in the laboratory in social groups with mothers and peers copulate normally as adults. Those males that have only limited exposure to peers fail, as youngsters, to display the typical pattern of double foot-clasp mounting, a pattern of juvenile behavior considered essential for the development of adequate adult sexual behavior (Goy and Goldfoot, 1974). Rearing males in isosexual groups results in adult sexual behavior that in most respects resembles that of heterosexually reared males (Berkovitch et al., 1988). These results suggest that social interaction during the early years of life in primates, especially social interaction with peers, provides experience necessary for species-typical social interactions in adulthood, rather than providing sexual experience per se. In the absence of such early social learning, the deficient sexual behavior seems to result from "response tendencies incompatible with being an effective partner" (Wallen et al., 1981:308). Although species differences likely exist, it appears, for the two primate species considered, that a male must learn species-typical forms of social interaction with peers while it is still prepubertal if it is to interact effectively and mate competently when it is an adult. Somewhat similar results were obtained earlier with certain strains of guinea pigs (Valenstein et al., 1955). In addition to demonstrating the importance of early contact with conspecifics for the development of normal male sexual behavior, the research on guinea pigs suggested that, in this species, specifically sexual experience might be required.

ADRENAL AND GONADAL HORMONES DURING ADRENARCHE AND PUBERTY

The chimpanzee is an especially useful model for the human in endocrine studies of development because the chimpanzee more closely resembles the human, in several areas related to sexual maturation, than any other species available for research. Several studies, for example, have demonstrated that only the chimpanzee and gorilla *(Gorilla gorilla)* exhibit an adrenarche similar to that preceding puberty in human children (Cutler et al., 1978; Winter et al., 1980; Collins et al., 1981; Smail et al., 1982). The relatively rapid rise in adrenal steroids that characterizes adrenarche, especially the rises in dehydroepiandrosterone (DHA) and in DHA sulfate (DHAS), was initially thought to be the stimulus for puberty in humans (Ducharme et al., 1976; Hopper and Yen, 1975; Sizonenko and Paunier, 1975). Subsequent research, however, failed to support this hypothesis, since the normal temporal relationship between adrenarche and puberty was dissociated in certain pathological conditions of children (Grumbach et al., 1978; Sklar et al., 1980). Whether adrenarche has some influence on the development of sexual behavior, independent of its relationship to puberty, is of some interest in that the adrenal androgen DHAS is the most abundant steroid in the circulation of adult humans (Baulieu et al., 1965).

Puberty has been defined in terms of several endocrine and other physiological and anatomical measures, including a rapid increase in gonadotropin and gonadal steroid levels (Martin et al., 1977; Winter et al., 1980; Hobson et al., 1981; Nadler et al., 1987), menarche in the female (Coe et al., 1979), and increased testis size in the male (Kraemer et al., 1982: Nadler et al., 1987). Research on puberty in the chimpanzee revealed a pattern of hormonal changes essentially identical to that in humans (Hobson et al., 1981). Gonadotropins are elevated immediately after birth and decline to low levels thereafter, until the time of puberty. Concomitant with the high levels of gonadotropins during the first few months of life are relatively high levels of testosterone in the male. Testosterone then declines to low levels that persist until 6–10 years of age, the approximate age of puberty in chimpanzees. The variability in age at onset of puberty reflects, in part, differences in the quality of rearing conditions. As in human children, follicle-stimulating hormone (FSH) levels of prepubertal chimpanzees are higher in females than in males; there is no sex difference in luteinizing hormone (LH) levels, however. Between the ages of 3 and 9 years, FSH was higher than LH in nine of eleven chimpanzee females, whereas LH was higher than FSH in nine of ten males (Nadler et al., 1987). The lower FSH levels in males have been ascribed to the negative feedback action of testosterone; ovarian activity of females is minimal during this period (Hobson et al., 1981). The same reasoning could account for the sex difference in the ratios of FSH:LH.

In contrast to the considerable research conducted on the endocrine development of chimpanzees, there have been few studies in which behavioral development was investigated in conjunction with the measurement of hormones. Two studies, however, reported significant correlations between various aspects of behavior and the levels of gonadotropins and testosterone (Kraemer et al., 1982; Nadler et al., 1987). In general, affiliative behavior declined in males with age and increasing hormone levels, and solitary, aggressive, and sexual behavior increased. The comparable relationships for females were not significant. As in other nonhuman primates and human children, males initiated play more frequently than females and engaged in play more frequently and for longer periods of time. The developmental differences in behavior of the males, however, occurred at an earlier age than the main developmental increases in hormones. This implies that the behavioral differences occurred independently of increases in testosterone or, more likely, as a result of relatively small increases prior to puberty. Because the sensitivity to negative feedback effects of steroids is greater in relatively young chimpanzees compared to older ones (Nadler et al., 1985a), behavioral sensitivity to hormonal increases at these ages may also be relatively high.

There were no significant correlations between age and/or body weight with the levels of DHA and DHAS, in contrast to the results on gonadotropins and testosterone (Nadler et al., 1987), supporting the view that adrenarche and puberty are independent developmental phenomena in the chimpanzee, as in the human (Smail et al., 1982). The failure to find significant correlations between the levels of cortisol and DHA/DHAS is consistent with data on humans suggesting that the secretion of these adrenal steroids is controlled by different endocrine mechanisms (Parker et al., 1983). Given the relatively high levels of circulating DHAS in chimpanzees and humans, and no known function of DHAS, we sought to determine if the levels of DHA or DHAS were correlated with any of the behavioral categories recorded in the chimpanzees. The absence of significant correlations between any of the categories of behavior in male (or female) chimpanzees and DHA/DHAS levels provides no support for the view that these adrenal steroids influence behavioral development in this species. Clearly, however, further experimental studies are required before such a conclusion is accepted.

The research on adrenarche and puberty in chimpanzees demonstrates the value of the chimpanzee for investigating endocrine issues of interest in the human. The comparability of data on adrenal and gonadal hormone levels during development in chimpanzees and humans, together with significant correlations between behavioral development and gonadal, but not adrenal steroid levels in chimpanzees, supports the view that the former hormones, but not the latter, play a significant role in behavioral development of male chimpanzees (and men).

ESTRUS AND MALE SEXUAL BEHAVIOR

One of the earliest issues of interest in the study of sexual behavior of nonhuman primates was the degree to which the male and female restricted their mating to a limited period of the menstrual cycle, i.e., the relationship between estrus and mating. The primates were considered especially appropriate for extrapolating results to humans in that it was recognized early that primates, more than other mammals typically studied in the laboratory, seem to resemble humans in their sexual relations. In particular, the earliest reports suggested that nonhuman primates did not exhibit estrus, i.e., a restriction of female sexual arousal to about the time of ovulation, but rather mated throughout the female cycle (see Nadler, 1981; Nadler et al., 1986). The earliest studies, however, were conducted by workers with only limited knowledge of the species under investigation and therefore lacked appropriate experimental conditions.

When the first experimental studies of sexual behavior were conducted with chimpanzees (Yerkes and Elder, 1936; Yerkes, 1939; Young and Orbison, 1944), however, a different perspective was obtained. It was found that the chimpanzees, for the most part, mated during a restricted period of the menstrual cycle, associated with maximal anogenital swelling of the female, i.e., the phase of elevated estrogen levels (Graham, 1981; Nadler et al., 1985b). This led the investigators to propose that estrus was indeed characteristic of chimpanzees, since they appeared, on this basis, to satisfy the definition of estrus (Heape, 1900). Since some of the chimpanzees also mated at other times in the cycle, however, the issue was clouded. Yerkes (1939) reported that mating temporally dissociated from the presumptive time of ovulation took place under certain specific conditions, such as when (1) either the male or female was immature and/or inexperienced sexually, (2) the male and female were unfamiliar with each other, (3) the male was dominant over the female and the female was timid or fearful, and (4) the animals had no opportunity to see each other prior to initiation of a behavioral test. Under these conditions, mating could occur at any time during the cycle, as well as during menstruation, gestation, and lactation. There was the firm conclusion, moreover, that such mating was due primarily to the influence of the male, since it appeared that males intimidated or coerced females to mate at those times when the female was not maximally swollen. The characterization that emerged was that female chimpanzees resembled other mammals in that they were most sexually aroused during a limited phase of their cycle, indicative of a significant hormonal (estrogenic) influence on their behavior. In addition, however, they could adjust their behavior to that of the males when threatened with physical injury; i.e., they demonstrated some emancipation from hormonal regulation of their sexual behavior. The male chimpanzees in these studies clearly preferred to mate during the period of maximal female anogenital swelling, but they also initiated mating at other times, under the conditions described above. Subsequent research on chimpanzees in other captive settings generally supported this view of male sexual behavior (Allen, 1981; Wallis, 1982), although evidence was also presented that females sometimes initiated mating independently of cycle phase (Lemmon and Allen, 1978).

Laboratory research on the other great apes supported Yerkes' (1939) hypothesis regarding the differential regulation of mating by males and females, in terms of both species-typical patterns of mating and the behavior of individual animals (Nadler, 1981; Nadler et al., 1986). When western lowland gorillas *(G. g. gorilla)* were tested in traditional laboratory pair tests, it was primarily the females that initiated mating, and copulation was limited to a relatively restricted 1–4 day period of the female cycle closely associated with maximal tumescence of the female's perineal labia (Nadler, 1975, 1976). In other words, female sexual initiation was associated with periovulatory mating at the species level. As with the chimpanzees, however, mating also occurred at other phases of the cycle, as a result of the male's initiative and the female's accomodation (Nadler and Miller, 1982; Nadler et al., 1983). When orang-utans *(Pongo pygmaeus)* were tested under similar conditions, a very different pattern of interaction was observed, but one that also conformed to Yerkes' hypothesis (Nadler, 1977a). Males initiated essentially all the copulations in a forcible manner, and mating took place on a near-daily basis; i.e., male initiation accounted for nonperiovulatory mating. Among pygmy chimpanzees *(Pan paniscus)* tested in a group setting, mating was initiated primarily by the male and was differentially successful according to the female's dominance status in the group. The most subordinate female mated during a greater proporiton of the cycle and at lower levels of anogenital swelling than either of the more dominant females and less often refused sexual solicitations by the male (Dahl, 1987). In all the great apes, therefore, evidence suggests that mating occurs more frequently under test conditions in captivity than it does in the wild, and in all species it appears that such mating, dissociated from reproductive function, is due to the male's initiative.

Yerkes (1939) had proposed that support for his hypothesis that males rather than females accounted for mating dissociated from reproductive function required that the males be prevented from intimidating or coercing the females to mate. Nearly 50 years after this proposal, such studies were conducted with common chimpanzees, gorillas, and orang-utans. These investigations used a new pair-test paradigm called the restricted-access test (RAT) with female choice to differentiate it from the traditional laboratory "free-access" test (FAT). In the RAT with gorillas and orang-utans, the test was initiated with the male and female in separate cage compartments separated by a doorway that was sufficiently large to permit passage by the female, but not by the male. For the less sexually dimorphic chimpanzee, the female was able to regulate sexual access by pressing a lever that opened the door to the male's compartment. The results of these tests for all three species supported Yerkes' hypothesis. For eight of nine pairs of chimpanzees tested, mating occurred less frequently during the entire cycle and was more closely associated with the presumptive time of ovulation in the RATs, in comparison to the FATs (Nadler et al., 1989). The results were even more dramatic for the orang-utans, in which there was frequently a complete reversal of the male's and female's roles in initiating copulation (Nadler, 1988b). Whereas in the FATs orang-utan males initiated mating forcibly on almost every day of the cycle, in the RATs the males, during midcycle, frequently reclined on their backs and solicited the females with penile erections. The males, so aggressive when they could physically coerce the females to copulate, became quite passive when prevented from doing this and in fact, allowed the females to mount them and deliver the thrusts to ejaculation. These data, therefore, support the view that female apes initiate mating primarily at the time in the cycle that is conducive to conception, whereas males, in addition, under certain conditions initiate mating that serves no direct reproductive function.

Although studies using the RAT implicated male apes as primarily responsible for nonperiovulatory mating, the basis for the males' behavior was not clear. For each species, however, examination of data on mating in the wild suggests that the conditions of the FAT are a relevant factor, especially the introduction and restriction of the female in a confined space with the more dominant male (Nadler, 1987, 1989). In gorillas, orang-utans, and common chimpanzees under natural conditions, sexual interaction, complete or incomplete, frequently occurs upon reunion of a male and female or following the establishment of proximity *after a period of separation.* Since separation and reunion occur daily during the FATs, these test conditions may simulate natural conditions that stimulate male sexual initiative and thereby account for the increased mating that occurs. On the other hand, these relationships may be entirely coincidental or their significance exaggerated, and what is primarily responsible for relatively frequent mating by males may be relatively continuous and high levels of male sexual arousability (Whalen, 1968).

One approach to assessing the merits of these two hypotheses is to remove or attenuate the social/sexual stimulation of the male associated with the introduction of the female at the beginning of the FAT. This can be accomplished by using the RAT with male choice, i.e., by placing the female in proximity and in visual contact with the male but preventing physical contact until the male lever-presses to gain access to the female. If it is specifically the stimulus of close physical contact that stimulates sexual initiative by the male in the FAT, then preventing such contact while permitting visual access should result in lever-pressing and mating of normal frequency and distribution in the cycle. On the other hand, if male sexual arousability is relatively high continuously, then mating should occur more frequently irrespective of cycle phase. Preliminary results on one pair of common chimpanzees supported the view that it is the conditions of FAT testing that stimulate male sexual initiative (Hartney, 1989). When required to lever-press for access to the female, the male copulated only during the latter portion of the phase of maximal anogenital swelling, i.e., only during the periovulatory period. If similar data were obtained with additional pairs, it would suggest that sexual activity unrelated to reproduction occurs among the great apes (1) in situations that have special significance in terms of sexually stimulating the males and (2) when males are able to coerce females to copulate because of their dominance over the females and the females' inability to avoid and/or escape from the males. It may be that the latter condition of female vulnerability is, in itself, sexually stimulating to the male or becomes so as a result of experience (Shields and Shields, 1983; Nadler, 1988a).

Very much the same problem of male sexual initiative confounding interpretation of female sexual motivation during the menstrual cycle has been encountered in research on humans, among whom cyclic influences, if any, are certainly more subtle than in the apes (Cavanaugh, 1969; Bancroft, 1978; Sanders and Bancroft, 1982). The results of one study of midcycle increases in female-initiated sexual behavior and noncyclic patterns of male-initiated sexual behavior and copulations (Adams et al., 1978) closely resemble the results with the great apes. Other carefully conducted studies, on the other hand, have not obtained such data (Schreiner-Engel et al., 1981; Sanders and Bancroft, 1982), consistent with the view that human sexual behavior is influenced by a multitude of factors that readily overwhelm any influences that hormones in the female might exert. If, however, it is mainly the male who prevents detection of cyclic influences on sexual behavior of women, then cyclic influences should be more apparent in the sexual activity of lesbian pairs. Exactly this finding was recently reported (Matteo and Rissman, 1984), reviving yet again the question of sexual motivation during the menstrual cycle in women and the influence of the male on the frequency and distribution of sexual activity.

GONADECTOMY AND HORMONE REPLACEMENT THERAPY

Earlier we discussed the role of testosterone during psychosexual differentiation and pubertal development. Once sexual differentiation is complete, testosterone plays a different role, as an activating agent, in that it brings to expression those behaviors characteristic of the male of the species. The importance of testosterone for the display of male sexual behavior and its mechanism of action is not nearly as clear as one might expect. This was seen above in consideration of the correlations between testosterone and the awakening of sexual activity at about the time of puberty and is especially true if one examines the effects on sexual behavior of castration in the adult male. McGill (1978), in examining the effects of castration in mice, quoted W. C. Young (1961), who wrote, "There is probably no other phase of the problem of hormones and mating behavior in which our progress has been so negligible." In dealing with the same problem, Beach (1948) cited Aristotle, who reported that castrated bulls "continue to unite sexually with a cow" and Steinach, who, a few thousand years after Aristotle, noted that horses and rodents mate for some time after being castrated as adults.

Beach (1947) proposed that with increased development of the forebrain there was a decreased dependency on hormones for the display of sexual behavior and an increased participation of higher cognitive functioning. The argument and other relevant literature have been summarized by Hart (1974). If the hypothesis is correct, one would expect that castration in the human would have a lesser effect on sexual behavior than castration in other mammals. The effects of castration in men, however, are reported to vary from "no loss of sexual capacities and responsiveness" to "decrease and total loss" (Money, 1961). The length of time that males of various species maintain the capacity to display the ejaculatory response after castration has, in fact, developed into something of a contest. Records vary from 147 days for the rat to 64 months for the dog. As McGill and Haynes (1973) pointed out, however, the latter record exceeds the 2 year life expectancy of B6D2F mice; a mouse of that genotype exhibited the ejaculatory response 14 months after castration. It is clear that the Beach hypothesis regarding corticalization of function and emancipation from hormonal control requires some modification.

It is also apparent that the way in which retention of ejaculatory function is measured and reported may be critical. We found that three of ten adult male rhesus monkeys ejaculated and five achieved intromission 1 year after castration, but the percentage of tests in which each male ejaculated varied from 0 to 66% (Phoenix et al., 1973). Others have reported that the ejaculatory response persists from 5 to 109 weeks in castrated rhesus males (Michael and Wilson, 1974). Although different aspects of sexual behavior may persist for months or years after castration, serum levels of testosterone drop rapidly. There was a significant decrease in the levels of testosterone in the systemic circulation of rhesus males 30 min after castration, and after 13 weeks only one male showed detectable levels of testosterone as measured by gas chromatography (Resko and Phoenix, 1972). After 1 year, testosterone as measured by radioimmunoassay varied from 0.2 to 0.6 ng/ml. There was no significant correlation, however, between levels of sexual performance and testosterone levels before or 1 year after castration.

One week of daily injections of TP (1 mg/kg) given to rhesus males 1 year after castration restored to precastration levels whatever aspects of behavior were affected (Phoenix et al., 1973). In another study, when castrated rhesus males were given injections of 2.0 mg/day of TP, mounting and intromission were restored to intact levels, but ejaculation was not fully restored (Michael and Wilson, 1974). Large individual differences in performance were observed in both studies; these overshadow other differences that could be expected in view of differences in testing procedures. Behavioral thresholds of testosterone in castrated rhesus varied among males, with values from 50 μg/day to 3.2 mg/day (Michael et al., 1984).

Treatment of castrated cynomolgus *(Macaca fascicularis)* males with TP produced results similar in most respects to those obtained with castrated rhesus monkeys except that the threshold level for the ejaculatory response was approximately eight times higher for cynomolgus males (Zumpe and Michael, 1985). It was suggested that the different thresholds may be related to the fact that the rhesus, unlike the cynomolgus monkey, is a seasonal breeder and may therefore be more sensitive to small increases in serum testosterone when levels are low. Seasonality, therefore, may be considered an important factor in selecting a primate animal model for the study of human reproductive physiology and behavior, although other considerations appear to have determined model selection in the past. Availability to researchers and prior information about particular primate species appear to have been among the most important considerations for animal model selection to date.

Knowledge that testosterone will restore sexual behavior in the castrated male is of interest both theoretically and practically, but the mechanism by which the hormone carries out its effects remains a question. It was suggested from research on the rat that testosterone had its effect on sexual behavior by aromatization to estradiol. Hence dihydrotestosterone (DHT) that could not be aromatized did not restore sexual behavior in the castrated rat (Feder, 1971; McDonald et al., 1970; Whalen and Luttge, 1971). However, DHT did restore sexual behavior in male rhesus monkeys that had been castrated 3 years earlier. The propionate ester of DHT (DHTP) was not as effective as TP in that it took about twice as many daily injections of DHTP to increase performance to levels of control males (Phoenix, 1974a). Treatment with DHTP, however, was carried out 3 years after castration, whereas TP treatment was given 1 year after castration. This difference in the interval between castration and hormone treatment, therefore, may have accounted for some of the difference in response sensitivity to DHTP. A somewhat similar study of the relative effectiveness of TP and DHTP in restoring sexual behavior of castrated cynomolgus males found significant increases in ejaculation with both TP and DHTP, but DHTP was effective only at higher doses (Michael et al., 1986). However, DHTP alone and in combination with estradiol benzoate failed to restore sexual behavior to precastration levels in three castrated stumptailed macaques *(Macaca arctoides)* (Schenck and Slob, 1986). The effectiveness of DHTP on sexual behavior of castrated or hypogonadal men has not been assessed, nor has the neural mechanism of action of the gonadal hormones that induce sexual behavior or libido been elucidated.

We have limited our discussion of the activating role of testosterone to sexual behavior, although a number of other forms of behavior and physiology that are not directly sexual in nature may be affected by the hormone. Aggression, for example, has been associated with testicular secretions since the writings of the ancient Greeks, but the specific connections between aggression and sexual behavior are not always clear. Increases of aggression among Japanese macaques *(Macaca fuscata)* during the breeding season have been well documented (Eaton, 1978), and fighting among males for access to females is commonly recognized in many species. Testosterone administration increased aggression in an all-male group of cynomolgus monkeys, but social status was a critical factor. Aggression was much greater in dominant males than it was in subordinate ones (Rejeski et al., 1988). We remain a long way from understanding the relationship of testosterone to sex and aggression in monkeys as well as in men.

Testosterone and DHT were also known to increase body weight of rhesus males (Phoenix, 1974a; Michael and Wilson, 1974; Phoenix and Chambers, 1986b), but no attempt was made to determine whether the increase in weight was associated with an increase in muscle. The effect of steroids on muscle building has been of greater interest to young American males in the 1980s than the question of what the hormones might do to their sexual prowess. It has been reported that approximately 6.6% of twelfth grade male students in the United States use or have used anabolic-androgenic steroids (Buckley et al., 1988). Twenty-seven percent of the students interviewed said they took steroids to improve their physical appearance. Some athletes take steroids (primarily testosterone) to improve athletic performance. Anecdotal reports suggest an increase in uncontrollable aggression and mood changes as common side effects. The use of testosterone by young men for whatever reason is relatively new, but the use of testicular secretions (testosterone) by old men is not new and the reason clear.

HORMONES AND SEXUAL BEHAVIOR IN AGING MALES

One hundred years ago, C. E. Brown-Sequard (1974), at the age of 72, injected himself with an aqueous solution of spermatic vein blood, semen, and the juice of crushed dog and guinea pig testicles. His description of increased muscular strength and improved biological and mental function may best be described as rejuvenation. The good doctor sought youth from the testicular products; today's youth, it would seem, seek the gold at athletic events. Nevertheless, research continues into the role of testosterone and other testicular products in the decline of sexual behavior associated with old age.

Whether serum testosterone levels decline in old men is still not fully resolved (Harman, 1983; Vermeulen, 1983; Nieschlag et al., 1982), but the sexual vigor of youth most certainly declines with age (Kinsey et al., 1948). We undertook a program of research using rhesus males, in part, as a comparative study of sexual behavior and the gonadal hormones in aging males and, in part, for what we could learn from a nonhuman primate that would contribute to our understanding of sexual behavior and aging in men (Phoenix, 1977; Phoenix and Chambers, 1986a). Old rhesus males like old mice and men experience a decline in sexual activity. The lower level of sexual behavior in old rhesus males, however, occurs without a statistically significant decline in testosterone levels, bound or free (Chambers et al., 1981), or a change in diurnal patterns of serum testosterone, DHT, or LH (Chambers and Phoenix, 1981; Chambers et al., 1982).

Treatment of old intact males with TP did not increase the level of sexual behavior displayed (Phoenix and Chambers, 1988), and treatment of old long-term castrated males with TP increased their level of performance only to that observed in intact males of comparable age (Chambers and Phoenix, 1983). Together with the data on hormone levels in aging rhesus males, these behavioral data suggest that target tissue sensitivity, rather than hormone level alterations, accounts for the age-related decline in sexual activity. The large individual differences among rhesus males, also reported among men, make it difficult to assign an age at which the decline in vigor begins, but the best estimate would probably be between 10 and 15 years of age.

Neither a change in environment (Phoenix and Chambers, 1984) nor a change in partner (Chambers and Phoenix, 1982), both reported to improve sexual performance among some old men, proved effective among old rhesus. Treatments with various drugs, such as apomorphine, deprenyl, and yohimbine, were also ineffective in increasing levels of sexual performance. It should be noted that different dosages and treatment durations might possibly be effective. What has been reported to be effective in old rats, however, did not prove to be so in old rhesus males (Chambers and Phoenix, 1989). On the other hand, when old rhesus males were paired with specially selected *preferred females,* their sexual behavior increased to levels they displayed as young males or to that of young males in general (Chambers and Phoenix, 1984). Repeated attempts have failed to identify the particular characteristics of a female that render her a preferred sexual partner. Preliminary attempts to identify vaginal odor as a significant variable have been unsuccessful (Phoenix et al., 1986). The finding that old rhesus males are able to perform sexually with a preferred partner is, however, a provocative one. It indicates that low levels of performance displayed with randomly selected females are not due to a physical disability associated with old age.

The number of subjects used in experimental studies of sexual behavior in nonhuman primates is very low. Studies that employ rodents in the same kind of experiments rarely use fewer than eight animals per group, and the subjects are usually drawn from an inbred strain and/or consist of littermates (Chambers and Phoenix, 1986). Because individual differences are so extensive among nonhuman primates, it is ironic (and less than ideal science) that researchers use so few subjects. Hundreds of hours of repeated testing may improve reliability of measurement for a small sample, but it does nothing to increase the generality of results to the larger population. Under the circumstances, it is remarkable that reports on sexual behavior of nonhuman primates are in such close agreement.

SEXUAL SELECTION AND INTERSPECIES DIFFERENCES
IN SEXUAL BEHAVIOR

One of the more interesting observations linking structure and function in reproductive biology of the great apes is the relationship between genital and gonadal morphology and behavior (Short, 1981). Although closely related in a number of ways, the common chimpanzee and gorilla differ substantially in penile size and visibility and in their testis weight/body weight ratios. The male chimpanzee has a relatively long, pink penis that is quite conspicuous when erect against a background of black hair. In contrast, the gorilla's small black penis is nearly invisible against black hair. Prior to mating, the chimpanzee performs a complex courtship display in which the erect penis plays a conspicuous role (Fig. 6–1), whereas the male gorilla shows little or no courtship behavior (Fig. 6–2). Short (1981) attributed these differences to sexual selection; an example of "form reflects function." The orang-utan was initially considered to be more similar to the gorilla than to the chimpanzee. Short noted, however, that, although the orang-utan's penis was smaller than the chimpan-

Fig. 6–1 Male common chimpanzee performs a bipedal swagger with penile erection as part of an active courtship display to solicit sexual presenting by the female. The pink penis is highly conspicuous against the background of black hair. (Photo: Ronald Nadler.)

zee's and generally inconspicuous because of the organ-utan's long hair, it was not known whether it played some role in courtship. In fact, it was only 10 years ago that courtship behavior by male orang-utans was first observed, both in captivity (Nadler, 1982) and in the wild (Shurmann, 1982). The courtship behavior of the orang-utan, however, differs from that of the chimpanzee. Whereas the chimpanzee performs an active bipedal swagger when courting, the orang-utan (in restricted-access tests) generally reclines, with an erect penis directed toward the female, and rather passively waits for the female to approach (Fig. 6–3). The orang-utan's penis is also longer than Short originally reported (Dahl, 1988). There is a direct relationship, therefore, between the size and conspicuousness of the penis in the great apes and the extent to which the males perform a courtship display (Table 6.1).

The testis weight/body weight ratio of the chimpanzee is many times greater than that of the gorilla and orang-utan, a relationship that Short (1981) as well as Harcourt (1981) attributed to differences in hourly rates of mating. When hourly rates of mating are used for comparison, however, the gorilla greatly exceeds the orang-utan (Harcourt, 1981) even though orang-utans have a larger

Fig. 6–2 Male lowland gorilla exhibits a tight-lipped, stiff-legged stance as a cue for the female to present sexually. The penis is small, black, and inconspicuous. (Photo: Ronald Nadler.)

testis weight/body weight ratio. On the other hand, if the measure used for comparison is the number of days of copulation in the menstrual cycle that normally precedes ovulation (Nadler et al., 1986; Nadler, 1987, 1988a), i.e., preovulatory copulation, then there is a direct relationship with the testis weight/body weight ratio (Table 6–1). Under natural conditions, preovulatory copulation is determined by the duration of estrus, 10–14 days or longer for the chimpanzee, about 5 days for the orang-utan, and about 2 days for the gorilla. Since males must be fertile at the end of estrus when ovulation occurs, larger testes, with presumably greater fertilizing capacity, are required for longer durations of estrus. That preovulatory copulation is an appropriate measure for assessing testis function is suggested by the finding in men that the reduction in sperm content with repeated ejaculations is cumulative over days (Freund, 1962). An alternate interpretation is that testis size is essentially dichotomous, i.e., that differences in testis size are related to differences in the potential for sperm competition between males of multimale and one-male mating systems (Harcourt et al., 1981).

Harcourt (1981) noted that several aspects of the reproductive behavior of great apes differ in relation to the number of males that simultaneously compete for access to females *at the time of estrus,* i.e., the extent of intermale competition for estrous females. Intermale competition for estrous females is greatest in the multimale mating system of the common chimpanzee, wherein essentially all the adult males of the community may copulate with all the

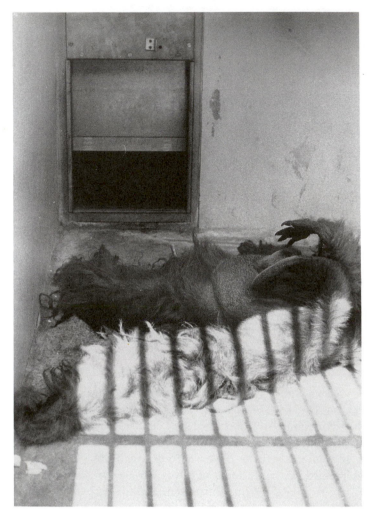

Fig. 6–3 Male Sumatran orang-utan reclines on its back with a penile erection ("penile display") as part of a relatively passive courtship display to solicit mounting by the female during a restricted-access test with female choice. Although generally inconspicuous amidst long body hair, the penis is quite conspicuous when it is erect and displayed in this manner. (Photo: Ronald Nadler.)

estrous females. Intermale competition for estrous females is least apparent among gorillas, in which only the leading male copulates with the parous females of the group. The comparison between the male gorilla and orang-utan with respect to intermale competition for estrous females is not entirely clear because of the limited data available for the orang-utan on this dimension of behavior. Since the estrous female orang-utan may have more than a single male available, due to the overlapping of male home ranges, intermale com-

Table 6–1 Reproductive behavior and anatomy of the great apes in relation to the number of males simultaneously competing for sexual access to estrous females

	Chimpanzee	Orang-utan	Gorilla
Number of males competing for estrous females	Many	None, one, or two	None, possibly one
Male sexual initiative	Pronounced	Moderate	Low
Male courtship display	Conspicuous, complex	Conspicuous, simple	Inconspicuous, simple
Penile visibility	High	Moderate	Low
Preovulatory copulation (days)	10–14	4–6	2–3
Testis weight/body weight ratio*	0.269	0.048	0.017

Source: Nadler et al. (1989).

*Does not take into account allometric considerations.

petition is ranked higher in orang-utans than in gorillas (Table 6–1). It can be seen in Table 6–1 that the extent of male sexual initiative and courtship behavior is directly related to intermale competition at estrus. The interpretation is that in a competitive mating system, such as the chimpanzee's, males that compete for females by courting and initiating sexual interactions will contribute more genes to subsequent generations than males that do not perform competitively. In the one-male mating system of the gorilla, on the other hand, intermale competition is minimal or absent altogether, and primarily females initiate mating. The orang-utan seems to be intermediate to the other great apes; i.e., male orang-utans initiate copulation more frequently than male gorillas but less frequently than male chimpanzees. Male orang-utans also exhibit a courtship display (described above), whereas the male gorilla does not.

As was noted above, the testis weight/body weight ratio in the great apes is directly related to the extent of preovulatory copulation during the menstrual cycle, and both of these measures are directly related to intermale competition for estrous females. Interpretation of the differences in duration of preovulatory copulation is made in terms of the duration of estrus, from which preovulatory copulation derives. A relatively long period of estrus provides a female in a multimale group, such as the chimpanzee's, a period of time (about 2 weeks) to select the male, from among the males in the community, with which the female will consort and mate at the end of estrus, during the periovulatory phase of the cycle (Nadler, 1977b). In fact, although consortships accounted for only 2% of all copulations *observed* in one study of chimpanzees, such consortships appeared to account for 50% of the conceptions (Tutin and McGinnis, 1981). A female in a one-male mating system such as the gorilla's, by contrast, has already selected the male with which she will mate at ovulation when she joined the group. A relatively short period of estrus, therefore, is sufficient in this species, i.e., about 2 days. Since the female orang-utan may have more than a single male available at the time of estrus, the species' intermediate ranking on intermale competition is consistent with its intermediate duration of estrus, i.e., about 5 days (Galdikas, 1981).

An obvious question at this point is whether an analysis of reproductive parameters for humans, comparable to those described above for the great apes, would have implications for proposing a human mating system. Two problems must be addressed before undertaking such an analysis. The first is a logical problem involving the difficulty of inferring cause from effect. In the present context, we may not infer a particular mating system from a group of reproductive parameters merely because the parameters are consistent with that system. We may, however, exclude a particular cause when its known effect is absent. We may exclude a particular type of mating system when the parameters that are consistently associated with it are lacking or when incompatible parameters are present. The second problem relates to the definitions of the parameters. The parameters of reproduction in the great apes may not be defined with sufficient precision to permit them to be falsified in humans. We may not be able reliably to recognize the difference between parameters of a multimale vs. a one-male mating system when they are encountered outside the original context of great ape reproduction. What, for example, is meant by "pronounced" male sexual initiative and what is "moderate" or "low" male sexual initiative in the human context? In discussing relationships among the great apes, it seems reasonable to use such relative terms. It may not be reasonable to extend such descriptors to humans. As a result of possible ambiguities with respect to categorizing the various parameters of reproduction in humans, we may have more confidence about the attributes of certain parameters than about others. This analysis depends, moreover, on specifying some criterion for suggesting that a particular pattern of human behavior likely reflects a significant contribution by biology, i.e., when that behavioral pattern is found in the majority of human societies for which data are available. This implies that such an impact of biology may not be apparent in any given society, including our own.

As was described above, male sexual initiative and courtship behavior are two parameters that vary significantly between multimale and one-male systems. Both forms of behavior are pronounced and conspicuous in the multimale mating system of the chimpanzee, whereas assertive female sexual initiative and the lack of a male courtship display characterize the one-male mating system of the gorilla. Courtship behavior and sexual initiative of modern men are probably least informative, among the parameters discussed, in attempting to characterize a human mating system, because such behavior varies so considerably by society and culture. Nonetheless, among the approximately 190 societies considered by Ford and Beach (1951) in their cross-cultural assessment of human sexual behavior, the majority are characterized by the belief (common in our own society) "that only men should take the initiative in seeking and arranging a sexual affair" (p. 101). Despite this formal proscription, however, women in most other societies, like those in our own, "do actively seek sexual liaison with men, even though they may not be supposed to do so" (pp. 101–102). In some societies, moreover, it is the women who primarily initiate sexual relationships. This would suggest that, among modern humans, courtship and the initiation of sexual interactions with individuals of the opposite sex are carried out by both men and women.

In contrast to the perspective presented immediately above is the evidence (cited earlier) obtained primarily from studies within our own society pertaining to the frequency and distribution of sexual behavior during the menstrual cycle. Since this research focuses on mating during the menstrual cycle, it may constitute a more relevant comparison to the data on great apes than the more general consideration of sexual initiative in the cross-cultural literature (Nadler, 1986). A common interpretation proposed for the lack of cyclicity in human sexual behavior is that the pattern during the cycle reflects an inordinate influence of male sexual initiative (e.g., James, 1971). This proposal, therefore, is essentially identical to Yerkes' (1939) original interpretation of the comparable data on chimpanzees. If similar data were available on a broader spectrum of human beings, they would suggest that the sexual initiative of men was significantly more pervasive during the cycle than the initiative of women.

The available data on courtship and initiation of sexual interactions in the extant members of our species, therefore, are varied. The cross-cultural data do not support the conclusion that our ancestors had a multimale mating system, such as the chimpanzee's, and by similar reasoning are inconsistent with a one-male system, such as the gorilla's. To the extent that our options are restricted to the three mating systems described for the great apes (Table 6–1), humans, by exclusion, most closely resemble the orang-utan with respect to courtship and initiation of mating. The data from the menstrual cycle studies of sexual behavior, by contrast, suggest that sexual initiative by human males is indeed "pronounced," in the sense described above for chimpanzees. If these data were truly representative of our species, then this degree of male sexual initiative would be clearly incompatible with a one-male mating system, such as the gorilla's. The case for the orang-utan, however, is more complicated. The field data on orang-utans indicate that this species is polygynous, but they are not entirely clear with respect to "the *number of males* simultaneously competing for and available to estrous females" (Harcourt, 1981:308, emphasis added). Since the orang-utan's mating system appears to differ considerably from the multimale mating system of the chimpanzee, and is similar in some respects to the gorilla's, it is considered for the present discussion to be some variant of a one-male system. Given the forcible initiation of copulation by male orang-utans under certain circumstances, in the wild as well as in captivity, it is probably inappropriate, on the basis of male sexual initiative, to exclude the possibility that humans might have a one-male mating system of this type. The analysis of courtship and sexual initiation among humans, therefore, is clearly incompatible only with the polygynous, one-male mating system of the gorilla.

Short (1981) concluded that the human penis was sufficiently large and conspicuous to serve a function in courtship similar to that of the chimpanzee's, i.e., that the human penis evolved its size as a result of being attractive to females. There is little evidence in the cross-cultural literature, however, to suggest such a function (Ford and Beach, 1951). The cross-cultural evidence suggests, on the contrary, that physical appearance of women is more important to men than physical appearance of men is to women. Although the penis is generally uncovered in more societies than is the vulva, it is probably a less significant mechanism for attracting modern women than is speech. It could be proposed, moreover, that regardless of size, the human penis is not as conspicuous as that of the chimpanzee, since its color does not contrast sharply with the surrounding hair (however, see Short, 1981, for a contrary view). On the other hand, the human penis is clearly more conspicuous than that of the gorilla, suggesting by exclusion that the human may be most similar to the orang-utan with respect to size and visibility of the penis. As regards structure, however, "the penis of *Pongo* is remarkably similar to that of *Pan* ... and unlike that of *Hylobates, Gorilla* or *Homo*" (Dahl, 1988:141).

As was suggested above, Short's hypothesis regarding the adaptation of the testes to accommodate species-typical frequencies of copulation in the great apes is most effectively supported when preovulatory copulation (actually, ejaculation) is used as the parameter of copulatory frequency. Preovulatory copulation, as defined for the great apes, is equal to the number of days of copulation with ejaculation that typically precedes ovulation, i.e., the duration of estrus that is characteristic of the species. The cross-cultural literature indicates that there is considerable variability in copulatory frequency among human societies as well as considerable differences among adults of different ages (Ford and Beach, 1951). Among young, healthy adults in the majority of human societies, however, copulation once daily or nightly is typical. This pattern and daily frequency of mating are different from those of the great apes, in that the latter generally ejaculate several times per day during a more limited period of the cycle. Interpreting the relationship between copulatory rate and testis size for great apes and humans, therefore, is not a straightforward matter.

It was proposed above for the great apes that the duration of estrus, which determines the extent of preovulatory copulation under natural circumstances, is directly related to opportunities for female choice (Nadler, 1977b; Harcourt, 1981). According to this interpretation, the pattern of mating in humans may be viewed as an extreme example. Since the human female can mate throughout the cycle, she could have continuous opportunity for choice of a partner, behavior that would be adaptive in a multimale mating system, such as the chimpanzee's. If female choice is viewed in a broader sense, such as subserving female reproductive investment in general, other interpretations of the human mating pattern can be proposed, such as strengthening or maintaining (1) a pair-bond in a one-male mating system or (2) affiliative relations with several or more males in multimale mating system. Whatever interpretation is made of the human mating pattern from the female perspective, however, it must be compatible with the fertilizing capacity of the human testis.

The testis weight/body weight ratio of the human (0.079) (Short, 1981) is less than one-third that of the chimpanzee (0.269) (Table 6–1), suggesting that the human testis is not adapted for the high rates of preovulatory copulation characteristic of a competitive, multimale mating system like the chimpanzee's. When assessed in terms of the potential for sperm competition, moreover, the human testis is more compatible with a one-male mating system (Harcourt et al., 1981). Since testis size in the human is more than four times greater than that of the gorilla, however, it is unlikely that humans are adapted to a one-male mating system like that of the gorilla. Based on testis size, therefore, the human most closely resembles the orang-utan among the great apes. This relationship between human and orang-utan may not be very meaningful, however, given the differences between the species in pattern of mating during the cycle and daily frequency of copulation.

Based on this analysis of reproductive parameters, a human mating system would least resemble the one-male mating system of the gorilla, among the great apes. The extent to which the human reproductive parameters differ from those of the chimpanzee (multimale) and the orang-utan (one-male) is debatable. Depending on the database used for comparison, sexual initiative of men could be considered either moderate or pronounced. Similarly, depending on the particular data used to evaluate the orang-utan, male initiative in this species could be considered pronounced or moderate. The forcible initiation of copulation by some males in the wild and by many in captivity is certainly pronounced. The more cooperative initiation of copulation by resident males in the wild and by males tested in the laboratory under conditions of female choice is more appropriately termed moderate. Whether or not men perform some act comparable to a courtship display that incorporates a conspicuous penis is largely a matter of individual interpretation (see Short, 1981). Because a penile display is incorporated into the sexual solicitation of females in both multimale (chimpanzee) and one-male (orang-utan) mating systems, however, this issue is moot. That humans typically mate during an extended period of the menstrual cycle, following as well as prior to ovulation, suggests that our mating is less restricted in the cycle than that of either chimpanzees or orang-utans. As was noted above, however, both chimpanzees and orang-utans mate more frequently during the cycle in laboratory tests wherein the female's options for regulating mating are compromised; this results in near-daily mating for orang-utans. Although one might assert that the data from the wild are more appropriate for assessing a biological influence, another might respond that the laboratory studies reveal the full extent of behavioral capacities and potential and that behavioral patterns of orang-utans in the wild, in any event, might have been "different in the recent evolutionary past" (Dahl and Nadler, 1990).

The final parameter under consideration, and probably the most objective one with which to evaluate a human mating system, is the testis weight/body weight ratio. Human testes, when assessed in cross-species perspective, are clearly of insufficient size and capacity to be included among the primates with multimale mating systems (Harcourt et al., 1981). This being the case, one would conclude that the human mating system is basically one-male and most closely resembles the orang-utan's, among the great apes. If one extends the analysis to other closely related primates, however, one must consider the one-male mating system of the monogamous lesser apes, gibbon *(Hylobates sp.)* and siamang *(Hylobates syndactylus).* Although data on the reproductive parameters for the lesser apes are not extensive, laboratory research comparable to that conducted on the great apes is underway with gibbons (Nadler and Dahl, 1989). These studies are particularly interesting in that they permit a critical test of Harcourt's hypothesis regarding importance of intermale competition for estrous females. Since the gibbon has a one-male mating system, it should resemble the gorilla if mating system rather than social structure determines the various parameters of reproductive function under consideration. It is noteworthy, moreover, that the gibbon and gorilla are the only hominoids other than humans in which there is a well-defined glans penis (Dahl, 1988). The testis weight/body weight ratio for the gibbon (0.084) (Schultz, 1938), moreover, is considerably greater than that for the gorilla (and is very close to that of humans). The data on testis size, therefore, are consistent with the suggestion that the human mating system is not multimale, in the sense that others found in primates are. This suggests, by exclusion, that the human would have some type of one-male mating system.

Short (1981) concluded that humans may have been "polygynous, or perhaps promiscuous, but certainly not monogamous" (p. 337), in the sense of the lesser apes, on the basis of our greater degree of sexual dimorphism. The solution he proposed is that the human mating system is most likely an example of serial monogamy. Since the fertile life of men exceeds that of women, men would increase their reproductive success by periodically competing for new, younger reproductive females. The intermale competition for such females would result in sexual selection comparable to, but less extreme than, that found in the polygynous gorilla and orang-utan. The current analysis is consistent with this characterization but does not exclude other possibilities.

One of the most significant aspects of human sexual behavior is the variety of mating relationships found cross-culturally. This makes it clear that humans can adapt to a variety of relationships, which accounts for the difficulty in specifying any one as typical of the species. If flexibility is seen as the most significant feature of our mating preferences, then one might propose that the orang-utan best exemplifies the human prototype. Although both the chimpanzee and gorilla exhibit alterations in species-typical patterns of mating when tested in the laboratory, these are not nearly as pronounced as that of the orang-utan, in which different test conditions can produce a complete reversal of male and female sexual initiation and grossly different patterns of sexual interaction during the menstrual cycle (Nadler, 1981, 1988b). Although we may exclude certain possibilities, we must be satisfied with a certain degree of ambiguity concerning the issue of a human mating system until additional data related to these issues are obtained. Further research on the great apes, in particular, may offer the best opportunity for deriving the human mating system, as was so eloquently expressed by Short (1981:340), "since they, our glass, can modestly discover to ourselves that which we yet know not of."

CONCLUSIONS

Although the original organizational hypothesis was based on research with rodents and was of some academic interest, it was the result of research with monkeys and the effects of prenatal testosterone on play and aggression as well as sexual behavior that gained the attention of clinicians. The research showed that the action of a hormone during fetal development may be very different from its action in adulthood. The implications for human psychosexual differentiation were clear, but details of the interaction of testosterone with conditions of rearing remain to be delineated.

The data suggest that early social experience (during the first year or so of life) may be essential for competent sexual performance in adulthood. Although Bingham (1928) originally concluded that it was sexual experience per se that was required, later research has suggested an alternate interpretation, i.e., that it is social experience rather than specifically sexual experience that is critical. This hypothesis receives its strongest support from research on the common chimpanzee that demonstrated competent sexual interaction by male chimpanzees that had been reared by humans from the first few days of life (Lemmon, 1971). It appears from this research and from comparable research on rhesus monkeys (e.g., Wallen et al., 1981) that nonhuman primates must have the opportunity to socialize with others, although not necessarily conspecifics, to learn broadly appropriate patterns of social interaction. Critical to this learning process is the modulation and integration of dominance/aggressive and submissive behavior into their social behavior repertoires.

The study of adrenarche and puberty in chimpanzees revealed close similarities between this species and humans in terms of hormone patterns and certain aspects of behavior. The behavioral changes during development were correlated with the circulating levels of FSH and testosterone and with the size of the testes but were not correlated with the levels of the adrenal androgens, dehydroepiandrosterone (DHA) or DHA sulfate. The activation of male sexual behavior, therefore, appears to result primarily from the pubertal increase in gonadal hormones and not the increase in adrenal hormones associated with adrenarche.

The research on estrus and mating during the menstrual cycle is an area of special comparative significance; it was one of the first in which the nonhuman primate was specifically proposed as a suitable model for the human, and the problem of interest is still unresolved. It appears from the results of the two types of tests conducted on the great apes in the laboratory, combined with data from the wild, that males are responsive to the hormonal fluctuations of the female during the menstrual cycle but are not as responsive as the female. Behavioral and nonbehavioral stimuli emanating from the female, presumably as a function of elevated estrogen levels, make the female especially sexually attractive to the male at about the time of ovulation (Beach, 1976). Males are also sexually attracted to females at other times, under certain conditions. The common features of the laboratory conditions under which such nonreproductive sexual behavior took place were that (1) the male was dominant over the female, (2) the female could not avoid or escape from the male, and (3) the male and female were paired after a period of separation. Unfamiliar and relatively unfamiliar females, moreover, stimulate sexual responses by males under natural conditions. The physical proximity of an unfamiliar or relatively unfamiliar and *vulnerable* female appears to represent a potent sexual stimulus to the male great apes.

There is little evidence to suggest that men are preferentially attracted to women at about the time of ovulation and considerable evidence to suggest that women are continuously attractive to men (e.g., Short, 1981). As was noted above, it was proposed that female vulnerability, perhaps analogous to that of the female great apes in the laboratory free-access tests, may stimulate certain types of sexual aggression among humans (Shields and Shields, 1983). Although the conditions that increase male sexual arousal in the male great apes appear to have a biological basis, personal experience is likely more relevant among humans.

The short-term effects of castration on the display of sexual behavior vary considerably across individuals, but all evidence indicates that over time there is a decline in sexual behavior. Appropriate treatment with testosterone restores sexual performance to precastration levels in most individuals. Whether in mice, in monkeys, or in men, however, testosterone does not cause sexual behavior, it merely permits its display.

There is a decline in the sexual behavior of old rhesus males, although there is no significant decline in serum levels of testosterone, and treatment with additional testosterone does not increase levels of performance. Not all investigators agree that serum levels of testosterone decline in healthy old men, but they do agree that the frequency of sexual activity declines. The cause of the decline in sexual behavior of old male monkeys and men remains illusive, but the youthful performance of old male monkeys with *preferred females,* in contrast to randomly selected ones, may be instructive.

The discussion of sexual selection in the great apes suggested that the number of males that simultaneously compete for estrous females accounts for the characteristics of several parameters of reproduction (Harcourt, 1981). A multimale mating system with high intermale competition for estrous females, such as the common chimpanzee's, is associated with pronounced male sexual initiative, courtship behavior, a relatively large and conspicuous penis (adapted for courtship), a relatively prolonged period of estrus and mating prior to ovulation (facilitating female choice), and relatively large testes to accomodate the prolonged period of preovulatory copulation (Table 6–1). The one-male mating system of the gorilla, by contrast, is characterized by little or no intermale competition of estrous females, little male sexual initiative and essentially no courtship, a small and inconspicuous penis, a short period of estrus and preovulatory copulation, and relatively small testes. The orang-utan appears to be intermediate to the chimpanzee and gorilla, in terms of both its mating system and the various parameters of reproductive function. When evaluated along similar lines, the human male most closely resembles the polygynous orang-utan, although a form of serial monogamy has also been suggested for the human (Short, 1981).

ACKNOWLEDGMENTS

Preparation of this chapter was supported in part by NIH grants RR-00165 and RR-00163 from the Division of Research Resources to the Yerkes Regional Primate Research Center and the Oregon Regional Primate Research Center, respectively; by NSF grant BNS87-08406 to R.D.N., and by NIH grant AG-01608 to C.H.P. The Yerkes and Oregon Research Centers are fully accredited by the American Association for Accreditation of Laboratory Animal Care.

REFERENCES

Adams, D. B., A. R. Gold, and A. D. Burt (1978). Rise in female-initiated sexual activity at ovulation and its suppression by oral contraceptives. *New England Journal of Medicine 299:*1145–1150.

Allen, M. L. (1981). Individual copulatory preference and the "strange female effect" in a captive group-living male chimpanzee *Pan troglodytes. Primates 22:*221–236.

Arnold, A. P., and S. M. Breedlove (1985). Organizational and activational effects of sex steroids on brain and behavior: a reanalysis. *Hormones and Behavior 19:*469–498.

Bancroft, J. (1978). The relationship between hormones and sexual behaviour in humans. Pp. 493–519 in J. B. Hutchinson, ed., *Biological Determinants of Sexual Behaviour.* Chicester: John Wiley & Sons.

Baulieu, E.-E., C. Corpechot, F. Dray, R. Emiliozzi, M.-C. Lebeau, P. Mauvais-Jarvis, and P. Robel (1965). An adrenal-secreted "androgen": Dehydroisoandrosterone sulfate. Its metabolism and a tentative generalization on the metabolism of other steroid conjugates in man. *Recent Progress in Hormone Research 21:*411–500.

Beach, F. (1947). Evolutionary changes in the physiological control of mating behavior in mammals. *Psychological Review 54:*297–315.

――――― (1948). *Hormones and Behavior.* New York: Paul B. Hoeber.

――――― (1976). Sexual attractivity, proceptivity, and receptivity in female mammals. *Hormones and Behavior 7:*105–138.

Bercovitch, F. B., M. M. Roy, K. K. Sladky, and R. W. Goy (1988). The effects of iso-sexual rearing on adult sexual behavior in captive male rhesus macaques. *Archives of Sexual Behavior 17:*381–388.

Bingham, H. C. (1928). Sexual development in apes. *Comparative Psychology Monographs 5:*1–165.

Brown-Sequard, C. E. (1974). The effects produced on man by subcutaneous injections of a liquid obtained from the testicles of animals. Pp. 21–22 in C. S. Carter, ed., *Hormones and Sexual Behavior* (reprinted from *Lancet 2:*105–106 1889). Stroudsburg, PA: Dowden, Hutchinson and Ross, Inc.

Buckley, W. E., C. E. Yesalis, K. E. Friedl, W. A. Anderson, A. L. Streit, and J. E. Wright (1988). Estimated prevalence of anabolic steroid use among male high school seniors. *Journal of the American Medical Association 260:*3441–3445.

Cavanagh, J. R. (1969). Rhythm of sexual desire in women. *Medical Aspects of Human Sexuality 3:*29–39.

Chambers, K. C., D. L. Hess, and C. H. Phoenix (1981). Relationship of free and bound testosterone to sexual behavior in old rhesus males. *Physiological Behavior 27:*615–620.

Chambers, K. C., and C. H. Phoenix (1981). Diurnal patterns of testosterone, dihydro-testosterone, estradiol, and cortisol in serum of rhesus males: relationship to sexual behavior in aging males. *Hormones and Behavior 15:*416–426.

———— (1982). Sexual behavior in old male rhesus monkeys: influence of familiarity and age of female partners. *Archives of Sexual Behavior 11:*299–308.

———— (1983). Sexual behavior in response to testosterone in old long-term-castrated rhesus males. *Neurobiology of Aging 4:*223–227.

———— (1984). Restoration of sexual performance in old rhesus macaques paired with a preferred female partner. *International Journal of Primatology 5:*287–297.

———— (1986). Testosterone is more effective than dihydrotestosterone plus estradiol in activating sexual behavior in old male rats. *Neurobiology of Aging 7:*127–132.

———— (1989). Apomorphine, (−)-deprenyl, and yohimbine fail to increase sexual behavior in rhesus males. *Behavioral Neuroscience 103:*816–823.

Chambers, K. C., J. A. Resko, and C. H. Phoenix (1982). Correlation of diurnal changes in hormones with sexual behavior and age in male rhesus macaques. *Neurobiology of Aging 3:*37–42.

Coe, C. L., A. C. Connolly, H. C. Kraemer, and S. Levine (1979). Reproductive development and behavior of captive female chimpanzees. *Primates 20:*571–582.

Collins, D. C., R. D. Nadler, and J. R. K. Preedy (1981). Adrenarche in the great apes (abstract). *American Journal of Primatology 1:*344.

Cutler, G. B., Jr., M. Glen, M. Bush, G. D. Hodgen, C. E. Graham, and D. L. Loriaux (1978). Adrenarche: a survey of rodents, domestic animals, and primates. *Endocrinology 103:*2112–2118.

Dahl, J. F. (1987). Sexual initiation in a captive group of pygmy chimpanzees *Pan paniscus. Primate Report 16:*43–53.

———— (1988). External genitalia. Pp. 133–144 in J. H. Schwartz, ed., *Orang-Utan Biology.* New York: Oxford University Press.

Dahl, J. F., and R. D. Nadler (1990). The male external genitalia of the extant *Hominodidea. American Journal of Physical Anthropology 81:* 211–212.

Davenport, R. K., and C. M. Rogers (1970). Differential rearing of the chimpanzee: a project survey. Pp. 337–360 in G. H. Bourne, ed., *The Chimpanzee* Vol. 3. Basel: Karger.

Diamond, M. (1965). A critical evaluation of the ontogeny of human sexual behavior. *Quarterly Review of Biology 40:*147–175.

Ducharme, J.-R., M. G. Forest, E. DePeretti, M. Sempe, R. Collu, and J. Bertrand (1976). Plasma adrenal and gonadal sex steroids in human pubertal development. *Journal of Clinical Endocrinology and Metabolism 42:*468–476.

Eaton, G. G. (1978). Longitudinal studies of sexual behavior in the Oregon troop of Japanese macaques. Pp. 35–59 in T. E. McGill, D. A. Dewsbury, and B. D. Sachs, eds., *Sex and Behavior: Status and Prospectus* New York: Plenum Press.

Eaton, G. G., R. W. Goy, and C. H. Phoenix (1973). Effects of testosterone treatment in adulthood on sexual behavior of female pseudohermaphrodite rhesus monkeys. *Nature 242:*119–120.

Feder, H. H. (1971). The comparative actions of testosterone propionate and 5a-andros-

ten-17b-ol-3-one propionate on the reproductive behavior, physiology and morphology of male rats. *Journal of Endocrinology 51:*241–252.

Ford, C. S. and F. A. Beach (1951). *Patterns of Sexual Behavior.* New York: Harper
Brothers and Paul B. Hoeber.

Freund, M. (1962). Interrelationships among the characteristics of human semen and
factors affecting semen specimen quality. *Journal of Reproduction and Fertility
4:*143–159.

Galdikas, B. M. F. (1981). Orangutan sexuality in the wild. Pp. 281–300 in C. E. Graham, ed., *Reproductive Biology of the Great Apes: Comparative and Biomedical
Perspectives.* New York: Academic Press.

Gorski, R. A., and C. A. Barraclough (1963). Effects of low dosages of androgen on the
differentiation of hypothalamic regulatory control of ovulation in the rat. *Endocrinology 73:*210–216.

Goy, R. W., and D. A. Goldfoot (1974). Social factors affecting the development of
mounting behavior in male rhesus monkeys. Pp. 223–247 in W. Montagna and
W. A. Sadler, eds., *Reproductive Behavior.* New York: New York.

Goy, R. W., and J. A. Resko (1972). Gonadal hormones and behavior of normal and
pseudohermaphroditic nonhuman female primates. *Recent Progress in Hormone
Research 28:*707–733.

Goy, R. W., K. Wallen, and D. A. Goldfoot (1974). Social factors affecting the development of mounting behavior in male rhesus monkeys. Pp. 223–247 in W. Montagna and W. A. Sadler, eds., *Reproductive Behavior* New York: Plenum Press.

Grady, K. L., C. H. Phoenix, and W. C. Young (1965). Role of the developing rat testis
in differentiation of neural tissues mediating mating behavior. *Journal of Comparative and Physiological Psychology 59:*176–182.

Graham, C. E. (1981). Menstrual cycle of the great apes. Pp. 1–43 in C. E. Graham, ed.,
*Reproductive Biology of the Great Apes: Comparative and Biomedical Perspec-
tives.* New York: Academic Press.

Grumbach, M. M., G. E. Richards, F. A. Conte, and S. L. Kaplan (1978). Clinical disorders of adrenal function and puberty: an assessment of the role of the adrenal
cortex in normal and abnormal puberty in man and evidence for an ACTH-like
pituitary adrenal androgen stimulating hormone. Pp. 583–612 in V. H. T. James,
M. Serio, G. Guisti, and L. Martini, eds., *The Endocrine Function of the Human
Adrenal Cortex.* New York: Academic Press.

Hamilton, G. V. (1914). A study of sexual tendencies in monkeys and baboons. *Journal
of Animal Behavior 4:*295–318.

Hampson, J. L., and J. G. Hampson (1961). The ontogenesis of sexual behavior in man.
Pp. 1401–1432 in W. C. Young, ed., *Sex and Internal Secretions.* Baltimore: Williams and Wilkins Co.

Harcourt, A. H. (1981). Inter-male competition and the reproductive behavior of the
great apes. Pp. 301–318 in C. E. Graham, ed., *Reproductive Biology of the Great
Apes: Comparative and Biomedical Perspectives.* New York: Academic Press.

Harcourt, A. H., P. H. Harvey, S. G. Larson, and R. V. Short (1981). Testis weight, body
weight and breeding system in primates. *Nature 293:*55–57.

Harman, S. M. (1983). Relation of the neuroendocrine system to reproductive decline
in men. Pp. 203–219 in J. Meites, ed., *Neuroendocrinology of Aging.* New York:
Plenum Press.

Harris, G. W. (1964). The Upjohn lecture of the endocrine society: Sex hormones, brain
development and brain function. *Endocrinology 75:*627–648.

Hart, B. L. (1974). Gonadal androgen and sociosexual behavior of male mammals: a comparative analysis. *Psychological Bulletin 81:*383–400.

Hartney, A. T. (1989). *Sexual Behavior of the Chimpanzee during the Menstrual Cycle and during Pregnancy.* Unpublished undergraduate thesis, Emory University.

Heape, W. (1900). The "sexual season" of mammals and the relation of the "proestrum" to menstruation. *Quarterly Journal of Microscopical Science 44:*1–70.

Hobson, W. C., G. B. Fuller, J. S. D. Winter, C. Faiman, and F. I. Reyes (1981). Reproductive and endocrine development in the great apes. Pp. 83–103 in C. E. Graham, ed., *Reproductive Biology of the Great Apes.: Comparative and Biomedical Perspectives.* New York: Academic Press.

Hopper, B. R., and S. S. C. Yen (1975). Circulating concentrations of dehydroepiandrosterone and dehydroepiandrosterone sulfate during puberty. *Journal of Clinical Endocrinology and Metabolism 40:*458–461.

Imperato-McGinley, J., L. Guerrero, T. Gautier, and R. E. Peterson (1974). Steroid 5a-reductase deficiency in man: an inherited form of male pseudohermaphroditism. *Science 186:*1213–1215.

James, W. H. (1971). The distribution of coitus within the human intermenstruum. *Journal of Biosocial Science 3:*159–171.

Jost, A. (1953). Problems of fetal endocrinology: the gonadal and hypophyseal hormones. *Recent Progress in Hormone Research 8:*379–418.

Kempf, E. J. (1917). The social and sexual behavior of infra-human primates with some comparable facts in human behavior. *Psychoanalytical Review 10:*127–154.

Kinsey, C. A., W. B. Pomeroy, and C. E. Martin (1948). *Sexual Behavior in the Human Male.* Philadelphia: W. B. Saunders Co.

Kollar, E. J., W. C. Beckwith, and R. B. Edgerton (1968) Sexual behavior of the ARL colony chimpanzees. *Journal of Nervous and Mental Disease 147:*444–459.

Kraemer, H. C., J. R. Horvat, C. Doering, and P. R. McGinnis (1982). Male chimpanzee development focusing on adolescence: integration of behavioral and physiological changes. *Primates 23:*393–405.

Lemmon, W. B. (1971). Experiential factors and sexual behavior in male chimpanzees. Pp. 432–440 in *Medical Primatology 1970.* Basel: S. Karger.

Lemmon, W. B., and M. L. Allen (1978). Continual sexual receptivity in the female chimpanzee *(Pan troglodytes). Folia Primatologica 30:*80–88.

Martin, D. E., R. B. Swenson, and D. C. Collins (1977). Correlation of serum testosterone levels with age in male chimpanzees. *Steroids 29:*471–481.

Mason, W. A., R. K. Davenport, and E. W. Menzel (1968). Early experience and the social development of rhesus monkeys and chimpanzees. Pp. 440–480 in G. Newson and S. Levine, eds., *Early Experience and Behavior.* Springfield, IL: Charles C. Thomas.

Matteo, S., and E. F. Rissman (1984). Increased sexual activity during the midcycle portion of the human menstrual cycle. *Hormones and Behavior 18:*249–255.

McDonald, P., C. Beyer, F. Newton, B. Brien, R. Baker, H. S. Tan, C. Sampsom, P. Kitching, and R. Greenhill (1970). Failure of 5a-dihydrotestosterone to initiate sexual behaviour in castrated male rat. *Nature 227:*964–965.

McGill, T. E. (1978). Genotype-hormone interactions. Pp. 161–187 in T. E. McGill, D. A. Dewsbury and B. D. Sachs, eds., *Sex and Behavior: Status and Prospectus.* New York: Plenum Press.

McGill, T. E., and C. M. Haynes (1973). Heterozygosity and retention of ejaculatory reflex after castration in male mice. *Journal of Comparative and Physiological Psychology 84:*423–429.

Michael, R. P., R. W. Bonsall, and D. Zumpe (1984). The behavioral thresholds of tes-
    tosterone in castrated male rhesus monkeys *(Macaca mulatta). Hormones and
    Behavior 18:*161–176.

Michael, R. P., and M. Wilson (1974). Effects of castration and hormone replacement
    in fully adult male rhesus monkeys *(Macaca mulatta). Endocrinology 95:*150–
    159.

Michael, R. P., D. Zumpe, and R. W. Bonsall (1986). Comparison of the effects of tes-
    tosterone and dihydrotestosterone on the behavior of male cynomolgus monkeys
    *(Macaca fascicularis). Physiology and Behavior 36:*349–355.

Missakian, E. A. (1969). Reproductive behavior of socially deprived adult male rhesus
    monkeys *(Macaca mulatta). Journal of Comparative and Physiological Psychol-
    ogy 69:*403–407.

Money, J. (1961). Sex hormones and other variables in human eroticism. Pp. 1383–1400
    in W. C. Young, ed., *Sex and Internal Secretions.* Baltimore: Williams and Wil-
    kins, Co.

Nadler, R. D. (1975). Sexual cyclicity in captive lowland gorillas. *Science 189:*813–814.

———— (1976). Sexual behavior of captive lowland gorillas. *Archives of Sexual Behav-
    ior 5:*487–502.

———— (1977a). Sexual behavior of captive orang-utans. *Archives of Sexual Behavior
    6:*457–475.

———— (1977b). Sexual behavior of the chimpanzee in relation to the gorilla and
    orang-utan. Pp. 191–206 in G. H. Bourne, ed., *Progress in Ape Research.* New
    York: Academic Press.

———— (1981). Laboratory research on sexual behavior of the great apes. Pp. 191–238
    in C. E. Graham, ed., *Reproductive Biology of the Great Apes: Comparative and
    Biomedical Perspectives.* New York: Academic Press.

———— (1982). Reproductive behavior and endocrinology of orangutan. Pp. 231–248
    in L. E. M. de Boer, ed., *The Orangutan. Its Biology and Conservation.* The
    Hague: Dr. W. Junk.

———— (1986). Great ape sexual behavior: human implications? Pp. 45–49 in P.
    Kothari, ed., *Proceedings of the 7th World Congress of Sexology.* Bombay: Vakil
    and Sons, Ltd.

———— (1987). Sexual initiation in common chimpanzee, gorilla and orang-utan. *Pri-
    mate Report 16:*35–42.

———— (1988a). Sexual aggression in the great apes. Pp. 154–162 in R. A. Prentky and
    V. L. Quinsey, eds., *Human Sexual Aggression: Current Perspectives, Annals of
    the New York Academy of Science.* Vol. 528. New York: New York Academy of
    Sciences.

———— (1988b). Sexual and reproductive behavior. Pp. 105–116 in J. H. Schwartz,
    ed., *Orang-utan Biology.* New York: Oxford University Press.

———— (1989). Sexual initiation in wild mountain gorillas. *International Journal of
    Primatology 10:*81–92.

Nadler, R. D., D. C. Collins, L. C. Miller, and C. E. Graham (1983). Menstrual cycle
    patterns of hormones and sexual behavior in gorillas. *Hormones and Behavior
    17:*1–17.

Nadler, R. D., R. W. Cooper, C. Roth-Meyer, E. Borreau, and G. Affre (1985a) Hor-
    mone responses to clomiphene citrate in young chimpanzees. *Journal of Medical
    Primatology 14:*117–132.

Nadler, R. D., and J. F. Dahl (1989). Reproductive biology at the Yerkes Regional Primate Research Center and the nature of animal welfare extremism. *Laboratory Primate Newsletter 28*:13–15.

Nadler, R. D., J. F. Dahl, M. E. Wilson, K. G. Gould, and D. C. Collins (1989). Reproductive behaviour and physiology of the chimpanzee. Pp. 25–29 in R. M. Eley, ed., *Comparative Reproduction in Mammals and Man.* Nairobi: Institute of Primate Research, National Museums of Kenya.

Nadler, R. D., C. E. Graham, R. E. Gosselin, and D. C. Collins (1985b). Serum levels of gonadotropins and gonadal steroids, including testosterone, during the menstrual cycle of the chimpanzee *(Pan troglodytes). American Journal of Primatology 9*:273–284.

Nadler, R. D., J. G. Herndon, and J. Wallis (1986). Adult sexual behavior: hormones and reproduction. Pp. 363–407 in G. Mitchell and J. Erwin, eds., *Comparative Primate Biology,* vol. 2A. New York: Alan R. Liss, Inc.

Nadler, R. D., and L. C. Miller (1982). Influence of male aggression on mating of gorillas in the laboratory. *Folia Primatologica 38*:233–239.

Nadler, R. D., J. Wallis, C. Roth-Meyer, R. W. Cooper, and E.-E. Baulieu (1987). Hormones and behavior of prepubertal and peripubertal chimpanzees. *Hormones and Behavior 21*:118–131.

Nieschlag, E., U. Lammers, C. W. Freischem, K. Langer, and E. J. Wickings (1982). Reproductive function in young fathers and grandfathers. *Journal of Clinical Endocrinology and Metabolism 55*:676–681.

Nissen, H. W. (1954). Genetic, psychological and hormonal factors in the establishment and maintenance of the patterns of sexual behavior in mammals. Unpublished manuscript on file in the University of Kansas Library.

Norman, R. L., and H. G. Spies (1986). Cyclic ovarian function in a male macaque: additional evidence for a lack of sexual differentiation in the physiological mechanisms that regulate the cyclic release of gonadotropins in primates. *Endocrinology 118*:2608–2609.

Page, D. C., R. Mosher, E. M. Simpson, E. M. C. Fisher, G. Mardon, J. Pollack, B. Mcgillvray, A. De La Chapelle, and L. G. Brown (1987). The sex-determining region of the human Y chromosome encodes a finger protein. *Cell 51*:1091–1104.

Parker, L. N., E. T. Lifrak, and W. D. Odell (1983). A 60,000 molecular weight human pituitary glycopeptide stimulates adrenal androgen secretion. *Endocrinology 113*:2092–2096.

Phoenix, C. H. (1974a). Effects of dihydrotestosterone on sexual behavior of castrated male rhesus monkeys. *Physiology and Behavior 12*:1045–1055.

———— (1974b). Prenatal testosterone in the nonhuman primate and its consequences for behavior. Pp. 19–32 in R. C. Friedman, R. M. Richart, and R. L. Vande Wiele, eds., *Sex Differences in Behavior.* New York: John Wiley & Sons.

———— (1977). Factors influencing sexual performance in male rhesus monkeys. *Journal of Comparative and Physiological Psychology 91*:697–710.

Phoenix, C. H., and K. C. Chambers (1982). Sexual behavior in adult gonadectomized female pseudohermaphrodite, female, and male rhesus macaques *(Macaca mulatta)* treated with estradiol benzoate and testosterone propionate. *Journal of Comparative and Physiological Psychology 96*:823–833.

———— (1984). Sexual behavior and serum hormone levels in aging rhesus males: effects of environmental change. *Hormones and Behavior 18*:206–215.

———— (1986a). Aging and primate male sexual behavior. *Proceedings of the Society for Experimental Biology and Medicine 183*:151–162.

———— (1986b). Threshold for behavioral response to testosterone in old castrated male rhesus macaques. *Biology of Reproduction 35*:918–926.

———— (1988). Testosterone therapy in young and old rhesus males that display low levels of sexual activity. *Physiology and Behavior 43*:479–484.

Phoenix, C. H., K. C. Chambers, J. N. Jensen, and W. Baughman (1984). Sexual behavior of an androgenized female rhesus macaque with a surgically constructed vagina. *Hormones and Behavior 18*:393–399.

Phoenix, C. H., R. W. Goy, A. A. Gerall, and W. C. Young (1959). Organizing action of prenatally administered testosterone propionate on the tissues mediating behavior in the female guinea pig. *Endocrinology 65*:369–382.

Phoenix, C. H., R. W. Goy, and J. A. Resko (1968). Psychosexual differentiation as a function of androgenic stimulation. Pp. 33–49 in M. Diamond, ed., *Perspectives in Reproduction and Sexual Behavior.* Bloomington, IN: Indiana University Press.

Phoenix, C. H., J. N. Jensen, and K. C. Chambers (1983). Female sexual behavior displayed by androgenized female rhesus macaques. *Hormones and Behavior 17*:146–151.

———— (1986). Stimulus qualities of a preferred female partner and sexual behavior of old rhesus males. *Physiology and Behavior 38*:673–676.

Phoenix, C. H., A. K. Slob, and R. W. Goy (1973). Effects of castration and replacement therapy on sexual behavior of adult male rhesuses. *Journal of Comparative and Physiological Psychology 84*:472–481.

Pomerantz, S. M., R. W. Goy, and M. M. Roy (1986). Expression of male-typical behavior in adult female pseudohermaphroditic rhesus: comparisons with normal males and neonatally gonadectomized males and females. *Hormones and Behavior 20*:483–500.

Pomerantz, S. M., M. M. Roy, and R. W. Goy (1988). Social and hormonal influences on behavior of adult male, female, and pseudohermaphroditic rhesus monkeys. *Hormones and Behavior 22*:219–230.

Pomerantz, S. M., M. M. Roy, J. E. Thornton, and R. W. Goy (1985). Expression of adult female patterns of sexual behavior by male, female, and pseudohermaphroditic female rhesus monkeys. *Biology of Reproduction 33*:878–889.

Rejeski, W. J., P. H. Brubaker, R. A. Herb, J. R. Kaplan, and D. Koritnik (1988). Anabolic steroids and aggressive behavior in cynomolgus monkeys. *Journal of Behavioral Medicine 11*:95–105.

Resko, J. A., and C. H. Phoenix (1972). Sexual behavior and testosterone concentrations in the plasma of the rhesus monkey before and after castration. *Endocrinology 91*:499–503.

Riesen, A. H. (1971). Nissen's observations on the development of sexual behavior in captive-born, nursery-reared chimpanzees. Pp. 1–18 in G. H. Bourne, ed., *The Chimpanzee,* Vol. 4. Basel: S. Karger.

Riesen, A. H., and E. F. Kinder (1952). *The Postural Development of Infant Chimpanzees.* New Haven, CT. Yale University Press.

Rogers, C. M., and R. K. Davenport (1969). Effects of early restricted rearing on sexual behavior of chimpanzees. *Developmental Psychology 1*:200–204.

Saunders, D., and J. Bancroft (1982). Hormones and the sexuality of women in the menstrual cycle. *Clinical Endocrinology and Metabolism 11*:639–659.

Schenck, P. E., and A. K. Slob (1986). Castration, sex steroids, and heterosexual behavior in adult male laboratory-housed stumptailed macaques *(Macaca arctoides).* *Hormones and Behavior 20*:336–353.

Schreiner-Engle, P., R. C. Schiavi, H. Smith, and D. White (1981). Sexual arousability and the menstrual cycle. *Psychosomomatic Medicine 43:*199–214.

Schultz, A. H. (1938). The relative weight of the testes in primates. *Anatomical Record 72:*387–394.

Schurmann, C. L. (1982). Mating behavior of wild orang utan. Pp. 269–284 in L. E. M. de Boer, ed., *The Orang utan: Its Biology and Conservation.* The Hague: Dr. W. Junk.

Shields, W. M., and L. H. Shields (1983). Forcible rape: an evolutionary perspective. *Ethology and Sociobiology 4:*115–136.

Short, R. V. (1981). Sexual selection in man and in the great apes. Pp. 319–341 in C. E. Graham, ed., *Reproductive Biology of the Great Apes: Comparative and Biomedical Perspectives.* New York: Academic Press.

Sizonenko, P. C., and L. Paunier (1975). Hormonal changes in puberty. III. Correlation of plasma dehydroepiandrosterone, testosterone, FSH, and LH with stages of puberty and bone age in normal boys and girls and in patients with Addison's disease or hypogonadism or with premature or late adrenarche. *Journal of Clinical Endocrinology and Metabolism 41:*894–904.

Sklar, C. A., S. L. Kaplan, and M. M. Brumbach (1980). Evidence for dissociation between adrenarche and gonadarche: studies in patients with idiopathic precocious puberty, gonadal dysgenesis, isolated gonadotropin deficiency, and constitutionally delayed growth and adolescence. *Journal of Clinical Endocrinology and Metabolism 51:*548–556.

Smail, P. J., C. Faiman, W. C. Hobson, G. B. Fuller, and J. D. Winter (1982). Further studies on adrenarche in nonhuman primates. *Endocrinology 111:*844–848.

Sokolowsky, A. (1923). The sexual life of the anthropoid apes. *Urologic and Cutaneous Review 27:*612–615.

Thornton, J., and R. W. Goy (1986). Female-typical sexual behavior of rhesus and defeminization by androgens given prenatally. *Hormones and Behavior 20:*129–147.

Tutin, C. E. G., and R. P. McGinnis (1981). Sexuality of the chimpanzee in the wild. Pp. 239–263 in C. E. Graham, ed., *Reproductive Biology of the Great Apes: Comparative and Biomedical Perspectives.* New York: Academic Press.

Valenstein, E. S., W. Riss, and W. C. Young (1955). Experiential and genetic factors in the organization of sexual behavior in male guinea pigs. *Journal of Comparative and Physiological Psychology 48:*397–403.

Van Wagenen, G., and J. B. Hamilton (1943). The experimental production of pseudohermaphroditism in the monkey. Pp. 583–595 in T. Cowles, ed., *Essays in Biology in Honor of Herbert M. Evans.* Berkeley, CA: University of California Press.

Vermeulen, A. (1983). Androgen secretion after age 50 in both sexes. *Hormone Research 18:*37–42.

Wallen, K., D. A. Goldfoot, and R. W. Goy (1981). Peer and maternal influences on the expression of foot-clasp mounting by juvenile male rhesus monkeys. *Developmental Psychobiology 14:*299–309.

Wallis, J. (1982). Sexual behavior of captive chimpanzees *(Pan troglodytes):* pregnant vs. cycling females. *American Journal of Primatology 3:*77–88.

Wells, L. J., and G. Van Wagenen (1954). Androgen-induced female pseudohermaphroditism in the monkey (Macaca mulatta): anatomy of the reproductive organs. *Contributions to Embryology 35:*93–106.

Whalen, R. E. (1968). Sexual motivation. *Psychological Review 73:*151–163.

Whalen, R. E., and W. G. Luttge (1971). Testosterone, androstenedione and dihydro-testosterone: effects on mating behavior of male rats. *Hormones and Behavior.* 2:117–125.

Wilson, J. D., J. E. Griffin, F. W. George, and M. Leshin (1984). Recent studies on the endocrine control of male phenotypic development. Pp. 223–232 in M. Serio, M. Motta, M. Aznisi, and L. Martini, eds., *Sexual Differentiation: Basic and Clinical Aspects.* New York: Raven Press.

Winter, J. S. D., C. Faiman, W. C. Hobson, and F. I. Reyes (1980). The endocrine basis of sexual development in the chimpanzee. *Journal of Reproduction and Fertility* 28:131–138.

Yerkes, R. M. (1939). Sexual behavior in the chimpanzee. *Human Biology 11*:78–111.

Yerkes, R. M., and J. H. Elder (1936). Oestrus, receptivity and mating in the chimpanzee. *Compararative Psychology Monographs 13*:1–39.

Young, W. C. (1961). The hormones and mating behavior. Pp. 1173–1239 in W. C. Young, ed., *Sex and Internal Secretions.* Baltimore: Williams and Wilkins.

Young, W. C., R. W. Goy, and C. H. Phoenix (1964). Hormones and sexual behavior. *Science 143*:212–218.

Young, W. C., and W. D. Orbison (1944). Changes in selected features of behavior in pairs of oppositely-sexed chimpanzees during the sexual cycle and after ovariec-tomy. *Journal of Comparative Psychology 37*:107–143.

Zumpe, D., and R. P. Michael (1985). Effects of testosterone on the behavior of male cynomolgus monkeys *(Macaca fascicularis). Hormones and Behavior 19*:265–277.

Primates and the Origins of Aggression, Power, and Politics among Humans

BERNARD CHAPAIS

Any morphological characteristic, the human hand, for example, is the result of a long phylogenetic history. This is a widely accepted fact, documented both by paleontology and by comparative anatomy, which together make it possible to recount part of that history. Nonetheless, the idea that a species' behavioral traits are also the outcome of an evolutionary trajectory has been accepted only recently and with a good deal of reluctance. Although this idea was proposed long ago (e.g., Darwin 1871, 1872), it matured in the 1940s with the rise of modern ethology, which explicitly defined the evolution of behavior as one of its main research questions.

Behavior is the visible aspect of the activities of neurobiological structures fed by, and organizing, the information from the physical and social environment. Whether neurological processes generate facial expressions, emotions, learning mechanisms, cognitive processes, or patterns of social interactions, they did not appear from nowhere. They too have an evolutionary past. How, then, can we recount the history of behavior? Fossilized bones and teeth allow for some degree of inference about locomotion, manipulation, diet, sexual dimorphism, etc. This information, when combined with paleoecological data and evolutionary theory, permits one to make further inferences about group composition, home range, mating systems, etc. But the available evidence is scanty and the amount of inference limited. In the human case, archaeology provides further information. However, it begins yielding limited data at about 2 million years ago. Paleontology is thus silent about the evolution of whole areas of behavioral and mental processes (language, much of social behavior, etc.), and archaeology is limited by the time span it covers.

Another source of knowledge about the history of human behavior is the comparative study of primate behavior. Because monkeys, apes, and humans are closely related, it is possible to look for behavioral similarities based on common descent (homologies) and to gain important insights about the phylogenetic background of human behavior. This chapter attempts to characterize this background in a specific area of social behavior, power relations.

Exercising power is carrying out one's will despite resistance from others (see, e.g., Weber, 1947; Blau, 1964). Defined that way, human power encompasses various causal, structural, and consequential phenomena: the control of others through the use of physical sanctions (aggression and dominance) or through the allocation of material compensations, emotional gratifications, or symbolic rewards; the manipulation of information; deception; the existence of social stratifications (e.g., hierarchies); the dynamics of social exploitation; the formation of alliances and coalitions; warfare; etc. Obviously it is not assumed here that power represents a homogeneous category of phenomena (i.e., that power has a unitary cause or function), nor is it assumed that there is a unique phylogenetic path leading to the various expressions of power. Rather, this multidimensional category is defined in relation to one common denominator, the use of constraint (physical or psychological) in the pursuit of one's goal, in contrast to nonconstraining forms of interaction (tolerance, sharing, balanced reciprocity, noncompetitive cooperation, etc.).

In this chapter, then, data on nonhuman primates are used as a source of knowledge about the origins of human behavior. More specifically, the comparison aims at identifying some of the phylogenetic precursors of certain human traits. In this specific sense, primatologists can be thought of as historians working on a large-scale time span. A pertinent example is research on the teaching of artificial languages to apes. These studies have generated a stimulating discussion among linguists and comparative psychologists on the nature of language and its human specificities (Ristau and Robbins, 1982; Premack and Premack, 1983) and also have made significant contributions to the study of the origins and evolution of language (see, e.g., Parker and Gibson, 1979).

Few studies have specifically sought behavioral homologies between nonhuman primates and humans (but see van Hooff, 1972; Passingham, 1982; Wrangham, 1987a). More specifically, in the area of power relations, there have been only a few, rather limited, attempts to compare nonhuman primates and humans (e.g., Masters, 1976; Trudeau et al., 1981; Willhoite, 1976; Rajecki, 1983; Savin-Williams, 1987). Most of these studies focused on one type of power, aggressive coercion, and concentrated on just a few of its dimensions, most often the formal (ethological) aspect of aggressive behaviors and dominance interactions. This chapter considers both aggressive and nonaggressive forms of power, and the various dimensions along which they can be analyzed (basis of power, source, scope, domain, context, structure, and underlying psychological mechanisms). Before beginning a discussion of power, however, it will be useful to examine in more detail the foundations of the comparative method for studying the evolution of behavior.

PRIMATOLOGY AND THE EVOLUTIONARY HISTORY OF HUMAN BEHAVIOR

Genetically related species share a number of morphological traits, which they may have inherited from a common ancestor. If the number of species sharing a given trait is large (e.g., pentadactyly among primates), the trait is likely to have characterized an ancient common ancestor of these species, in which case it would be said to be primitive or conservative. In contrast, if the trait is shared by only a small number of related species (e.g., the one-toed foot of horses and zebras), it is likely to be derived from the last common, and more recent, ancestor of these few species only. Thus, among related species, the more widespread a trait is, the phylogenetically older it is likely to be.

The same reasoning applies to behavioral traits.[1] The fact that behavioral genotypes have an evolutionary past makes it possible to search for behavioral homologies among related species. For example, Lorenz (1941, cited by Hinde, 1970) compared species of ducks and geese as a way of establishing taxonomic relationships among them. Studies on the evolutionary history of behavior are few but are particularly interesting and intriguing. One good example is that of the fly *Hilaria sartor* (Kessel, 1955). In this species, the male courts the female by giving her a large, empty silk balloon that he has constructed. A comparison of the numerous existing species of the family to which *Hilaria sartor* belongs suggests a plausible scenario for the evolution of that complex behavior. Briefly, in some species, the male gives the female a live prey before copulating with her. In other species, the male hands the female a prey partially wrapped with silk threads. In still other species, the prey is completely wrapped but dry, the male having previously sucked its internal juices. Finally, in some species, this gift takes the form of an empty silk envelope. This example assumes that the behavioral similarities represent homologies and refers to logical relationships among the variants to infer that some forms must have preceded others (see Alcock, 1984, for other examples of behavioral homologies and evolutionary scenarios).

In contrast to the apparent simplicity of this example, the application of the concept of homology to behavior usually entails some formidable difficulties. Environmental variables have much more influence on behavior than on morphology, molding behavior throughout ontogeny. The resulting plasticity and variability of behavioral phenotypes generally makes it extremely difficult to study the phylogeny of behavior. In the above example, the phylogenetic reconstruction was greatly facilitated by the behavior being under rather strict genetic control. An additive model of genetic changes might account for the proposed evolutionary scenario. In the case of more complex behaviors or mental processes, however, an additive evolutionary model is unlikely to be adequate. A complex trait could have gone through a series of modifications (losses, transformations, or restructuring of components) because it was under the influence of specific selective pressures or because it was ontogenetically, physiologically, or mechanically constrained by the evolution of other traits affecting its expression. Such a complex trait might not be easily recognizable in some of the species being compared, and consequently the homology might not be identifiable.

Additional difficulties may complicate the search for behavioral homologies. A primary problem is that homologies can be confused with analogies (similarities resulting from the independent evolution of the trait under similar selective pressures). Analogies cannot be used for reconstructing the evolution of behavior, but they may suggest its adaptive functions (Lorenz, 1974). Nonetheless, distinguishing between homologies and analogies may be facilitated in some situations. For example, in the case of traits that are shared by a large number of species (primitive or conservative traits), the common descent hypothesis appears more parsimonious than the hypothesis of an independent and parallel evolution of the trait in every lineage.

With the present state of our knowledge, the existence of behavioral homologies cannot be fully demonstrated. The same is true, however, for morphological homologies, e.g., the chimpanzee hand and the human hand, yet the hypothesis of a homology in the case of the hand is widely accepted. This consensus is based on the facts that the trait is primitive; the degree of morphological, embryological, and ontogenetic resemblance is high; and alternative explanations (e.g., analogical similarity) are less plausible. It seems reasonable that our attitude towards behavioral homologies should depend on the same type of criteria (see Atz, 1970, for a critical review of the application of the concept of homology to behavior). It should also be mentioned that hypotheses on homologies are falsifiable. For example, new data could indicate that the processes or structural connections underlying supposedly homologous behaviors are in fact different in the species being compared.

In view of the especially high degree of behavioral plasticity found among primates, all the above-mentioned difficulties are particularly acute in the present case. In an attempt to minimize these problems, this review concentrates on the most fundamental and general principles characterizing power relations among nonhuman primates. By virtue of their generality, these principles are likely to represent phylogenetically old (primitive) traits. Comparing these principles with the human case, then, should make it possible to identify part of the primate legacy in human power relations, as well as those characteristics that are specifically human. The next section analyzes the concept of power and the various dimensions along which human–nonhuman comparisons will be made.

THE NATURE OF POWER

Utilizing power is only one way of satisfying one's needs. Thus it may be useful to present a framework integrating power relations into the context of need satisfaction and access to resources. This framework is somewhat simplified and is presented mainly for heuristic purposes.

Consider a group of individuals in need of some vital resources (food, water, mates, etc.). How will these resources be partitioned? At one extreme, the access can be noncompetitive, ranging from free and peaceful access to tolerance and active sharing. At the other extreme, partitioning can be determined by the structure of power relations and the dynamics of aggressive competition. Whether peace or competition will prevail depends largely on two dimensions: the abundance of resources and their defensibility. If resources are abundant, competition is not expected, and the individuals move about freely. In contrast, if resources are limited, competition may be advantageous. If resources that are limited are sufficiently concentrated, they may be defensible against other competitors, in which case power-based (e.g., aggressive) competition may ensue, with partitioning a function of the distribution of power among individuals. On the other hand, if limited resources are not defensible, being too dispersed or occurring in too large a quantity, scrambling (scramble competition), tolerance, and sharing may prevail.

In summary, free and peaceful access to resources is expected when resources are abundant, tolerance and sharing when they are limited but not defensible, and power-based competition when resources are both limited and defensible.[2] These general principles have a wide application. For example, because resource competition can take the form of competition for the space in which these resources are found, the above principles relate to the evolution of mating, reproductive, or feeding territories (see reviews by Brown and Orians, 1970; Wilson, 1974; Davies, 1978). Furthermore, because individuals may cooperate for finding, acquiring, or defending resources, the above principles also relate to the evolution of group living and social structure (see, e.g., Emlen and Oring, 1977; Wrangham, 1980, 1982, 1987b).

It is well beyond the scope of this chapter to analyze the occurrence of competition, tolerance, and sharing in relation to ecological conditions in primates and humans. This chapter concentrates rather on power as one mode of access to resources. A definition of power was given above, the capacity of an actor to carry out its own will despite resistance. There exist several classifications of power. For example, building on the work of French and Raven (1959) and Etzioni (1961) on human power relations, Bacharach and Lawler (1980) proposed to differentiate four *bases* of power, i.e., four categories of means, the control of which permits one to exercise power. *Coercive* power is based on the control of punishment and "can take the form of withdrawing or limiting rewards as well as administering costs" (p. 174). Thus coercive power includes aggression as well as other forms of sanctions, such as resource or emotional deprivation. *Remunerative* power is exercised through the control of rewards and differs from coercive power in that it implies increasing another's outcomes (mainly material resources). *Normative* power rests on the control of symbolic rewards (e.g., an award) and *knowledge* power on the control and manipulation of information.

In this chapter, these four bases are reorganized into two major categories of power: aggression-based power and dependence-based power. Let us consider, first, dependence-based power. Bacharach and Lawler's (1980) four categories of power have in common a fundamental feature, the dependence dimension of power relationships. Individual A has power over individual B inasmuch as B *depends* on A's resources, services, or emotional gratifications (coercive or remunerative power); symbolic rewards (normative power); or information (knowledge power). Thus, A's power over B stems from B's dependence on A and from A's capability to manipulate the outcomes valued by B. The greater B's dependence on A, the greater A's power over B. In this paper, the capacity to control others by manipulating the resources, services, information, emotional gratifications, and symbolic rewards on which they depend is subsumed under the label "dependence power."

It should be noted that because social dependence can be bilateral, power-dependence theory (Emerson, 1962; Blau, 1964) relates directly to the issue of social bargaining. Bacharach and Lawler (1980), for example, recognize four basic tactics of bargaining: B can reduce his dependence on A by either seeking alternatives outside the A–B relationship or by devaluing what A provides him; B can also try to increase A's dependence on him by either reducing A's alternatives outside the A–B relationship or by increasing the extent to which A values what B provides. The same tactics are available to A. Thus a state of dependence between A and B can be balanced or imbalanced. At one end of this continuum, dependence is asymmetrical, translating into power and exploitation. At the other end of the continuum, dependence is balanced and relates to exchange and reciprocity theory.

The second major basis for the exercise of power is aggressive coercion. A generally accepted definition of aggression is "any behavior directed toward the goal of harming or injuring another living being who is motivated to avoid such treatment" (Baron, 1977:7). Aggressive coercion, therefore, implies the threat, or the actual use, of aggression. Note, however, that the reciprocal proposition, that all aggression is necessarily coercive, can be debated.[3]

Thus far we have identified the two primary *bases* of power, namely, aggression and dependence. Five other dimensions of power can be usefully distinguished. Bacharach and Lawler (1980) define the *sources* of power as whatever permits one to control the bases of power. They identify four sources: structural position (boss, priest, etc.), personal characteristics (charisma, intelligence, etc.), expertise (medical, legal, etc.), and opportunity (e.g., a secretarial position that allows access to specific information). They further define the *domain* of power as "the number of units or individuals under the control of a superior" and the *scope* of power as "the range of behaviors or activities controlled for each unit, whether individuals or groups" (p. 38). Finally, in this chapter, power is analyzed along two other dimensions. The *context* of power refers to the social situation in which power is used (e.g., competition for resources), and the *structure* of power relations refers to the distribution of power among individuals (e.g., a dominance hierarchy or an alliance network). These dimensions (basis, source, domain, scope, context, and structure of power) provide a useful framework for comparing nonhuman primate and human power relations. The following analysis proceeds separately for aggression-based power and dependence-based power.

AGGRESSION-BASED POWER IN NONHUMAN PRIMATES

As was mentioned above, in situations where the availability of vital resources or social partners is limited, the use of aggression may be advantageous for gaining access to, or defending, these resources or partners. This is congruent with the empirical finding that primates generally use aggression during intragroup or intergroup competition for resources, dominance status (defined below), and social (mostly sexual) partners (see Table 7–1). Thus competition appears to be the major driving force behind aggression. Personal defense and the protection of certain conspecifics account for most other contexts of aggression.

The benefits that accrue from the use of aggression are counterbalanced, however, by the costs incurred in the course of fighting. Mathematical models that take into account the benefits (resources at stake) and costs of aggression (injuries and energy spent) provide satisfactory explanations for the evolution of behaviors that control the frequency or intensity of aggression (e.g., Maynard Smith, 1982; for discussions bearing specifically on the adaptive aspects of aggression in primates, see Popp and DeVore, 1979; Dunbar, 1988). One particularly effective way of limiting the damage of aggression is to submit to the opponent. Hence, among nonhuman primates, most aggressive acts occur in the context of so-called dominance/subordinance relationships. This type of relationship is typified by the occurrence of submissive behaviors by individual B in the absence of aggression by A. An approach or a threat by A suffices to elicit a retreat, a flight, or a more specific and ritualized behavior, such as the fear grimace, by B (Fig. 7–1). It is easy to see why aggressive interactions lead to the establishment of dominance relationships. In species forming small to medium-sized groups, whose members are capable of recognizing each other, as primates are, it is highly advantageous for the loser of a fight to avoid further

Fig. 7–1 Aggression is by far the major basis of power among nonhuman primates. Fights commonly result in the establishment of dominance/subordinance relations, and social alliances are a major determinant of rank in many species, such as Japanese macaques. Here, the female on the left displays a clear submissive signal, the "fear grimace," in response to an "open mouth threat" by the female on the right. The latter has an ally (in the background). (Drawing: Shona Teijeiro.)

fighting with the same opponent. Accepting a subordinate status for as long as the distribution of power remains unchanged appears to be the best strategy.

Source, Scope, and Domain of Aggression-Based Power

In nonhuman primates, the main *source* of aggressive coercion is an individual's dominance status. As we shall see below, a position in the dominance order can be acquired mainly in the context of dyadic fights, or it may express the relative power of the individuals' allies. The important point here is that dominance is primarily a matter of relative fighting ability, physical size, and coalition size. Other factors may affect the rank of an individual, but they do so mainly through their effects on the outcome of physical conflicts. For example, intelligence may affect the acquisition of rank, as in the use of an empty kerosene can by a male chimpanzee *(Pan troglodytes)* in competition for dominance (Goodall, 1986) or in the skilled manipulation of allies (de Waal, 1982; Nishida, 1983).

Much debate has surrounded the issue of the *scope* of dominance, i.e., the range of activities and consequences associated with dominance status. Much of the confusion has stemmed from the search for perfect correlations between dominance rank and specific activities, and from the unwarranted assumption that dominance alone might explain much of the distribution of interactions and resources among individuals. Because the concept of dominance failed in both respects, some workers have been tempted to reject it altogether ("the baby and the bath water" phenomenon; Bernstein, 1981) When dominance is analyzed as one of a number of integrated variables (sex, age, degree of relatedness, length of tenure, etc.) affecting the distribution of resources, however, its explanatory value appears important and essential.

In general, it can safely be said that dominance carries real benefits, at least in some circumstances (especially when resources are scarce). The scope of dominance-based power amounts to securing priority of access to physical resources (space, food, water, sleeping sites) and mates. As a consequence, dominant males sire more offspring, on average, than subordinate males, and dominant females may produce more surviving offspring than lower-ranking females (see Richards [1974] and Hinde [1978] for a discussion of the proximate correlates of dominance; see Bernstein [1981] and Dunbar [1988] for a general discussion of the concept; see Fedigan [1983], Silk [1987], and Harcourt [1987] for reviews on the relations between rank and reproductive success; and see de Waal [1986] for a discussion of the integration of dominance and social bonding).

Interestingly, the scope of dominance power in nonhuman primates is almost entirely restricted to obtaining prior access to incentives: high-ranking animals drink, eat, mate, etc., before subordinates, but they are not observed to exploit and take advantage of subordinates by forcing them to do something under the threat of aggression, other than giving way or following (as in the case of herding behavior displayed by male hamadryas baboons [*Papio hamadryas*]; Kummer, 1968). Another possible form of exploitation would be to use subordinates as social tools. For example, Dunbar (1988) discussed the use of infants by adult males among macaques (*Macaca* sp.) and baboons (*Papio* sp.) in terms of exploitation.[4] The forms of such exploitation, however, are very few. Nonhuman primate dominance, therefore, appears essentially as a nonexploitative form of coercive power.

The *domain* of dominance in nonhuman primates (i.e., the number of individuals under the control of a dominant animal) comprises all group members ranking below that individual. However, dominance power is exercised mainly on a one-to-one basis or between groups of allies. A dominant individual acting alone does not control subordinates as a group but rather in a sequential manner. Leadership in group movement cannot be invoked as an example of coercive group control by a single individual, because it is not coercive (Dunbar, 1988).

Structure of Aggressive Power: Dominance Orders

An ideal dominance relationship would be perfectly asymmetrical, the dominant directing no submission to the subordinate and the latter directing no aggression to the dominant. Frequently, however, this asymmetry is imperfect; the subordinate may counterattack the dominant and the latter may react with mild (nonritualized) types of submission. This phenomenon most often expresses the effect of social alliances; e.g., if individual A dominates B on the basis of an alliance with C, B can challenge A in the absence of C. We shall return to the topic of alliances in the next section.

A set of dominance/subordinance relationships among the members of a group can take the form of a linear dominance order or dominance hierarchy (A dominant to B dominant to C, etc.). Many factors affect the nature of these hierarchies, the most important being group composition, i.e., whether the group is composed of a single monogamous pair, a single one-male unit, several one-male units, or a multimale-multifemale unit, etc. A second factor is the pattern of group transfer. In the majority of species, females are phylopatric (i.e., they remain in their natal group as adults), whereas males emigrate and join neighboring groups. In such species, females of the same group are genetically related to each other, whereas adult males usually are not (except in the case of male relatives joining the same group; Meikle and Vessey, 1981). In some species, it is the females that leave their natal group and males are phylopatric (e.g., hamadryas baboons, chimpanzees), reversing the pattern of relatedness among adult groupmates. Finally, in still other species, both sexes emigrate (red colobus [*Colobus badius*], howler monkeys [*Alouatta* sp.], gorillas [*Gorilla gorilla*]; for a review, see Pusey and Packer, 1987). Thus patterns of transfer are a major determinant of the groups' genealogical structure. Kinship, in turn, strongly affects the dynamics of dominance relationships in primates, genetic relatedness being a primary determinant of alliances. With these factors in mind, we now turn to a brief overview of the nature of dominance orders involving females, males, and the two sexes combined.

Female–female dominance relations

One well-known female dominance system is the matrilineal (or nepotistic) type. It is found in species in which females are phylopatric and related, and it is common in cercopithecine species (macaques, baboons and vervets [*Cerco-pithecus aethiops*]; Kawamura, 1965; Sade, 1967; Missakian, 1972; Silk et al., 1981; de Waal, 1977; Paul and Kuester, 1987; Estrada et al., 1977; Horrocks and Hunte, 1983; Lee and Oliver, 1979; Dunbar, 1980; Hausfater et al., 1982; Johnson, 1987). In this system, rank is transmitted from mother to daughters so that a female is dominant over the females that are subordinate to her mother. This pattern results in a hierarchy of families. The mother often remains dominant to her daughter, although daughters sometimes outrank their mother (Chikazawa et al., 1979; Silk et al., 1981; Hausfater et al., 1982; Chapais, 1985). As for rank relations between sisters, younger ones generally outrank older sisters when all are mature (Kawamura, 1965; Datta, 1988), but this does not seem to be the case in all species (Paul and Kuester, 1987). As we will see, both the acquisition and maintenance of rank in such a system have much to do with the dynamics of alliances among kin and non-kin.

Not all female dominance hierarchies are matrilineal. In some species, daughters either outrank their mother on a regular basis (e.g., Hanuman langurs [*Presbytis entellus*]: Hrdy and Hrdy, 1976; mantled howler monkeys [*Alouatta palliata*]: Jones, 1980), or do not acquire their mother's rank (e.g., gorillas: Harcourt and Stewart, 1987). In these species, therefore, kin do not necessarily occupy adjacent ranks. Although such interspecific variation in the structure of female dominance relations is not well understood, it probably relates to a large extent to interspecific differences in demographic and life-cycle parameters (mortality and fecundity rates, rate of population increase, inter-birth interval, growth rate of individuals, etc.). These factors are indeed important determinants of the basic components of power, namely, the relative size of individuals of differing age and the relative size of kin alliances (for discussions of female dominance systems, see Chapais and Schulman, 1980; Hausfater et al., 1987; Dunbar, 1988).

Finally, it should be noted that the existence of well-defined female dominance hierarchies is far from being the rule among nonhuman primates. For example, female dominance relations are weakly developed or not apparent in several species of cercopithecines living in one-male groups (Cords, 1987) as well as in some species of colobines (Struhsaker and Leland, 1987). This probably reflects the prevalence of lower levels of resource competition in these species, ultimately related to feeding strategies and ecological conditions.

Male–male dominance relations

In species forming matrilineal dominance hierarchies, immature males may rank according to their mother's rank until they leave their natal troop (macaques: Koyama, 1967; Missakian, 1972), or they may free themselves of the influence of maternal rank even as immatures, ranking according to their relative age/size (baboons: Lee and Oliver, 1979; Johnson, 1987; Pereira, 1988).[5] As adults, males rank according to their size and fighting ability. In some species, rank further correlates with age and time spent in the nonnatal group (rhesus [*Macaca mulatta*] and Japanese macaques [*Macaca fuscata*]: Drickamer and Vessey, 1973; Norikoshi and Koyama, 1975), whereas, in other species, males can acquire a high rank soon after immigrating into a new group (baboons and vervets: Strum, 1982; Henzi and Lucas, 1980).

Among common chimpanzees, a species in which males are phylopatric, maternal influence is secondary and mature males are generally dominant over both young and very old males (Bygott, 1979). Maternal influence seems more important in the pygmy chimpanzee *(Pan paniscus),* a species also characterized by male phylopatry (Itani, 1987).

Again, the cause of such interspecific variation in the structure of male dominance relations is not well understood. Factors affecting the propensity of males to form alliances are certainly crucial (see below). For example, in baboons and rhesus macaques, the rank of nonnatal males may be affected by their alliances with adult females (Smuts, 1985; Chapais, 1986). In common chimpanzees, alliances between males seem to be a preponderant factor (de Waal, 1982; Nishida, 1983; Goodall, 1986).

Male–female dominance relations

Dominance relationships between males and females are strongly influenced by the degree of sexual dimorphism present in each species. In species with moderate or no sexual dimorphism, intersexual dominance relations are either ambiguous and infrequent (e.g., New World monogamous primates and gibbons [*Hylobates* sp.]; Smuts, 1987a; Leighton, 1987) or in favor of females (females are invariably dominant over males in prosimians; Richards 1987). In contrast, in species characterized by a marked degree of sexual dimorphism (e.g., baboons and great apes) males are invariably dominant to females, and there exist two distinct hierarchies. Finally, when the degree of sexual dimorphism is intermediate (e.g., macaques and vervets), males are generally dominant to females but the latter can form effective coalitions against males. In these species, therefore, the male and the female hierarchies are not always clearly separate. For example, in a group of rhesus monkeys, the alpha female was dominant over the highest-ranking males (Chapais, 1986).

Structure of Aggressive Power: Primate Alliances

An alliance is defined as any situation in which an individual benefits from the support of another individual against a third party. Such polyadic aggressive interactions always affect the outcome of conflicts, and in many species they are the key factor for understanding the dynamics of dominance relations and the structure of dominance orderings (Fig. 7–1). Six functional categories of alliances are described below.[6] In all cases, the following hierarchy is assumed: $A > B > C > D$.

Protective alliances

In a protective alliance, individual A consistently helps C against B, with the result that the beneficiary (C) outranks the target (B) and ranks just below its ally (A). Several studies on matrilineal dominance systems point to the role of this sort of protective intervention in the transmission of rank along the maternal line. In conflicts between any two females from different families, the female from the dominant family is (1) more likely to receive support than the female from the subordinate family and (2) has a larger number of potential allies (kin and non-kin) that rank higher than those of the subordinate female. These factors combine to give the higher-born female more successful or effective support (Cheney, 1977; Datta, 1983a; Horrocks and Hunte, 1983; Pereira, 1989; reviewed by Chapais, N.D.). Moreover, the core of the process of rank inheritance can be reproduced experimentally (Chapais, 1988a,b). Rank reversals were induced in a captive group of Japanese macaques by manipulating the structure of the alliance network. Any female could assume a dominant or a subordinate status in relation to any other female depending on the relative number and relative body size of each female's allies. Aggressive interventions were the key process inducing rank changes.

Consistent protection may have effects other than the establishment of rank relations among females. For example, instead of leaving their natal group, the sons of high-ranking females may remain in it and occupy a high rank in the male dominance hierarchy (Koford, 1963; Tilford, 1982). Two cases for which good data are available reveal that such exceptions resulted from the active support the natal males received from their family (Chapais, 1983a). As another example, estrous females forming a consortship with a male can be temporarily dominant over other males and females (Carpenter, 1942; Datta 1983a). Finally, females can outrank their mothers, as well as unrelated females, with the help of females or males ranking higher than the targets (Marsden, 1968; Chance et al., 1977; Gouzoules, 1980; Chapais, 1985).

Revolutionary alliances
In these situations, individuals B and C jointly attack and outrank A. Each of the two partners is individually subordinate to the target, but their combined force is greater than that of the target. The most spectacular cases are probably the contests for the alpha position observed in three different populations of common chimpanzees (in captivity: de Waal, 1982, 1984; in the wild: Nishida, 1983; Goodall, 1986). Revolutionary coalitions have also been observed among male langurs (Curtin, 1981) and male Barbary macaques (*Macaca sylvanus*: Witt et al., 1981) in captivity. Such alliances can also be induced experimentally. In Japanese macaques, the members of a subordinate family can jointly outrank single dominant females that are experimentally deprived of their allies (Chapais, 1988a).

Conservative alliances
In this type of alliance, individuals A and B support each other against subordinates who would otherwise be able to outrank them in dyadic conflicts or by forming a coalition. Individuals acquiring a high rank either on their own or through revolutionary alliances or consistent protection (see above) can later form a conservative alliance. The latter may ensure that they will be able to

maintain their rank when they become older and weaker, or after losing some of their initial allies. For example, in a group of rhesus macaques, the most dominant of the nonnatal males was also the oldest. He was obviously less fit physically than most of the lower-ranking and younger males. This male maintained intense affiliative relationships with many of the group's females and attacked other males jointly with these females. This pattern was interpreted as a conservative alliance (Chapais, 1986).

Conservative alliances seem to play an important role in the remarkable stability of matrilineal dominance hierarchies (Chapais and Schulman, 1980). Experiments carried out on Japanese macaques revealed that unrelated high-ranking females were mutually dependent for purposes of maintaining their rank above a subordinate family (Chapais, N.D.). Such bilateral alliances were also observed among unrelated female vervets (Hunte and Horrocks, 1986).

Resource-specific coalitions

This type of coalition has been observed in a single context and in one species, savanna baboons *(Papio cynocephalus)*. Two males may form a coalition against a higher-ranking male to gain access to a female in consort with the latter. The joint attack may result in one of the coalition partners mating with the female (Packer, 1977; Noe, 1986; Bercovitch, 1988). These coalitions have been interpreted in terms of reciprocal altruism by Packer (1977), but new evidence suggests that they rather express a form of cooperation (Bercovitch, 1988), with potential immediate benefits for either partner. Note that these coalitions do not affect the relative ranks of the participants.

Defensive coalitions

Here, B and C help each other against A, without challenging A's dominance status. For example, female kin frequently help each other against adult males and adult females (Massey, 1977; de Waal, 1977; Kurland, 1977; Kaplan, 1978; Watanabe, 1979, Berman, 1980; Silk, 1982; Chapais, 1983b; Datta, 1983a,b; Bernstein and Ehardt, 1985; Hunte and Horrocks, 1986). Interestingly, *non-kin* females, although they ally against adult males, do not do so against higher-ranking females (Cheney, 1983; Chapais, 1986). A possible reason is that if a female is already using the support of higher-ranking females for maintaining her rank in the female hierarchy (conservative alliances), she can hardly ally against these same females without jeopardizing her alliances with them. Empirical evidence indicates that females are ready to jeopardize such alliances only when aiding their relatives.

Xenophobic alliances

This category includes all instances in which individuals belonging to the same group jointly threaten or attack one or more members of another group. Such coalitions may develop in three contexts: territorial defense, the repelling of potential immigrants, and dominance interactions between groups. For example, common chimpanzees collectively defend their territory (Nishida and Hiraiwa-Hasegawa, 1987), and Goodall et al. (1979) described what appears to be a rare phenomenon: a territorial invasion between adjacent communities of common chimpanzees during which several members of the smaller community were killed (see also Nishida et al., 1985). Another type of xenophobic alliance takes place in one-male groups when outside males attempt to take over the group. In this context, the single reproductive male may be helped by another male in his attempt to repel the outsider (e.g., howler monkeys: Sekulic, 1983). Females may also form coalitions against male immigrants, in both unimale groups (e.g., patas, *Erythrocebus patas:* Hall, 1967; langurs: Hrdy, 1977) and multimale groups (e.g., Japanese macaques: Packer and Pusey, 1979). Females and males can also combine their forces in this context (e.g., langurs: Hrdy, 1977).

Aggression-Based Power in Nonhuman Primates: Summary

Aggression is a fundamental and widespread *basis* of power among nonhuman primates. Competition for food and mates encompasses most *contexts* of aggression and appears to be the major driving force behind aggression. The costs of aggression are considerably reduced through the formation of dominance relationships between individuals known to each other. Thus, the main *source* of aggressive power in nonhuman primates is dominance status. The *scope* of dominance status is limited to securing priority of access to resources and mates; i.e., dominant individuals usually do not actively exploit subordinates. The *domain* of dominance status includes all group members occupying lower ranks, but power is exercised on a one-to-one basis; i.e., it does not include coordinated group control. The basic *structure* of aggression-based power in primate societies is the dominance hierarchy. Dominance orderings are deeply affected by group composition, group transfer, and the resulting pattern of genetic relatedness among group members and by the degree of sexual dimorphism. In many species, power relations reflect the structure of aggressive *alliances.* Nonhuman primates form alliances to gain access to resources either directly (resource-specific coalitions), indirectly (through competition for rank: protective, revolutionary, and conservative alliances), or by defending access to territory and group (xenophobic alliances). Much of the structure of alliances can be accounted for by a few principles relating to the benefits and costs of combining forces. First, nonhuman primates ally when they need to do so. For example, males rarely ally with each other, or with females, against females. In both cases, larger males can overpower females by acting alone. Second, primates prefer and compete for allies that are higher-ranking than themselves. For example, high-ranking females form alliances against lower-ranking females in matrilineal hierarchies. Third, primates minimize the risks incurred when forming coalitions. For example, they do not ally unilaterally against higher-ranking targets, except when defending a relative.

DEPENDENCE-BASED POWER IN NONHUMAN PRIMATES

Do primates make use of power bases other than aggressive coercion? Can individual A exert control over the behavior of B by manipulating symbolic rewards, information, resources, services, or emotional gratifications that are valued by B? In other words, does social dependence exist in primates, and is it clearly asymmetrical and exploitative (as opposed to balanced)?

We can readily eliminate normative power, which implies symbolic transactions and the definitions of group norms. Although the roots of symbolic activity can be found in the behavior of great apes (see, e.g., Ristau and Robbins, 1982; Premack and Premack, 1983), no primatologist has ever reported the existence of symbol-granting institutions in primates. Normative power appears to be a uniquely human phenomenon.

What about power derived from the control of information? Theoretically, this could take two forms: the exploitation of others through the bargaining of information or through deception. Bargaining for information means exchanging information in return for resources, services or other information. One condition, therefore, is the possession of differential knowledge about the world among group members. For example, old individuals are more experienced than younger ones. They know more about the location of water and food sources, the behavior of predators, certain techniques of food extraction and tool use (for a review of tool use, see Passingham, 1982; Nishida, 1987), etc. Such differential knowledge, however, does not seem to translate into bargaining relationships, either symmetrical (reciprocal) or asymmetrical (exploitation). There are at least two reasons for this. First, it is easy for the individuals not possessing a piece of information to acquire it through observational learning, since the information cannot be easily controlled by its possessors. Second, a transaction involving informational components (vs. material ones) requires a system of referential communication, which is at best embryonic in primates (see, e.g., Seyfarth et al., 1980).

As for exploitation through deception, laboratory studies have revealed that common chimpanzees are capable of deceiving others by inhibiting behaviors that could lead conspecifics to discover the location of food (Menzel, 1974) or by withholding information when interacting with noncooperative human partners (Woodruff and Premack, 1979). Whiten and Byrne (1988) reviewed anecdotal evidence of spontaneous acts of deception among primates. They grouped their data into a number of functional classes: concealment (e.g., hiding from view or avoiding looking at a desirable object), distraction (e.g., looking or leading away), presentation of a neutral or affiliative image of oneself, etc. Alternative explanations, however, exist for many of the examples Whiten and Byrne describe, although some cases might be properly labeled as deception.[7] Thus, some species of nonhuman primates (mostly chimpanzees) seem capable of influencing others through deception.

Let us examine now the possibility of controlling others by manipulating valued resources, either physical or social. Primates are devoid of material possessions. They do not carry or conserve tools, food, or other resources. Nonetheless, captive male hamadryas baboons have been observed to respect the possession of food containers by other males (Sigg and Falett, 1985). Males can also defend females permanently (e.g., in hamadryas baboons; Kummer, 1968) or temporarily when they come into estrus (during consortships), but such control does not usually form the basis of transactions between males. One possible reason is that females cannot be controlled completely; they can exercise choice and refuse to copulate with a male (see Smuts, 1987, for a review of female choice of mating partners). Interestingly, de Waal (1982) reported a form of transaction between the second- and third-ranking males in a group of captive chimpanzees. The second-ranking male allowed his partner access to estrous females in return for his support against the alpha male. If the second-ranking male denied his partner access to the females, the third-ranking male supported the alpha male against his former partner. In this example, the benefits were bilateral, and power, therefore, appeared to be reciprocal. Only by measuring the degree of asymmetry of the transaction could we assess whether exploitation and dependence-based power were involved. We shall come back to this issue later.

This leads us to the possibility of exploitation through the bargaining of services. For example, it has been suggested that in many species of cercopithecines dominant females receive grooming from lower-ranking females in exchange for their aggressive support against third parties (Seyfarth, 1977, 1983; Seyfarth and Cheney, 1984). Another possibility relates to the existence of long-term relationships between unrelated males and females in baboons and macaques. In this situation, females are in a position to provide males with some specific services: they can refuse or accept mating, they can influence male dominance relations, and they can affect the speed at which males become integrated into a group. On the other hand, males can provide females with another type of service: they can protect them and/or their offspring against other males (Strum, 1983; Smuts, 1985; Chapais, 1986). A third example is the observation that female pygmy chimpanzees sometimes copulate with a dominant male before being allowed to eat food under his control (Kuroda, 1984).

These three examples may be interpreted in terms of transactions involving different currencies. Again, however, these transactions do not seem to be asymmetrical; services flow bilaterally in all cases. We are thus faced with two possibilities. Interdependence involving services may be asymmetrical, this asymmetry going undetected by primatologists, or social dependence may be generally balanced and reciprocal among nonhuman primates. In the latter case, it might be that primates lack some of the prerequisites for exploiting the dependence of others (i.e., for giving significantly less than they receive). In other words, social dependence, although it does exist, may not translate into dependence-based power in nonhuman primates. This second possibility appears the most likely on the basis of the available evidence.

Finally, dependence power might be exercised through the manipulation of emotional gratifications. For example, one could ask whether nonhuman primates manipulate others by withholding or rewarding them with affection (a common source of obedience between mothers and offspring in humans). Because primates do not communicate symbolically, such emotional gratifications would have to be conveyed behaviorally, e.g., through touching, cuddling, or grooming. Consequently, this type of transaction might be difficult to differentiate from dependence power involving services (behaviors). In any case, it does not seem to have been reported to date. In humans, another important source of emotional dependence stems from the capacity of others to increase an individual's self-esteem or to reduce guilt. This requires self-awareness, and at present, although great apes may be self-aware to some extent (Gallup, 1987), there is no evidence that they can control others by manipulating those individuals' self-esteem. Much of this type of control in humans is done using language, which primates also lack.

In summary, primates seem very limited in their use of dependence-based power. They do not manipulate symbolic rewards, nor do they bargain for information, although they may be capable of some forms of manipulation using deceptive tactics. They do not seem to manipulate others either by controlling material resources or emotional gratification. They do appear to be capable of using the dependence of others in relation to the services they can provide, but such dependence appears to be mutual and reciprocal, so that power and influence are exercised bilaterally.

EVOLUTIONARY BASIS OF HUMAN AGGRESSION

This review has proceeded beyond the particulars of species and ecological contexts to put forward a number of fundamental principles characterizing the dynamics of power relations among nonhuman primates. It seems reasonable to assume that these general principles describe phylogenetically conservative processes, processes that probably shaped aggression among the first hominids as well as among living primates. It then becomes possible to look at the survival and transformation of that evolutionary heritage among contemporary humans. We shall examine separately aggressive and dependence power in that perspective. Table 7–1 summarizes the comparison of aggression between nonhuman primates and humans.

Formal Aspect of Aggressive Coercion

The formal aspect of aggression refers to its behavioral manifestation. Nonhuman primate aggression takes the form of nonvocal threats, attacks, and displays as well as aggressive and submissive vocalizations. Morphologically similar components are found in humans' aggressive interactions (e.g., Chevalier Skolnikoff, 1973; Camras, 1980; Trudeau et al., 1981; Rajecki, 1983). Humans, however, have added the elements of verbal threat and aggression to the repertoire of vocal and nonvocal aggressive acts. Although the vocalizations of nonhuman primates may convey some information about the interaction itself (in addition to the emotional state of the subjects; see, e.g., Seyfarth et al., 1980; Gouzoules et al., 1984), this semantic component appears rather limited in a comparative perspective.

In contrast, ethologists studying human behavior have classified verbal statements into motivational/functional categories, many of which relate to aggres-

Table 7–1 Comparison of eight dimensions of aggressive coercion between nonhuman primates and humans

Dimensions	Components	Nonhuman primates	Humans
Forms	Threat (visual or vocal)	Yes	Yes
	Physical attack	Yes	Yes
	Aggressive display	Yes	Yes
	Verbal threat or aggression	No	Yes
Bases	Individual physical power	Yes	Yes
	Collective physical power	Yes	Yes
	Technological power (weapons)	No[a]	Yes
Sources	Dominance status	Yes	Yes
	Control of weapons	No	Yes
	Legitimate authority	No	Yes
Scope	Priority of access to resources	Yes	Yes
	Exploitation of subordinates	Rare[b]	Yes
Domain	Dominant controls all subordinates	Yes	Yes
	One-to-one basis	Yes	Yes
	Coordinated group control	No	Yes
Contexts	Competition for resources		
	Dyadic	Yes	Yes
	Resource-specific coalitions	Yes	Yes
	Crimes (theft, burglary, etc.)	No	Yes
	Competition for mates		
	Male–male competition	Yes	Yes
	Male aggression on female	Yes	Yes
	Rape	No[c]	Yes
	Female–female competition	Rare	Yes
	Female aggression on male	No	Yes
	Competition for associates	Yes	Yes
	Competition for dominance per se		
	Dyadic	Yes	Yes
	Polyadic (alliances)	Yes	Yes
	Protection of conspecifics		
	Kin	Yes	Yes
	Unrelated infant	Yes	Yes
	Sexual associate	Yes	Yes
	Social partner (friend)	Yes	Yes
	Policing	Yes	Yes
	Personal defense		
	Individual	Yes	Yes
	Cooperative	Yes	Yes
	Play derived aggression	Yes	Yes
	Redirection of aggression	Yes	Yes
	Psychopathic crimes	No[d]	Yes
	Male takeovers (in one-male groups)	Yes	No
	Infanticide	Yes	Yes
	Dyssocial aggression (see text)	No	Yes
	Intergroup		
	Xenophobia	Yes	Yes
	Territorial defense	Yes	Yes
	Intergroup aggression	Yes	Yes

(*Continued*)

Table 7–1 Comparison of eight dimensions of aggressive coercion between nonhuman primates and humans (*Continued*)

Dimensions	Components	Nonhuman primates	Humans
	Group invasion	No[e]	Yes
	Warfare	No	Yes
	Intergroup alliances	No	Yes
	Violent demonstrations	No	Yes
	Terrorism	No	Yes
Structure	Intragroup structure		
	Dominance orders	Yes	Rare[f]
	Revolutionary alliances	Yes	Yes
	Conservative alliances	Yes	Yes
	Protective alliances	Yes	Yes
	Intergroup structure (see above)	Weak	Yes
Psychological processes	Individual learning		
	Material reinforcement (resource)	Yes	Yes
	Social reinforcement (status)	Yes	Yes
	Modeling influences		
	Social facilitation	Yes	Yes
	Observational learning	Probable	Yes
	Instructional control	No	Yes
	Self-reinforcement		
	Moral justification	No	Yes
	Displacement of responsibility	No	Yes
	Dehumanization of victims	No	Yes
	Attribution of blame to victims	No	Yes

[a]Exceptions: in chimpanzees, one adult male using an empty kerosene can to frighten other males; in gorillas, males throwing vegetation during chest-beating displays (Schaller, 1963).
[b]For example, manipulation of infants by males in "agonistic buffering" (see text).
[c]Except, perhaps, in orang-utangs (MacKinnon, 1979).
[d]Except, perhaps, in the case of a female chimpanzee cannibalizing infants (Goodall, 1977).
[e]Exception in chimpanzees (Goodall et al., 1979; see text).
[f]Mostly in children (see text).

sion and coercion: command/order, ridicule/tease, protest/object, threaten/warn, accuse/blame, criticize/insult, etc. (Weigel and Johnson, 1981). The evolution of language, therefore, has brought about an extension of the expression of aggression and submission into the symbolic realm, not only increasing the ethological repertoire of aggressive behaviors but enlarging as well the range of activities under the control of dominant individuals (see below).

Component Bases of Aggressive Coercion

The human species makes ample use of the primitive basis of aggressive coercion, namely, individual and collective physical power. A new component, however, has been added, technological power (weapons). This capacity is at best incipient in primates. When utilizing tools as weapons (e.g., throwing rocks or branches), nonhuman primates use them against other species (Passingham, 1982). Cases of tool use in the context of intra- or intergroup aggression are extremely rare (see footnote a in Table 7–1). One important consequence of the evolution of technological aggression is that an individual's aggressive power ceases to be correlated with its physical strength. This fact has many implications: An armed individual can easily defeat a larger but unarmed opponent; sexual dimorphism is no longer the sole determinant of relative power, and aggression between the sexes can be less asymmetrical; dominance status may not correlate with mere physical power; small groups may defeat larger groups; etc. Thus the technological component of aggressive power not only has increased tremendously the amount of power available but has deeply affected the distribution of power among individuals and groups.

Sources of Aggressive Coercion

Among nonhuman primates, an individual's potential to use aggressive coercion is defined and limited by its dominance rank. Human coercive power, in contrast, stems from more diverse sources. In certain situations, humans' dominance relations also appear to emerge from spontaneous aggressive interactions among individuals (e.g., among preschool children; see below), but a more important source of aggressive coercion is legitimate authority. Humans define normative positions conferring authority and carrying considerable coercive power, including the right to apply physical constraints (e.g., imprisonment) and physical punishments. Examples of such normative positions are found in various military and religious organizations. It is noteworthy that in this context human aggression can take place through intermediaries: individual A can be aggressive towards B in obedience to C's orders. Among nonhuman primates, aggression is direct.

Furthermore, obedience not only occurs in response to the fear of physical sanctions (aggressive coercion) but may also occur in response to the promise of material rewards, to ideological (e.g., religious) beliefs, and to an expert's request. Milgram (1974) conducted famous experiments on obedience to authority, which illustrated dramatically the scope of "expert power." Subjects were told by an experimenter to administer shocks to an individual participating in a learning experiment conducted at Yale University. The victim acted in connivance with the experimenter. For example, at "315 volts, after a violent scream, the victim reaffirmed vehemently that he was no longer a participant. He provided no answers, but shrieked in agony whenever a shock was administered. After 330 volts he was not heard from . . ." (p. 23). In one series of experiments, 62.5% of the subjects complied with the experimenter's instructions and gave the maximum shock available (450 volts). We touch here on some of the links between aggressive coercion and dependence-based (in this case, expert) power. Although dependence power is not aggressively coercive per se, actors making use of dependence power can induce the individuals they control into utilizing aggression towards third parties.

In summary, humans have enriched the realm of aggression-based power by integrating a new form (verbal aggression) and a new basis (weapons) and by increasing the number of sources giving access to power, namely, normative positions commanding obedience through various means and legitimizing the use of aggression.

Scope and Domain of Aggressive Coercion

Humans have also extended dramatically the scope of aggression-based power. Among nonhuman primates, dominance status confers mainly priority of access to resources. In humans the scope of power has extended to various forms of active exploitation. Language has made it possible for dominant individuals to give orders to subordinates, using the threat of physical sanctions. Thus, by integrating the capacity for language (as well as some other cognitive processes such as intentional communication) into a system of dominance relationships, we have a system of boss–servant relations in which subordinates bring resources to, and accomplish various tasks for, the dominants. This may range from exploitation among siblings to large-scale slavery. At the same time, language has extended the domain of power, i.e., the number of units (individuals and groups) under the verbal control of dominant individuals.

Contexts of Aggressive Coercion

A comparison of the contexts of aggression in nonhuman primates and humans reveals striking similarities (Table 7–1). In both groups, aggression usually occurs in the contexts of competition for resources, sexual partners and status, personal defense, protection of conspecifics, play, policing, intergroup encounters, etc. These similarities point to the primate legacy. On the other hand, the comparison also reveals some important differences in kind and frequency. Several new contexts of aggression appear at the human level, including theft, rape, female aggression in sexual contexts, political crimes, war, terrorism, psychopathological crimes, dyssocial aggression (i.e., gratuitous aggressive acts performed to gain approbation from a reference group; Tinklenberg and Ochberg, 1981), etc. Some of these differences relate to the evolution of dependence-based power and can be better understood in light of that new development (see below). Others are discussed in the next section on the structure of aggressive coercion.

Structure of Aggressive Coercion

Nonhuman primates form small groups, the members of which interact with each other on a regular basis. As a result, aggressive interactions almost invariably lead to the establishment of dominance relations. Among humans, in contrast, large, anonymous groups are common, especially in agricultural and industrial societies. Consequently, dominance hierarchies cannot be a regular feature of such groups. Moreover, the existence of small human groups is not a sufficient condition for the formation of primate-like dominance orders. Aggression in humans is only one power basis among others (subsumed under dependence power). Accordingly, human beings are expected to form primate-like dominance hierarchies only in those situations where aggressive coercion is the only, or major, power basis available, i.e., in situations in which the control of material, behavioral, symbolic, informational, or emotional rewards does not interfere with relative physical power. Needless to say, such conditions are rarely met (see below).

Interestingly, however, primate-like dominance hierarchies (established on the basis of agonistic interactions) are observed among preschool children (e.g., Strayer and Strayer, 1976; Sluckin, 1980; Strayer and Trudel, 1984). Such hierarchies have also been reported for small groups of human adolescents (Savin-Williams, 1980, 1987). In the latter studies, dominance was assessed on the basis of verbal and nonverbal instances of "assertion of an individual over another" (verbal directives, verbal ridicule, physical assertiveness, object displacement, etc.). The results demonstrated the existence of clear dominance hierarchies. Other measures were also taken and correlated with the dominance order: athletic ability, age, pubertal maturation, leadership, intelligence, etc. For example, in three groups of male adolescents aged between 12 and 14 years, athletic ability, bed position (in relation to the adult counselor), and leadership correlated positively with the dominance order. Among older adolescent males, mental abilities, camp experience, and social skills gained in relative importance as correlates of dominance status (Savin-Williams, 1987).

In this set of studies, it was often not possible to determine whether a given correlate was a cause or a consequence of agonistic dominance. For example, athletic ability could well be a determinant of status, in which case physical assertiveness and verbal ridicule would be consequential correlates of a predominantly nonagonistic status. Clearly, therefore, the use of the word *dominance* in these studies extends the meaning of that concept well beyond its use for nonhuman primates. The interesting point, however, is the very existence of primate-like dominance structures among young children and the observed ontogenetic trend of a progressive integration of criteria other than physical strength in the determination of power and status.

Another aspect of the structure of coercive power relations in humans is found at the intergroup level. Humans have extended the dynamics of aggressive coercion well beyond the level of the social group. A major aspect of this phenomenon is warfare. War may be defined as armed combat between groups of people constituting separate territorial teams, political units, or societies (see Ferguson, 1984b, for a review). Although the causes of war are the object of much debate among anthropologists, materialist/ecological approaches seem to offer the best explanations to date. According to these analyses, war appears to be a form of intergroup competition for vital and scarce resources, trade routes, or healthier habitats (see contributions in Ferguson, 1984a). Other suggested reasons for war (e.g., the avenging of previous deaths, ideological conflicts, etc.) cannot be discarded but appear to be secondary justifications.

As defined above, war seems to be a specifically human phenomenon: Nonhuman primates only rarely make use of weapons in their intraspecific conflicts and they do not form symbolically recognized political units. Moreover, territorial invasion is extremely rare in primates. It is difficult, however, to dismiss the basic primate pattern of intergroup competition, territorial defense, and xenophobic reactions as a possible phylogenetic antecedent of intergroup conflict among humans. Intergroup conflicts in primates indeed always relate to territorial defense, access to vital resources, and competition for mates (for a review, see Cheney, 1987). Again, the human species has greatly enriched that basic phenomenon, and intergroup conflicts have become symbolically structured and justified. Moreover, as we shall see, the evolution of dependence-based power deeply transformed the dynamics of alliances. Human coalitions not only aim at maximizing the collective coercive power of their members but commonly are formed on the basis of shared values and goals. This considerably increased and diversified the sources of intergroup conflicts, concurrently extending the range of justifications and opportunities for warfare.

In summary, the phylogenetic roots of some aspects of the structure of aggression among humans can be found in the dominance and alliance behaviors of nonhuman primates at both the intragroup and the intergroup levels. However, nonaggressive forms of power have a pervasive influence on the structure of power relations in all human groups, as we shall see.

Psychological Processes Underlying Aggressive Coercion

Limited space precludes an appropriate treatment of the psychological foundation of aggressive coercion in this chapter. However, an examination of some of the major psychological theories of aggression (e.g., Bandura's [1983] social learning theory, and cognitive approaches such as Dodge's [1982] social information-processing model) reveals that some common psychological processes underlie the use of aggression in nonhuman primates and humans. For example, individual learning, attentional processes, social learning (e.g., the observation of adult models), and perhaps the capacity to attribute hostile intentions to others (see Dodge and Coie, 1987) play important roles in the development of aggression and the acquisition of status in nonhuman primates. Some higher cognitive processes may also be involved in the dynamics of primate coalitions (see, e.g., Cheney and Seyfarth, 1986; de Waal, 1982).

Other psychological processes requiring self-awareness and symbolic communication appear to be specifically human, however. Some of these processes (moral justification, attribution of blame; Bandura, 1983) are listed in Table 7–1. As factors instigating or maintaining aggressive responses, they have undoubtedly extended the realm of human aggression. At the same time, however, self-awareness and language have provided humans with ways of controlling aggression. These mechanisms are mentioned in the following discussion.

Evolutionary Basis of Human Aggression: Discussion

The preceding analysis strongly suggests that the human capacity to exercise aggressive coercion is a primitive (conservative) primate characteristic. Human aggression has a phylogenetic history and a biological basis. The primate legacy can be recognized in all the dimensions of this phenomenon, as summarized in Table 7–1 (some of these similarities, however, may conceal different causal mechanisms, as in the case of infanticide; see Hausfater and Hrdy, 1984). On the other hand, it is interesting to see how this primate core was modified by the new mental abilities (and related spin-offs) acquired by the human species (self-awareness, language, technology, etc.). It is as if the human brain had seized hold of the fundamentals of primate dominance and developed the full potential of this phenomenon, creating verbal aggression, technological aggression, normative positions legitimizing the use of aggression, new ways of exploiting subordinates, new contexts in which to use coercive power, and new structural levels of aggressive relationships.

In view of such complex integration of phylogenetically old influences with more recent acquisitions, the exact influence of the primate legacy on the functioning of human aggression is difficult, if not impossible, to assess. To take a crude analogy, it is as if one wanted to understand the functioning of a present-generation computer solely from knowledge about the working of a first-generation machine. It would probably be possible to identify certain structural similarities between the two computers, reflecting historical modifications "by descent" between their structures. It might well be the case, however, that the workings of the primitive features within the more recent computer are deeply affected by the operation of its new features. For this reason, data on nonhuman primates generate, at best, simplified models of human phenomena. Such models are useful primarily in that they may orient research in some unexpected directions (see Hinde, 1974, 1987). Primate models of human pathologies, for example, have repeatedly proven useful in biomedical sciences.

Although it is difficult to determine the exact influence of the primate legacy on human aggression, it is much easier to state what our evolutionary heritage does *not* imply. For example, the phylogenetic foundation of aggression does not indicate that aggression is innate and inevitable. Some well-known theories have indeed proposed that aggression stems from a death instinct, the consequences of which could be partly lessened through the discharge of aggressive energy or of nondestructive energy (e.g., through violent sports; the so-called cathartic effect of Freud, 1933), or again that aggressive energy accumulates within the organism at a constant rate until it is discharged (Lorenz, 1966). These theories of aggression are now widely rejected. Energy models of aggression cannot withstand the ethological evidence showing that animal aggression is stimulus- and context-specific (Hinde, 1970, 1974). The catharsis principle is contradicted by the results of both experimental studies (e.g., Mallick and McCandless, 1966; Ryan, 1970; Baron, 1977) and cross-cultural analyses (Sipes, 1973; Segall, 1983). Environmental influences on aggression have been amply demonstrated: The contexts and frequency of aggression vary across cultures (Sipes, 1973; Segall, 1983), across families (Patterson, 1982), and across immediate physical settings (e.g., space and quantity of toys available (Hartup, 1974; Gump, 1975; Smith and Connelly, 1980). Also, the effect of learning on aggression has been demonstrated more directly. For example, aggressive mod-

els (i.e., examples of aggression) are known to increase the subsequent frequency of aggression (Bandura, 1983). Finally, aggression can be controlled
using various means: teaching conflict-reducing techniques, increasing awareness of the harmful effects of aggression (increasing empathy), reorganizing the
environment, and reinforcing acts that are incompatible with aggression (e.g.,
sharing and cooperation; Donnerstein and Donnerstein, 1977; Brown and
Elliot, 1965; Slaby and Crowley, 1977; Baron, 1977; Zahavi and Asher, 1978;
for a general review of the development of human aggression, see Parke and
Slaby, 1983).

These and a host of other studies clearly establish that environmental factors
and learning mechanisms exert a considerable hold on the expression of
human aggression. Such strong learning influences, however, are perfectly
compatible with, and in fact require, the existence of a neurophysiological sub-
strate underlying aggression. Animals and humans are equipped with neural
and hormonal structures underlying aggressive responses (e.g., Simmel et al.,
1983; Svare, 1983; Moyer, 1976, 1987). This biological foundation can be conceived as a universal set of constraints affecting the probability that an individual will act aggressively in a number of situations (e.g., being the target of an
attack, seeing one's close kin receive aggression, being deprived of scarce
resources, feeling exploited by a nonreciprocator, being sexually jealous, etc.).
It should be noted that aggressive responses are only one of several possible
types of responses to some of these situations, with withdrawal and resignation,
psychosomatization, achievement, self-anesthetization with drugs, being
among the alternatives (Bandura, 1983).

Because these universal, phylogenetic constraints affecting the probability of
aggressive responses are operating in a multitude of sociocultural environments, they result in the wide-ranging cross-cultural diversity observed in the
expression of human aggression, from "nonaggressive" to highly aggressive
societies. For this reason, the existence of nonaggressive societies can hardly be
invoked as evidence to deny the existence of genetically based neurobiological
structures underlying human aggression (see, e.g., Segall, 1983). Identical
pianos played by different people produce different sounds, and the existence
of that variation does not mean that the pianos have no common potential.

EVOLUTION OF DEPENDENCE-BASED POWER

Dependence-based forms of power (i.e., controlling others through the manipulation of material, emotional, behavioral, informational, or symbolic rewards)
are only incipient among nonhuman primates. What is lacking at the nonhuman level but present among humans that accounts for our remarkable development of dependence-based power? Let us return briefly to the nature of
aggressive coercion. All animals possess mechanisms for perceiving, evaluating,
and reacting to physical pain. This makes sense in view of the fact that physical
suffering directly impinges on the survival of organisms. Exercising power
through aggression, therefore, is rooted in the basic need of all organisms to

avoid physical injuries. This need is so fundamental that the first power basis to appear during evolution was the control of physical harm. In that limited sense, any victim of aggression, or any subordinate, is *dependent* on the dominant's willingness not to inflict further physical injuries.

Theoretically, all the needs of an individual (i.e., all the sources of pleasure and displeasure it can feel) are subject to being manipulated and controlled by others. These needs include those to eat, to drink, to reproduce, to keep warm, to be protected, to receive affection, to obtain social approval, etc. The capacity of organisms to manipulate any one of these needs corresponds to a specific component of dependence power. For example, the capability of individual A to satisfy B's need for protection makes it possible for A to control the behavior of B on this basis. Theories on dependence-based power illustrate well the cognitive complexities involved in the bargaining of social needs. Consider the four basic bargaining tactics available to each partner, mentioned above. The cognitive processes involved in the evaluation of the costs and benefits of each tactic have little in common with those involved in aggressive coercion.

Many prerequisites for the exercise of dependence-based power can be recognized. The manipulation of social dependence requires in the first place that individuals be part of extensive and well-developed networks of social dependence. This implies that dependence power must have emerged with the evolutionary development of mutualistic interactions (reciprocity, sharing, exchange, cooperation), which are fundamental components of human social behavior. Some of the phylogenetic antecedents of mutualism can be observed in the behavior of nonhuman primates, as was shown above.

Second, the exercise of dependence power requires that individuals be differentially able to satisfy the needs of others through controlling and allocating food, mates, affection, security, protection, social approval, etc. Equal control would indeed foster symmetrical reciprocity, leaving no room for the exploitation of dependence. Differences in the abilities to satisfy the needs of others may stem from various sources: differences in dominance status, charisma, various skills (from tool making to athletic ability and subsistence activities), intelligence, etc.

Third, the dynamics of social bargaining took on entirely new dimensions of complexity when it became possible to communicate about the needs of individuals using a symbolic form of communication. Language indeed pervades and possibly accounts for all existing forms of dependence power.

Finally, other cognitive processes seem equally fundamental. For example, assessing the costs and benefits of alternative tactics implies various degrees of reasoning, and self-awareness is a prerequisite for the manipulation of guilt and social approval.

It would be rather speculative to attempt to define the causal and evolutionary connections between all the above prerequisites of dependence power. In any case, the evolutionary developments of language, interdependence, self-awareness, etc. deeply transformed the dynamics of power relations during the hominization process. The number of *bases* of power increased (consider the variety of needs that can be manipulated). The *sources* of power also diversified. From physical dominance they expanded to include expertise, charisma, intelligence, knowledge, etc. The *domain* and *scope* of power enlarged as well. People and groups could now be manipulated by a single individual having charisma, moral authority, or expert knowledge, and rebelling against expertise or moral and charismatic leaders is probably more difficult, in general, than rebelling against a dominant individual in the context of power relations based on physical strength. We shall now examine some of the implications of dependence power at the level of the structure of power relations.

Dependence Power and Social Hierarchies

With regard to dominance, the most powerful individuals in a hierarchy are the ones having the greatest physical power (whether individual or collective). In the context of dependence power, however, the most powerful individuals (the ones most able to control the behaviors of others) are those with the greatest number of individuals depending on them for the satisfaction of needs. In a hunter-gatherer society, skilled hunters have more persons counting on them for meat than do unskilled hunters. The potential power of the former is, therefore, greater. Charismatic leaders (whether political or religious) have power over all the individuals they can influence. Experts in various fields (e.g., medical doctors) may exercise power over all the individuals depending on their knowledge. Company executives have power over all the people whose wages they control.

There are indeed many sociocultural scales of dependence-based power, in fact as many as there are areas of social dependence. High position in any one of these hierarchies confers prestige, contributes to self-esteem, and is actively sought. Humans, then, compete not so much for physical dominance but for numerous other, dependence-based, sources of power. Successful people are often the object of admiration, trust, respect, and envy (see Barkow, 1975, for a cross-cultural review of prestige seeking). Why is this so? Why are people attracted to high-ranking individuals, bestowing prestige upon them? One interpretation is that admiring and contributing to the self-esteem of high-ranking individuals may be the best way to rise in that hierarchy or otherwise to obtain resources, services, information, symbolic rewards, or emotional gratifications associated with high rank. In other words, providing power holders with self-esteem may be a strategy of rank acquisition and need satisfaction. This interpretation is supported by the observation that, by withholding their approval and gratifications when they feel exploited, lower-ranking individuals can influence power holders.

According to the present interpretation, competing for high rank within any scale of dependence power would aim at gaining influence over others. Barkow (1975) dismissed power as the ultimate goal of prestige striving and suggested instead that people, when seeking prestige, ultimately strive for self-esteem. The present perspective integrates both power and self-esteem in a single scheme: power would be the ultimate goal of social competition, self-esteem and prestige the proximate goals. This pattern may be homologous to the attractiveness of high rank among nonhuman primates. Affiliating (e.g., grooming) with the highest-ranking individuals seems to increase the probability of receiving various benefits in return (Seyfarth, 1983).

In contrast to dominance-based power, dependence power generates a great number of social hierarchies. Moreover, any individual usually belongs to more than one social scale at a time, and his status can be a combined product of his position in each of these. As we shall now see, dependence power revolutionized the dynamics of alliances as well.

Dependence Power and Alliances

Models of alliance formation in humans are numerous. Some of their basic principles are outlined here to demonstrate that they relate more to dependence power than to aggressive coercion (for reviews, see Murnighan, 1978; Bacharach and Lawler, 1980). In their simplest form, these models define the coalitions expected within triads of individuals or groups. Similar principles apply to larger groups. They begin with a given distribution of resources among the three units, e.g., individuals. Caplow (1968) defined eight such categories of resource distribution. For example, when individual A controls more resources than B, and B more than C, three distributions are possible: the sum of the resources controlled by B and C can be greater than, equal to, or less than those controlled by A. On the basis of such initial distribution patterns, the models calculate the interest of each of the three individuals in a given coalition (B and C against A, A and B against C, or A and C against B) and then define the probability of each coalition's occurrence.

Models of coalition formation in humans differ substantially in their basic assumptions. Caplow's (1968) model assumes that individuals form coalitions to maximize their rank, i.e., the number of individuals subordinate to the coalition. This model is applicable to the coalitional behavior of nonhuman primates. Another model (Gamson, 1961) assumes that the goal of coalition partners is not to maximize rank but to obtain the maximal amount of pooled resources, according to the equity principle (each partner's share of resources is proportional to his initial contribution). A third model (Komorita and Chertkoff, 1973) is similar to the previous model but differs in that bargaining intervenes in the distribution of resources (some partners aim at an equal share, others at an equitable one). Other models of coalition formation further integrate ideological considerations by assuming that partners seek to minimize ideological divergences (Murnighan, 1978). Each of these models may define different optimal coalitions (A–B, B–C, or A–C) for a given pattern of resource distribution.

Most of these models, therefore, are not applicable to the behavior of nonhuman primates. This is true because humans are strongly dependent on various resources, services, techniques, etc., and are further capable of pooling the latter for purposes of increasing their influence on others. As a result, human alliances take various forms: consumer associations, lobbies, political parties, religious sects, corporate businesses, worker unions, professional corporations, parents' associations, etc., each vying to exercise power over specific target groups.

CONCLUSIONS

The human species not only has retained the capacity for aggressive coercion but has greatly extended the range of that power basis through all of its dimensions. Aggression has become verbal, technological, exploitative, socially legitimized, and highly coordinated at both the intra- and the intergroup levels. Humans have further added to that primitive power basis the extremely diversified area of dependence power, creating multiple scales of social competition and multidimensional networks of alliances. Furthermore, aggressive and dependence power intermingle in many ways. Aggression draws new sources of expression in obedience to normative authority, and, reciprocally, normative authority can rely on aggressive coercion when other tactics of bargaining prove ineffective.

Considering only these evolutionary events, we must conclude that social inequalities expanded tremendously during the hominization process, and we can rightly ask how human relationships have remained bearable. The answer is twofold. First, it is important to realize that it is the potential for the utilization of power that has increased, not necessarily its actual employment. As was mentioned above, aggressive competition is advantageous only in certain circumstances, such as in situations of resource scarcity. Second, and perhaps most importantly, the evolution of the potential for power has been counterbalanced by the parallel and unprecedented evolution of mutualistic interactions. Following Wrangham (1982), we can define two categories of mutualism: noninterference mutualism and interference mutualism. When engaging in noninterference mutualism, individuals cooperate without intentionally harming their conspecifics, as in cooperative hunting. Such activities entail a high degree of social cohesion and reciprocity and are hardly compatible with too high a degree of social exploitation. The fact is that humans engage in a great variety of such cooperative activities, which inhibit or limit significantly the expression of power among cooperators.

The second type of mutualism, which is also common among humans, involves the formation of coalitions for competing against other groups. Again, a high degree of exploitation among the members of any coalition is incompatible with the degree of social cohesion required when opposing another group. Intergroup competition is known to increase intragroup cohesion (Stein, 1976). Thus the evolution of both categories of mutualism among humans, while creating new opportunities for the exploitation of dependence, simultaneously increased the degree of social cohesion within the cooperative units. The human species, therefore, appears to be original in its combination of a formidable potential for inequalities and social conflicts with a tremendous capacity for mutualism and peaceful coexistence.

The nonhuman primate legacy lies within us, interacting with new mental abilities and new types of power. If we knew nothing about the behavior of other primates, we might be forever pondering the origins of our potential for aggressive and dependence power and remain at a loss to explain their all too numerous manifestations.

ACKNOWLEDGMENTS

I am grateful to Daniel Pérusse for his helpful comments on the manuscript. The preparation of this chapter was facilitated by grants from the Natural Sciences and Engineering Research Council of Canada and the Fonds FCAR of the Province of Québec.

NOTES

1. Behavior is an aspect of the phenotype, and as such it expresses the interplay between a genetic basis and environmental influences. Although we have not yet identified precisely the genetic basis of any human behavior (except perhaps for certain pathologies), we infer the activity of genes from the mere existence of neurophysiological mechanisms responsible for behavior (in the same manner that we do not doubt the existence of the hand's genetic makeup, although we have not yet identified the specific genes involved).

2. Note, however, that, even when limited resources are defensible, they may be shared. This can occur in situations where reciprocation of access is advantageous (e.g., when resources are located or obtained randomly by one or a few individuals at a time, as may be the case with hunting).

3. One could argue, for example, that defensive or reactive aggression is carried out with no intention of controlling the behavior of others but rather with that of relieving a perceived or actual threat (see, e.g., Dodge and Coie, 1987). Strictly speaking, therefore, defensive aggression would not be coercive. To distinguish between coercive and noncoercive aggression, we need to understand the motivational basis of aggression. This issue is, however, unresolved. Aggression and its variously defined subcategories (defensive, instrumental, hostile, etc.) probably represent motivationally heterogeneous categories of behavior; i.e., aggression performed in one context may reflect the interplay between two or more motivational bases (Attili and Hinde, 1986). In these circumstances, it is difficult to differentiate between coercive and noncoercive aggression. For that reason, the present comparison of aggressive coercion between nonhuman primates and humans considers all types and contexts of aggression.

4. Infants may be carried by males in the course of agonistic contests with other males. Possible explanations are that this behavior reduces the tension between the males (agonistic buffering hypothesis; Deag and Crook, 1971; but see Taub, 1980) or that the infant's use is "an implicit threat that escalation by the opponent will result in the infant's mother intervening on his behalf" (Dunbar, 1988:255). In this example, exploitation is made possible by the discrepancy in size between the dominant and the subordinate.

5. This interspecific difference may result from differences in the degree of sexual dimorphism, which is considerably more pronounced in baboons than in rhesus monkeys and Japanese macaques (Pereira, 1988).

6. Species for which we have good data on alliances are few in number because of the

relatively late interest of primatologists in these phenomena and the quality of data required to carry out meaningful analyses (i.e., good observability, individual recognition of group members, knowledge of age and degrees of relatedness among animals). Well-studied species are mostly Old World monkeys and apes: macaques and baboons (roughly a dozen species), vervets, common chimpanzees, and to a lesser extent gorillas, langurs, etc. The present account is, thus, biased for these species.

7. The psychological mechanisms underlying deception in nonhuman primates remain to be assessed. Whiten and Byrnes (1988) favor high-level cognitive explanations. However, mechanisms based on individual or social learning may provide satisfactory explanations for many of the reported cases (see Baldwin, 1988).

REFERENCES

Alcock, J. (1984). *Animal Behavior.* Sunderland, MA: Sinauer.

Attili, G., and R. A. Hinde (1986). Categories of aggression and their motivational heterogeneity. *Ethology and Sociobiology 7:*17–27.

Atz, J. W. (1970). The application of the idea of homology to behavior. Pp. 53–74 in L. R. Aronson, et al., eds. *Development and Evolution of Behavior.* San Francisco: W. H. Freeman.

Bacharach, S. B., and E. J. Lawler (1980). *Power and Politics in Organizations.* San Francisco: Jossey-Bass.

Baldwin, J. D. (1988). Learning how to deceive. *Behavioral and Brain Sciences 11:*245–246.

Bandura, A. (1983). Psychological mechanisms of aggression. Pp. 1–40 in R. G. Geen and E. I. Donnerstein, eds. *Aggression: Theoretical and Empirical Reviews* Vol. 1. New York: Academic Press.

Barkow, J. H. (1975). Prestige and culture: a biosocial interpretation. *Current Anthropology 16:*553–572.

Baron, R. A. (1977). *Human Aggression.* New York: Plenum Press.

Bercovitch, F. B. (1988). Coalitions, cooperation and reproductive tactics among adult male baboons. *Animal Behaviour 36:*1198–1209.

Berman, C. M. (1980). Early agonistic experience and rank acquisition among free-ranging infant rhesus monkeys. *International Journal of Primatology 1:*153–170.

Bernstein, I. (1981). Dominance: the baby and the bathwater. *Behavioral and Brain Sciences 4:*419–457.

Bernstein, I. S., and C. L. Ehardt (1985). Agonistic aiding: kinship, rank, age and sex influences. *American Journal of Primatology 8:*37–52.

Blau, P. M. (1964). *Exchange and Power in Social Life.* New York: John Wiley & Sons.

Brown, J. L., and G. H. Orians (1970). Spacing patterns in mobile animals. *Annual Review of Ecological Systems 1:*239–262.

Brown, P., and R. Elliot (1965). Control of aggression in a nursery school class. *Journal of Experimental Child Psychology 2:*103–107.

Bygott, J. D. (1979). Agonistic behavior, dominance, and social structure in wild chimpanzees of the Gombe National Park. Pp. 405–427 in D. A. Hamburg and E. R. McCown, eds. *The Great Apes.* Menlo Park, CA: Benjamin/Cummings.

Camras, L. (1980). Animal threat displays and children's facial expression: a comparison. Pp. 121–136 in D. R. Omark, F. F. Strayer, and D. G. Freedman, eds. *Dominance Relations.* New York: Garland STPM Press.

Caplow, T. (1968). *Two Against One.* Englewood Cliffs, NJ: Prentice Hall.

Carpenter, C. R. (1942). Sexual behavior of free-ranging rhesus monkeys, *Macaca mulatta.* I: Specimens, procedures and behavioral characteristics of estrus. *Journal of Comparative Psychology 33:*113–142.

Chance, M.R.A., G. Emory, and R. Payne (1977). Status referents in long-tailed macaques *(Macaca fascicularis):* precursors and effects of a female rebellion. *Primates 18:*611–632.

Chapais, B. (1983a). Matriline membership and male rhesus reaching high rank in their natal troop. Pp. 171–175 in R. A. Hinde, ed. *Primate Social Relationships: An Integrated Approach.* Oxford, England: Blackwell.

——— (1983b). Dominance, relatedness and the structure of female relationships in rhesus monkeys. Pp. 208–219 in R. A. Hinde, ed. *Primate Social Relationships: An Integrated Approach.* Oxford, England: Blackwell.

——— (1985). An experimental analysis of a mother-daughter rank reversal in Japanese macaques *(Macaca fuscata). Primates 26:*407–423.

——— (1986). Why do adult males and females affiliate during the birth season? Pp. 173–200 in R. G. Rawlins and M. Kessler, eds., *The Cayo Santiago Macaques.* New York: SUNY Press.

——— (1988a). Rank maintenance in female Japanese macaques: experimental evidence for social dependency. *Behaviour 104:*41–59.

——— (1988b). Experimental matrilineal inheritance of rank in Japanese macaques. *Animal Behaviour 36:*1025–1037.

——— (N.D.) Role of alliances in the social inheritance of rank among female primates. In A. H. Harcourt and F.B.M. de Waal, eds., *Coalitions and Competition in Animals and Humans.* Oxford, England: Oxford University Press (in press).

Chapais, B., and S. R. Schulman (1980). An evolutionary model of female dominance relations in primates. *Journal of Theoretical Biology 82:*47–89.

Cheney, D. L. (1977). The acquisition of rank and the development of reciprocal alliances among free-ranging immature baboons. *Behavioral Ecology and Sociobiology 2:*303–318.

——— (1983). Extrafamilial alliances among vervet monkeys. Pp. 278–286 in R. A. Hinde, ed. *Primate Social Relationships: an Integrated Approach.* Oxford, England: Blackwell.

——— (1987). Interaction and relationships between groups. Pp. 267–281 in B. Smuts, et al., eds., *Primate Societies.* Chicago: University of Chicago Press.

Cheney, D. L., and R. M. Seyfarth (1986). The recognition of social alliances among vervet monkeys. *Animal Behaviour 34:*1722–1731.

Chevalier Skolnikoff, S. (1973). Facial expression of emotion in nonhuman primates. Pp. 11–89 in P. Ekman, ed., *Darwin and Facial Expressions.* New York: Academic Press.

Chikazawa, D., T. Gordon, C. Bean, and I. Bernstein (1979). Mother-daughter dominance reversals in rhesus monkeys *(Macaca mulatta). Primates 20:*301–305.

Cords, M. (1987). Forest guenons and patas monkeys: male-male competition in one-male groups. Pp. 98–111 in B. B. Smuts, et al., eds., *Primate Societies.* Chicago: University of Chicago Press.

Curtin, R. A. (1981). Strategy and tactics in male gray langur competition. *Journal of Human Evolution 10:*245–253.

Darwin, C. (1871). *The Descent of Man and Selection in Relation to Sex.* London: J. Murray.

—— (1872). *The Expression of Emotions in Man and Animals.* Chicago: University of Chicago Press [1965].

Datta, S. (1983a). Relative power and the acquisition of rank. Pp. 93–103 in R. A. Hinde. ed., *Primate Social Relationships: An Integrated Approach.* Oxford, England: Blackwell.

—— (1983b). Relative power and the maintenance of dominance. Pp. 103–112 in R. A. Hinde, ed., *Primate Social Relationships: An Integrated Approach.* Oxford, England: Blackwell.

—— (1988). The acquisition of dominance among free-ranging rhesus monkey siblings. *Animal Behaviour 36:*754–779.

Davies, N. K. (1978). Ecological questions about territorial behavior. Pp. 317–350 in J. R. Krebs and N. B. Davies, eds., *Behavioral Ecology: An Evolutionary Approach.* Oxford, England: Blackwell.

Deag, J. H., and J. M. Crook (1971). Social behavior and "agonistic buffering" in the wild Barbary macaque (*Macaca sylvanus* L.). *Folia Primatologica 15:*183–200.

Dodge, K. A. (1982). Social information processing variables in the development of aggression and altruism in children. Pp. 280–302 in C. Zahn-Waxler, M. Cummings, and M. Radke-Yarrow, eds., *The Development of Altruism and Aggression: Social and Sociobiological Origins.* New York: Cambridge University Press.

Dodge, K. A., and J. D. Coie (1987). Social-information-processing factors in reactive and proactive aggression in children's peer groups. *Journal of Personality and Social Psychology 53:*1146–1158.

Donnerstein, M., and E. Donnerstein (1977). Modeling in the control of interracial aggression: the problem of generality. *Journal of Personality 45:*100–116.

Drickamer, L. C., and S. Vessey (1973). Group changing in free-ranging male rhesus monkeys. *Primates 14:*359–368.

Dunbar, R. I. M. (1980). Determinants and evolutionary consequences of dominance among female gelada baboons. *Behavioral Ecology and Sociobiology 7:*253–265.

—— (1988). *Primate Social Systems.* Ithaca, NY: Cornell University Press.

Dunbar, R. I. M., and E. P. Dunbar (1976). Contrasts in social structure among black and white colobus monkey groups. *Animal Behaviour 24:*84–92.

Emerson, R. M. (1962). Power-dependence relations. *American Sociological Review 27:*31–41.

Emlen, S. T., and L. W. Oring (1977). Ecology, sexual selection, and the evolution of mating systems. *Science 197:*215–223.

Estrada, A., R. Estrada, and F. Ervin (1977). Establishment of a free-ranging colony of stumptail macaques *(Macaca arctoides):* social relations I. *Primates 18:*647–676.

Etzioni, A. (1961). *A Comparative Analysis of Complex Organizations.* New York: Free Press.

Fedigan, L. (1983). Dominance and reproductive success in primates. *Yearbook of Physical Anthropology 26:*91–129.

Ferguson, R. B. (1984a). *Warfare, Culture and Environment.* New York: Academic Press.

—— (1984b). Introduction: studying war. Pp. 1–57 in R. B. Ferguson, ed., *Warfare, Culture and Environment.* New York: Academic Press.

French, J. R., and B. H. Raven (1959). The bases of social power. In D. Cartwright, ed., *Studies in Social Power.* Ann Arbor, MI: University of Michigan Press.

Freud, S. (1933). *New Introductory Lectures on Psychoanalysis.* New York: W.W. Norton.

Gallup, G. G. (1987). Self awareness. Pp. 3–6 in G. Mitchell and J. Erwin, eds., *Comparative biology,* Vol. 2B. New York: Alan R. Liss, Inc.

Gamson, W. A. (1961). A theory of coalition formation. *American Sociological Review 26:*373–382.

Goodall, J. (1977). Infant killing and cannibalism in free-ranging chimpanzees. *Folia Primatologica 28:*259–282.

———(1986). *The Chimpanzees of Gombe: Patterns of Behavior.* Cambridge, MA: Harvard University Press.

Goodall, J., A. Bandora, C. Bergman, C. Busse, H. Matama, E. Mpongo, A, Pierce, and D. Riss (1979). Intercommunity interactions in the chimpanzee population of Gombe National Park. Pp. 13–54 in D. Hamburg and E. R. McCown, eds., *The Great Apes.* Menlo Park, CA: Benjamin/Cummings.

Gouzoules, H. (1980). A description of genealogical rank changes in a troop of Japanese monkeys *(Macaca fuscata). Primates 21:*262–267.

Gouzoules, S., H. Gouzoules, and P. Marler (1984). Rhesus monkey (*Macaca mulatta*) screams: representational signaling in the recruitment of agonistic aid. *Animal Behaviour 32:*182–193.

Gump, P. V. (1975). Ecological psychology and children. Pp. 75–126 in M. Hetherington, ed., *Review of Child Development Research,* Vol. 5. Chicago: University of Chicago Press.

Hall, K. R. L. (1967). Social interactions of the adult males and adult females of a patas monkey group. Pp. 261–280 in S. A. Altmann, ed., *Social Communication Among Primates.* Chicago: University of Chicago Press.

Harcourt, A. H. (1987). Dominance and fertility among female primates. *Journal of Zoology, 213:*471–487.

Harcourt, A. H., and K. Stewart (1987). The influence of help in contests on dominance rank in primates: hints from gorillas. *Animal Behaviour 35:*182–190.

Hartup, W. W. (1974). Aggression in childhood: developmental perspectives. *American Psychologist 29:*336–341.

Hausfater, G., J. Altmann, and S. A. Altmann (1982). Long-term consistency of dominance relations among female baboons *(Papio cynocephalus). Science 217:*752–755.

Hausfater, G., S. J. Cairns, and R. N. Levin (1987). Variability and stability in the rank relations of nonhuman primate females: analysis by computer simulation. *American Journal of Primatology 12:*55–70.

Hausfater, G., and S. B. Hrdy (eds.) (1984). *Infanticide: Comparative and Evolutionary Perspectives.* Hawthorne, NY: Aldine.

Henzi, S. P., and J. W. Lucas (1980). Observations of the intertroop movement of adult vervet monkeys *(Cercopithecus aethiops). Folia Primatologica 33:*220–235.

Hinde, R. A. (1970). *Animal Behavior: A Synthesis of Ethology and Comparative Psychology,* 2nd ed. New York: McGraw-Hill.

———(1974). *Biological Bases of Human Social Behavior.* New York: McGraw-Hill.

———(1978). Dominance and role: two concepts with dual meaning. *Journal of Social and Biological Structures 1:*27–38.

———(1987). Can nonhuman primates help us understand human behavior? Pp. 413–420 in B. B. Smuts, et al., eds., *Primate Societies.* Chicago: University of Chicago Press.

Hooff, J.A.R.A.M. van (1972). A comparative approach to the phylogeny of laughter and smiling. Pp. 209–241 in R. A. Hinde, ed., *Non-Verbal Communication.* Cambridge, England: Cambridge University Press.

Horrocks, J., and W. Hunte (1983). Maternal rank and offspring rank in vervet monkeys: an appraisal of the mechanisms of rank acquisition. *Animal Behaviour* 31:772–782.

Hrdy, S. B. (1977). *The Langurs of Abu.* Cambridge, MA: Harvard University Press.

Hrdy, S. B., and D. B. Hrdy (1976). Hierarchical relations among female hanuman langurs. (Primates: Colobinae: *Presbytus entellus*). *Science 193*:913–915.

Hunte, W., and J. Horrocks (1986). Kin and non-kin interventions in the aggressive disputes of vervet monkeys. *Behavioral Ecology and Sociobiology 20*:257–263.

Itani, J. (1987). Inequality versus equality for coexistence in primate societies. Pp. 75–104 in D. McGinness, ed., *Dominance, War and Aggression.* New York: Paragon House Publishers.

Johnson, J. A. (1987). Dominance rank in juvenile olive baboons *Papio anubis:* the influence of gender, size, maternal rank and orphaning. *Animal Behaviour 35*:1694–1708.

Jones, C. B. (1980). The functions of status in the mantled howler monkey, *Alouatta palliata* Gray: intraspecific competition for group membership in a folivorous Neotropical primate. *Primates 21*:389–405.

Kaplan, J. R. (1978). Fight interference and altruism in rhesus monkeys. *American Journal of Physical Anthropology 49*:241–250.

Kawamura, S. (1965). Matriarchal social ranks in the Minoo-B troop: a study of the rank system of Japanese monkeys. Pp. 105–112 in S. A. Altmann, ed., *Japanese Monkeys: A Collection of Translations.* Atlanta: S. A. Altmann.

Kessel, E. L. (1955). Mating activities of balloon flies. *Systematic Zoology 4*:97–104.

Koford, C. B. (1963). Rank of mothers and sons in bands of rhesus monkeys. *Science 141*:356–357.

Komorita, S. S., and J. Chertkoff (1973). A bargaining theory of coalition formation. *Psychological Review 80*:149–162.

Koyama, N. (1967). On dominance rank and kinship of a wild Japanese monkey troop in Arashiyama. *Primates 8*:189–216.

Kummer, H. (1968). *Social Organization of Hamadryas Baboons.* Chicago: University of Chicago Press.

Kurland, J. A. (1977). *Kin Selection in the Japanese Monkey.* Basel: S. Karger.

Kuroda, S. (1984). Interaction over food among pygmy chimpanzees. Pp. 301–324 in R. L. Sussman, ed., *The Pygmy Chimpanzee: Evolutionary Biology and Behavior.* New York: Plenum.

Lee, P. C., and J. I. Oliver (1979). Competition, dominance and the acquisition of rank in juvenile yellow baboons *(Papio cynocephalus). Animal Behaviour 27*:576–585.

Leighton, D. R. (1987). Gibbons: territoriality and monogamy. Pp. 135–145 in B. B. Smuts, et al., eds., *Primate Societies.* Chicago: University of Chicago Press.

Lorenz, K. (1966). *On Aggression.* New York: Harcourt Brace & World.

—— (1974). Analogy as a source of knowledge. *Science 185*:229–234.

MacKinnon, J. (1979). Reproductive behaviour in wild orangutan populations. Pp. 256–273 in D. A. Hamburg and E. R. McCown, eds., *The Great Apes.* Menlo Park, CA: Benjamin/Cummings.

Mallick, S. K., and B. R. McCandless (1966). A study of the catharsis of aggression. *Journal of Personality and Social Psychology 4*:591–596.

Marsden, H. M. (1968). Agonistic behavior of young rhesus monkeys after changes induced in the social rank of their mothers. *Animal Behaviour 16*:38–44.

Massey, A. (1977). Agonistic aids and kinship in a group of pigtail macaques. *Behavioral Ecology and Sociobiology 2:*31–41.

Masters, R. D. (1976). The impact of ethology on political science. Pp. 197–233 in A. Somit, ed., *Biology and Politics.* The Hague: Mouton.

Maynard Smith, J. (1982). *Evolution and the Theory of Games.* Cambridge, England: Cambridge University Press.

Meikle, D. B., and S. H. Vessey (1981). Nepotism among rhesus monkey brothers. *Nature 194:*160–161.

Menzel, E. W. (1974). A group of young chimpanzees in a one-acre field. Pp. 83–153 in A. M. Schrier and F. Stollnitz, eds., *Behavior of Nonhuman Primates,* Vol. 5. New York: Academic Press.

Milgram, S. (1974). *Obedience to Authority: An Experimental View.* New York: Harper and Row.

Missakian, E. A. (1972). Genealogical and cross-genealogical dominance relations in a group of free-ranging rhesus monkeys *(Macaca mulatta)* on Cayo Santiago. *Primates 13:*169–180.

Moyer, K. E. (1976). *The Psychobiology of Aggression.* New York: Harper and Row.

———— (1987). The biological basis of dominance and aggression. Pp. 1–34 in D. McGinness, ed., *Dominance, War and Aggression.* New York: Paragon House Publishers.

Murnighan, J. K. (1978). Models of coalition behavior: game theoretic, social psychological and political perspectives. *Psychological Bulletin 85:*1130–1153.

Nishida, T. (1983). Alpha status and agonistic alliance in wild chimpanzees *(Pan troglodytes schweinfurthii). Primates 24:*318–336.

———— (1987). Local traditions and cultural transmission. Pp. 462–474 in B. B. Smuts, et al., eds., *Primate Societies.* Chicago: University of Chicago Press.

Nishida, T., and M. Hiraiwa-Hasegawa (1987). Chimpanzees and bonobos: cooperative relationships among males. Pp. 165–177 in B. B. Smuts, et al., eds., *Primate Societies.* Chicago: University of Chicago Press.

Nishida, T., M. Hiraiwa-Hasegawa, T. Hasegawa, and Y. Takahata (1985). Group extinction and female transfer in wild chimpanzees in the Mahale Mountains. *Zeitschrift fur Tierpsychologie 67:*284–301.

Noe, R. (1986). Lasting alliances among adult male savannah baboons. Pp. 381–392 in J. G. Else and P. C. Lee, eds., *Primate Ontogeny, Cognition and Social Behaviour.* Cambridge, England: Cambridge University Press.

Norikoshi, K., and N. Koyama (1975). Group shifting and social organization among Japanese monkeys. In S. Kondo, M. Kawai, A. Ehara, and S. Kawamura, eds., *Proceedings from the Symposia of the Fifth Congress of the International Primatological Society.* Tokyo: Japan Science Press.

Packer, C. (1977). Reciprocal altruism in olive baboons. *Nature 265:*441–443.

Packer, C., and A. E. Pusey (1979). Female aggression and male membership in troops of Japanese macaques and olive baboons. *Folia Primatologica 31:*212–218.

Parke, R. D., and R. G. Slaby (1983). The development of aggression. Pp. 547–641 in P. H. Mussen, ed., *Handbook of Child Psychology,* Vol. 4. New York: John Wiley & Sons.

Parker, S. T., and K. R. Gibson (1979). A developmental model for the evolution of language and intelligence in early hominids. *Behavioral and Brain Sciences 2:*367–408.

Passingham, R. (1982). *The Human Primate.* Oxford, England: Freeman.

Patterson, G. R. (1982). *Coercive Family Processes.* Eugene, OR: Castilia Press.

Paul, A., and J. Kuester (1987). Dominance, kinship and reproductive value in female Barbary Macaques *(Macaca sylvanus)* at Affenberg, Salem. *Behavioral Ecology and Sociobiology 21:*323–331.

Pereira, M. E. (1988). Agonistic interactions of Juvenile savanna baboons I. Fundamental features. *Ethology 79:*195–217.

——— (1989). Agonistic interactions of Juvenile savanna baboons II. Agonistic support and rank acquisition. *Ethology 80:*152–171.

Popp, J. L., and I. DeVore (1979). Aggressive competition and social dominance theory: synopsis. Pp. 317–338 in D. A. Hamburg and E. R. McCown, eds., *The Great Apes.* Menlo Park,, CA: Benjamin/Cummings.

Premack, D., and A. Premack (1983). *The Mind of an Ape.* New York: W.W. Norton.

Pusey, A. E., and C. Packer (1987). Dispersal and philopatry. Pp. 250–266 in B. B. Smuts, et al., eds., *Primate Societies.* Chicago: The University of Chicago Press.

Rajecki, D. W. (1983). Animal aggression: implications for human aggression. Pp. 189–211 in R. G. Geen and E. I. Donnerstein, eds. *Aggression: Theoretical and Empirical Reviews,* Vol. 1. New York: Academic Press

Richards, S. (1974). The concept of dominance and methods of assessment. *Animal Behaviour 22:*914–930.

Richards, A. (1987). Malagasy prosimians: female dominance. Pp. 25–33 in B. B. Smuts, et al., eds., *Primate Societies.* Chicago: University of Chicago Press.

Ristau, C., and D. Robbins (1982). Language in the great apes: a critical review. Pp. 147–255 in J. Rosenblatt, R. A. Hinde, C. Beer, and M. C. Busnel, eds., *Advances in the Study of Animal Behavior,* Vol. 12. New York: Academic Press.

Ryan, E. D. (1970). The cathartic effect of vigorous motor activity on aggressive behavior. *Research Quarterly 41:*542–551.

Sade, D. S. (1967). Determinants of dominance in a group of free-ranging rhesus monkeys. Pp. 99–114 in S. A. Altmann, ed., *Social Communication among Primates.* Chicago: University of Chicago Press.

——— (1972). A longitudinal study of social behavior of rhesus monkeys. Pp. 378–398 in R. Tuttle, ed., *The Functional and Evolutionary Biology of Primates.* Chicago: Aldine.

Samuels, A., J. B. Silk, and J. Altmann (1987). Continuity and change in dominance relations among female baboons. *Animal Behaviour 35:*785–793.

Savin-Williams, R. (1980). Dominance and submission among adolescent boys. Pp. 217–229 in D. R. Omark, F. F. Strayer, and D. G. Freedman, eds., *Dominance Relations.* New York: Garland STPM Press.

——— (1987). Dominance systems among primate adolescents. Pp. 131–173 in D. McGinness, ed., *Dominance, War and Aggression.* New York: Paragon House Publishers.

Schaller, G. B. (1963). *The Mountain Gorilla: Ecology and Behavior.* Chicago: University of Chicago Press.

Segall, N. H. (1983). Aggression in global perspective: a research strategy. Pp. 1–43 in A. P. Goldstein and M. H. Segall, eds., *Aggression in Global Perspective.* New York: Pergamon Press.

Sekulic, R. (1983). Male relationships and infant deaths in red howler monkeys *(Alouatta seniculus). Zeitschrift fur Tierpsychologie 61:*185–202.

Seyfarth, R. M. (1977). A model of social grooming among adult female monkeys. *Journal of Theoretical Biology 65:*671–698.

——— (1983). Grooming and social competition in primates. Pp. 182–190 in R. A.

Hinde, ed., *Primate Social Relationships: An Integrated Approach.* Oxford, England: Blackwell.

Seyfarth, R. M., and D. L. Cheney (1984). Grooming, alliances and reciprocal altruism in vervet monkeys. *Nature 308:*541–543.

Seyfarth, R. M., D. L. Cheney, and P. Marler (1980). Vervet monkey alarm calls: semantic communication in a free-ranging primate. *Animal Behaviour 28:*1070–1094.

Sigg, H., and J. Falett (1985). Experiments on respect of possession and property in hamadryas baboons *(Papio hamadryas). Animal Behaviour 33:*978–984.

Silk, J. B. (1982). Altruism among female *Macaca radiata:* explanations and analysis of patterns of grooming and coalition formation. *Behaviour 79:*162–188.

———— (1987). Social behavior in evolutionary perspective. Pp. 318–329 in B. B. Smuts, et al., eds., *Primate Societies.* Chicago: University of Chicago Press.

Silk, J. B., A. Samuels, and P. Rodman (1981). Hierarchical organization of female *Macaca radiata* in captivity. *Primates 22:*84–95.

Simmel, E. C., M. E. Hahn, and J. K. Walters (1983). *Aggressive Behavior: Genetic and Neural Approaches.* Hilsdale, NJ: Lawrence Erlbaum Associates.

Sipes, R. G. (1973). War, sports and aggression: an empirical test of two rival theories. *American Anthropologist 75:*64–86.

Slaby, R. G., and C. G. Crowley (1977). Modification of cooperation and aggression through teacher attention to children's speech. *Journal of Experimental Child Psychology 23:*442–458.

Sluckin, A. M. (1980). Dominance relationships in preschool children. Pp. 159–176 in D. R. Omark, F. F. Strayer, and D. G. Freedman, eds., *Dominance Relations.* New York: Garland STPM Press.

Smith, P. K., and K. J. Connolly (1980). *The Ecology of Preschool Behavior.* New York: Cambridge University Press.

Smuts, B. B. (1985). *Sex and Friendship in Baboons.* Chicago: Aldine.

———— (1987a). Gender, aggression and influence. Pp. 400–412 in B. B. Smuts, et al., eds., *Primate Societies.* Chicago: University of Chicago Press.

———— (1987). Sexual competition and mate choice. Pp. 385–399 in B. B. Smuts, et al., eds., *Primate Societies.* Chicago: University of Chicago Press.

Stein, A. (1976). Conflict and cohesion: a review of the literature. *Journal of Conflict Resolution 20:*143–172.

Strayer, F. F., and J. Strayer (1976). An ethological analysis of social agonism and dominance relations among preschool children. *Child Development 47:*980–989.

Strayer, F. F., and M. Trudel (1984). Developmental changes in the nature and function of social dominance among young children. *Ethology and Sociobiology 5:*279–295.

Struhsaker, T. T., and L. Leland (1987). Colobines: infanticide by adult males. Pp. 83–97 in B. Smuts, et al., eds., *Primate Societies.* Chicago: University of Chicago Press.

Strum, S. C. (1982). Agonistic dominance in male baboons: an alternative view. *International Journal of Primatology 3:*175–202.

———— (1983). Use of females by male olive baboons *(Papio anubis). American Journal of Primatology 5:*93–109.

Svare, B. B. (1983). *Hormones and Aggressive Behavior.* New York: Plenum Press.

Taub, D. M. (1980). Testing the agonistic buffering hypothesis, 1. The dynamics of participation in the triadic interaction. *Behavioral Ecology and Sociobiology 6:*187–197.

Tilford, B. (1982). Seasonal rank changes for adolescent and subadult natal males in a

free-ranging group of rhesus monkeys. *International Journal of Primatology*
3:483–490.

Tinklenberg, J. R., and F. M. Ochberg (1981). Patterns of adolescent violence: a California sample. Pp. 121–140 in D. A. Hamburg and M. B. Trudeau, eds., *Biobehavioral Aspects of Aggression.* New York: Alan R. Liss, Inc.

Trudeau, M. B., E. Bergmann-Riss, and D. A. Hamburg (1981). Towards an evolutionary perspective on aggressive behavior: the chimpanzee evidence. Pp. 27–40 in D. A. Hamburg and M. B. Trudeau, eds., *Biobehavioral Aspects of Aggression.* New York: Alan R. Liss, Inc.

Waal, F. B. M. de (1977). The organization of agonistic relations within two captive groups of Java monkeys *(Macaca fascicularis). Zeitschrift fur Tierpsychologie* 44:225–282.

————— (1982). *Chimpanzee Politics.* New York: Harper and Row.

————— (1984). Sex differences in the formation of coalitions among chimpanzees. *Ethology and Sociobiology* 5:239–255.

————— (1986). The integration of dominance and social bonding in primates. *Quarterly Review of Biology* 61:459–479.

Walters, J. (1980). Interventions and the development of dominance relationships in female baboons. *Folia Primatologica* 34:61–89.

Watanabe, K. (1979). Alliance formation in a free-ranging troop of Japanese macaques. *Primates* 20:459–474.

Weber, M. (1947). *The Theory of Social and Economic Organization.* New York: Oxford University Press.

Weigel, R. W., and R. P. Johnson (1981). An ethological classification system for verbal behavior. *Ethology and Sociobiology* 2:55–66.

Whiten, A., and R. W. Byrne (1988). Tactical deception in primates. *Behavioral and Brain Sciences* 11:233–273.

Willhoite, F. H. (1976). Primates and political authority: a biobehavioral perspective. *American Political Science Review* 70:1110–1126.

Wilson, E. O. (1975). *Sociobiology: The New Synthesis.* Cambridge, MA: Harvard University Press.

Witt, R., C. Schmidt, and J. Schmidt (1981). Social rank and Darwinian fitness in a multimale group of Barbary macaques (*Macaca sylvanus* Linnaeus, 1758). *Folia Primatologica* 36:201–211.

Woodruff, G., and D. Premack (1979). Intentional communication in the chimpanzee: the development of deception. *Cognition* 7:333–362.

Wrangham, R. W. (1980). An ecological model of female-bonded primate groups. *Behaviour* 75:262–300.

————— (1982). Mutualism, kinship and social evolution. Pp. 269–289 in King's College Sociobiology Group, eds., *Current Problems in Sociobiology.* Cambridge, England: Cambridge University Press.

————— (1987a). The significance of African apes for reconstructing human social evolution. Pp. 51–71 in W. G. Kinzey, ed., *The Evolution of Human Behavior: Primate Models.* New York: SUNY Press.

————— (1987b). Evolution of social structure. Pp. 282–296 in B. B. Smuts, et al., eds., *Primate Societies.* Chicago: University of Chicago Press.

Zahavi, S., and S. R. Asher (1978). The effect of verbal instructions on preschool children's aggressive behavior. *Journal of School Psychology* 16:146–153.

8

Kinship

DONALD STONE SADE

I have been asked to write on whether studies on the primates other than humans have revealed anything about human kinship, not about whether or in what way human behavior and primate behavior share common elements, nor about the processes of hominization that transformed an ape-like biogram into the human biogram during the past few million years.

Now it is a truism in ethology that every species is unique. Even when the form of a display of one species is nearly identical to the homologous display in a related species, one does not assume, except as a working hypothesis, that the meaning of the display is the same until that has been demonstrated by empirical research on the second species. It is hypotheses, whether inductive or deductive, that pass between studies, not the empirical findings regarding any particular species. Hypotheses are tested against each species' uniqueness, then rejected or corrected and revised and passed on again with improved generality. This process is easy when applied within a series of closely related species, similar in their morphologies and their ecological relations.

When the human species is added to the series, however, special difficulties arise that are of a kind distinct from those encountered when only nonhuman species are compared. This is because the human brain has evolved capacities—for abstraction, for integrating past and present, for projection, and for the symbolic representation and enhancement of reality—that are unparalleled, even if adumbrated, elsewhere in the animal kingdom. These capacities work on every aspect of human behavior and social intercourse.

On the other hand, the unique capacities of the human brain work within constraints imposed by motivational mechanisms that are more similar to those of our close relatives than are the mechanisms underlying the higher mental functions. This explains the ease with which the primate ethologist develops an empathy for his subjects and the ease with which anthropomorphisms enter the primatological literature. So, we have both: a distinction between the human species and others that is of a kind and degree not found elsewhere and a common heritage of behavioral propensities that potentially permit studies of humans and other species to be integrated within a common science.

Now social anthropologists have emphasized the former to the neglect of the latter, whereas those primate ethologists who have speculated on the nature of the human species have considered the latter and ignored the former. Yet both have used the same words, including the term "kinship," to label their respective but nonidentical phenomena. To establish the common processes in human and nonhuman behavior, the distinctions between them should first be clearly stated. Only then can the actual identities be established. Now let us turn to the assigned topic of this essay.

What have studies of nonhuman primates (hereafter "alloprimates")[1] taught us about human kinship? The easy answer is "nothing," and this essay could end here, but perhaps some explanation for such an abrupt dismissal of the question is required. The trivial reason for the lack of influence of alloprimate studies on anthropology is that social anthropologists, the main students of human kinship systems, have been largely uninterested in the literature on alloprimates or have used the findings of the ethologists metaphorically rather than analytically. But there are less trivial and in fact profound reasons for the disinterest of the social anthropologists. Social anthropology is primarily the study of the symbolized abstractions of relations and interactions rather than of behavior. Although the symbolopoietic[2] process has invested every aspect of human behavior, it is the symbols, not the behaviors, that are the subject of social anthropology. In contrast, although the more the alloprimates (and other animals) are studied the more self-aware and cognitively complex they seem, there are no real findings that suggest even the inklings of the symbolic generalization of status that is the very hallmark of human social organization.

KINSHIP DEFINED

Social anthropology has emphasized two aspects of human kinship systems in categorizing social structure. The first aspect is the way in which categories of relatives are defined and named. Particular attention has been focused on the names given to parallel cousins and to cross cousins. "Parallel cousins" are the offspring of same-sexed siblings; "cross cousins" are the offspring of opposite-sexed siblings. Whether they are given the same label and whether that label is the same or different from the label for siblings of the same sex has seemed to some investigators a crucial characteristic for distinguishing social structures (see Murdock, 1949). The origin of this attention to kinship terminology derives from the great Lewis Henry Morgan's studies of the Iroquois Indians in New York State (Morgan, 1870). A second emphasis has been on the importance of systems of descent. When descent is considered a crucial factor, social structure is defined by the manner in which property is inherited, in the paternal line, in the maternal line, in both, or by some other arrangement. "Property" is here meant in the broadest possible sense and includes not only physical objects or land but also membership in groups such as clans or sibs, the requirement or privilege to acquire a mate (or mates) from a particular other group, names, rights, privileges, and expected ways of behaving. This emphasis on lineality in the transmission of "property" is not so narrowly associated with a single name but rather with the work of the British social anthropologists, perhaps most notably Radcliffe-Brown and his students. Perhaps this was because much influential early work in British social anthropology was done in African societies, in which lineality of descent is strongly emphasized.

Readers who are unacquainted with the intricacies of the work of the social anthropologists[3] will naturally refer mentally to the kinship system within which they themselves are culturally embedded. Upon reflection, many North American readers may discover that their cross and parallel cousins are distinguished from siblings by kinship terms but not from each other (Eskimo terminology). Descent of property is largely ambilineal, yet family names still generally descend through the line of fathers, reflecting the formerly more strongly patrilineal social structure of their European ancestors. Although today residence is usually neolocal, the now rapidly declining custom for the wife to take the husband's name may reflect the past condition in which women moved to their husband's holdings upon marriage (patrilocal residence).[4] The functional correlates of the modern North American kinship system are the emphasis on the autonomy and isolation of the nuclear family within the larger society and the diminished importance of the larger kinship groupings that formerly had controlling interest in the lives of individuals. The consequence of being embedded in this system is that in our minds kinship has become equivalent to genealogy.

GENEALOGY

A "genealogy" or pedigree is a record of all the descendants or ancestors of a particular individual. A genealogical diagram shows the biological relation between any individual and any other included within it irrespective of how the particular category of relationship might be named in the society's system of kinship terminology. Elicitation of genealogical information from informants and constructing genealogies for a community being studied is a major part of the work of the social anthropologist in the field. The culturally defined kinship relations and groups are then mapped onto the biological relations, which provide a universal framework for cross-cultural comparison. A genealogy, however, does not necessarily or usually correspond to a corporate kin group or to any group recognized by society at large or that might be demonstrable as a behavioral unit through a sociometric procedure.

Long-term observations on several populations of monkeys, first at the Japan Monkey Center (the early work is in English translation in Altmann, 1965) and later at Cayo Santiago, Puerto Rico (summarized in Sade et al., 1985) provided the first pedigrees for free-ranging alloprimates. Since matings are promiscuous among both Japanese macaques *(Macaca fuscata)* and the rhesus monkeys *(Macaca mulatta)* of Cayo Santiago, paternity is unknown. The pedigrees are therefore traced entirely through females. A population newly placed under observation includes a number of females whose genealogical relations are unknown. Therefore, each becomes the founding female of a distinct pedigree, which expands in numbers of individuals as descendant generations are added through continued observation (See Figure 8–1). Plates diagraming 97 pedigrees for rhesus monkeys are found in Sade et al. (1985:98 ff.).

Various terms such as *family, matriarchy, matriline,* and *lineage* were applied to these pedigrees, but all of these have specific meanings in social anthropology and refer to particular culturally determined patterns of descent or kin group membership that are distinct from simple biological relatedness. I believe that I was the first to use the neutral term *genealogy* in reference to an alloprimate social structure (Sade, 1965), and I did so deliberately to avoid any connotation of kinship except biological descent. In retrospect, perhaps *pedigree* would have been a better choice. This is because some colleagues seem unable to understand that *genealogy* means a collection of individuals related by descent from a single ancestor as well as meaning the process of studying pedigrees. A difficulty with *pedigree* is that it has more of the connotation of the line ancestors of an individual. In contrast, *genealogy* more often connotes the line of descendants of an individual. The latter connotation is actually more accurate for an alloprimate society, since relatedness must always be traced through descending generations by observation. In any case, *genealogy* works well enough for me, and I will use it in this chapter.

GENEALOGY AND BEHAVIOR

The importance of kinship (in the sense of genealogy) in studies of alloprimate social organization was first established by work at the Japan Monkey Center

Fig. 8–1 A portion of a rhesus monkey genealogy. An old female, with her latest infant at the breast, grooms or sits with three of her older offspring. (Photo: James Loy.)

(Altmann, 1965) and at Cayo Santiago, Puerto Rico (the early studies include Koford, 1963; Sade, 1965, 1967, 1968, 1972; Missakian, 1973). Many later studies on these and other species that extend these early findings are summarized in Gouzoules and Gouzoules (1987). In brief, it was discovered that:

1. Social groups were composed of members of several genealogies.
2. The members of the distinct genealogies formed interactional clusters within the social groups.
3. The members of the distinct genealogies were ranked in a status hierarchy of agonistic dominance.
4. Mating between close relatives, at least between mother and son, appeared to be infrequent or did not occur at all.

Taken together, these findings seemed to provide many of the elements from which human kinship systems were constructed. This seemed especially so in that the variety of known kinds of social organization among the alloprimates increased as wild populations of other species came under controlled observation. These species, especially the common chimpanzee (*Pan troglodytes;* Goodall, 1968, 1986) and the hamadryas baboon (*Papio hamadryas;* Kummer, 1968), revealed additional elements not apparent in macaque societies. Alliances of brothers among chimpanzees and stable matings between a male and several females among hamadryas baboons seemed to add to the alloprimate repertory of protokinship types. These alloprimates, although all members of the same infraorder as humans (Anthropoidea), are all more or less promiscuous in their mating systems. One might argue that the hamadryas baboon is male-polygynous rather than promiscuous. Yet the mating bond seems entirely based on male herding (Kummer, 1968) and therefore is a poor model for a human polygamous family. Curiously, given the emphasis that has been placed on the nuclear family (a more or less permanently mated adult male and female and their offspring) in studies of human kinship, the alloprimates whose social groups most closely resemble this condition, the Callitrichidae,[5] have been almost completely ignored by anthropologists (but see Knox, 1989).

ROBIN FOX

The intriguing parallels between the genealogically based interactional subgroups of some alloprimate societies and the kin-based formal groupings of human societies have been most sensibly considered (from the point of view of social anthropology) in a paper by Robin Fox (1975), one of the few who have sought to broaden the scope of social anthropology beyond its single-minded attention to the cultural.

Many of the early studies on the alloprimates were reviewed in this insightful article, which followed a companion piece published a few years before (Fox, 1972). The two papers should really be read together. Although many detailed field studies of primate behavior have since been published,[6] I doubt that Fox's major arguments would be altered except through enrichment by the addition of abundant new detail. Nor do I think that a better statement of the relation of studies on primate behavior to studies in traditional social anthropology would be easy to achieve.

Needless to say, the present summary must omit many of Fox's insightful comments. Nor is this the place to quibble about points with which I might disagree (especially regarding his review of brain evolution); that would detract from the first purpose of this chapter. For present purposes, Fox's arguments can be summarized as follows. Human kinship systems all are built on three principles, alliance, descent, and exogamy.

Descent has been defined earlier. In a restricted sense, *alliance* means "a relatively permanent assignment of mates" (Fox, 1975). Exogamy is the rule that marriages must be between rather than within kinship groups, however those groups may be defined within a particular culture. The rule may specify that marriages must be between certain kinship classes, such as cross cousins while prohibiting marriage between classes of equal biological degree of relatedness, such as parallel cousins. Thus *exogamy* should not be taken to be equivalent to the genetic concept of "outbreeding." It is a social and not a biological concept.

According to Fox (1972, 1975), exogamy creates alliances between different descent groups through the exchange of women (and associated property in the sense discussed earlier) by marriage. Reviewing the literature on alloprimate behavior, Fox found that each of the three principles was present in one or another species, but that in no species were they combined. "Descent" was a prominent organizing principle in the societies of macaques studied by the Japanese and at Cayo Santiago. "Alliance" (in the restricted sense) seemed to dominate the societies of species whose social groups consisted of a single adult male more or less stably mated to several females (especially the hamadryas baboon, but also many of the monkeys of the genus *Cercopithecus;* as was mentioned earlier, the monogamous Callitrichidae seem to have escaped the attention of anthropologists). The possibility of an inhibition of mating between mother and son suggested at least the incipient tendency for incest avoidance, considered an essential ingredient of exogamy by several early anthropologists. (These are cited in Fox [1972, 1975]. Fox thought, however, that the essential features of kinship systems were present among alloprimates without the necessity of the incest taboo.) In no species, however, were all three of these elements combined.[7] Therefore, according to Fox, the essence of the problem of accounting for the origin of human kinship systems from alloprimate beginnings is to explain how the essential elements of kinship became united into a single system. "That the descent system itself should determine the allocation of mates will probably remain the uniquely human distinction" (Fox, 1975:29). He saw the linguistic labeling of kinship categories as an elaboration on a system of kinship relations essentially already in place and therefore not an essential problem in accounting for the origins of kinship. His comment on the origin of new societies of Japanese monkeys through group fission is a good summary of the detail of his thinking (Fox's data come from Koyama, 1970):

> Most of the males from group A eventually ended up in group B and vice versa. Thus, we have a picture . . . of two "related" groups, each consisting of ranked matrilineages,

which exchanged males over a period. Had these males been involved in some system of alliance, in my terms, that is, had they ended up as relatively permanent mates in a matriline of the other group, then there would have been little difference, except at the symbolic level, between this and an "Iroquoian" system of two moieties each composed of ranked matrilineages, and a rule of moiety exogamy plus matrilocal residence—all very human, and yet little more than a naming system away from the Japanese monkeys. If groups A and B were called "Eaglehawk" and "Crow," and the various lineages "Snake," "Beaver," "Antelope," etc., then a picture emerges of a protosociety on a clan-moiety basis which would have delighted Morgan and McLennan (and Bachofen), but depressed Westermarck and Maine for sure. For one thing, it would have completely bypassed the nuclear family, and would not even require an incest taboo, much less a "pair bond" (Fox, 1975:28).

Given all this, why the skepticism of the first paragraph of the present essay? The rest of the paper addresses this point.

BEHAVIOR

Behavior as the ethologist means it is distinct from *behavior* as the social anthropologist often uses the term. The latter often means "expected relationship," whereas the former means physical movement of some part of the individual's anatomy or some other direct biological result of neural activity, such as the secretion of a scent.

When social anthropologists elicit the names of kinship categories and the rules of marriage, residence, and descent from their informants, they are not really describing behavior but rather discovering the cognitive map of expected kinship relationships that is held in the mind of the informant. Should the anthropologist have several or many informants, a composite picture of "cultural knowledge" is built up that becomes an idealized statement about expected kinship relationships. How close the actual relationships expressed in the behavior of individuals approach the ideal will be a statistical matter, and some approximations may be poor because some freedom for individual choice may be present, and individuals may find means to manipulate the rules. Yet the study of actual interactions is much more difficult than eliciting the conceptual ideal from informants, and it would be fair to say that more social anthropology consists of the latter than of the former. Studies on the societies of the alloprimates, on the other hand, consist only of the former.[8]

Status in alloprimate societies derives directly from interaction (and direct observation by animals of the interactions of other animals). The interactional clusters observed in many alloprimate groups result from attachments or bonding, not from "kinship." The cross-generational correlations of dominance status that seems such a prominent aspect of the social organization of the Old World monkeys do not derive from "membership" in a genealogical subgroup in the sense of membership in a human clan, moiety, sib, lineage, family, or other kin group. The fact that interactions between two individuals can appear to result from prior interactions between other individuals does mean that a status among alloprimates has a history. The history is entirely behavioral, however, not cultural (in the sense of a history of ideas[9] about status). Observed kinship systems among humans are end products of cultural history and do not derive directly from interactional processes.

Now, indeed, behavior may be determined by kinship relations in humans. That is, expected behaviors may be enforced upon individuals by training or coercion and admonishment or punishment offered for violation of the rules. However, the direction of causality is reversed between alloprimate and human societies. In human societies, behavior is determined by the predefined statuses of the individuals and is enforced by third parties. In alloprimate societies, the statuses emerge out of the behavioral interactions of the individuals themselves.[10] Napoleon Chagnon (1988) has recently emphasized the opposite, that humans can manipulate kinship terms to achieve individual behavioral and reproductive goals. Manipulation of the behavior of third parties according to social status is a recently fashionable topic among students of the alloprimates and has been most convincingly illustrated in a captive colony of chimpanzees by de Waal (1982). However, it is easy to fall into metaphor when reflecting on the analogies between alloprimate and human societies. A true understanding of the relations between the behavioral networks of the alloprimates and the symbolic social networks of humans requires careful distinctions, not reasonable analogies.

It is true that the adoption of inclusive fitness theory by certain social scientists has led to observations of behavior (in the sense of the ethologist) constrained by kinship (in the sense of genealogy or degree of relatedness). The famous *Axe Fight,* filmed by Napoleon Chagnon, in which a biological brother comes to the rescue while classificatory "brothers" hold back, and the attention to differential infanticide of natural and adopted children by Daly and Wilson (1983) are cases in point. In no sense, however, did inclusive fitness theory enter the human social sciences via studies of primate behavior. Rather, it was through the influence of particular biologists, such as Alexander, Wilson, Hamilton, and Trivers, themselves students of social insects or other nonprimate animals, on particular anthropologists, including DeVore, Chagnon, Irons, Hrdy, and Dickemann, some of whom happened to have studied alloprimates.

FROM STATUS TO EIDOLON

A distinct terminology is needed to distinguish behavioral status that results from interaction and symbolized status that is defined by cultural rules. Earl Count proposed the term *eidolon* for the latter. Eidolon "means definition of status such that within a society it connotes certain associated activities, certain demand-rights and duties, privilege-rights and no-demand rights, powers and liabilities, immunities and no-powers" (Count, 1958:1074). It is the "definition of status" that is the key to understanding the distinction between human and alloprimate society and that should be the focus of research. Individual humans occupy both face-to-face statuses derived from direct behavioral interactions[11] and eidola that they acquire by descent or through rites of passage.

In reference to the quotation from Robin Fox given earlier, an eidolon is much more than a named status, and an Iroquoian kinship system is therefore much more than a Japanese macaque society with a naming system added. The names or labels are themselves merely symbols of status generalizations. In the difference between status and eidolon is the essence of the difference between alloprimate and human society.

By *relation* hereafter in this chapter I mean a mechanistic or proximal causative connection, whatever the direction of causation in each case. What is the relation between status and eidolon? Everyone knows the difficulties that result from the "boss" being behaviorally subordinate to a formal inferior, but there seems to be no well-developed theory that could be used by both primate behaviorists and social anthropologists to explore the process. This question, however, suggests where research should be focused if the question that motivated this essay is of continued interest. Although the transition stages in the evolution[12] of kinship systems exist nowhere in nature and we cannot study them, we may wish to study the shared primitive behavioral characteristics of the alloprimates that may resemble the ancestral conditions out of which, through the process of hominization, our own species-specific social characteristics evolved. If so, attention should be directed to the following questions.

How elaborate, among alloprimates, are cognitive maps of the social network? There are a number of hints that alloprimate individuals can recognize the relative statuses between other individuals, an essential ability upon which to elaborate status generalization. Breuggeman (1973) noted "infant return" among rhesus monkeys, in which an unrelated monkey would pick up a lost infant and carry it to its proper mother. Dasser (1988) experimentally revealed that Java macaque *(Macaca fascicularis)* females could develop an abstract concept of "mother–infant affiliation." Walters (1980) argued that maturing young in a baboon *(Papio cynocephalus)* troop could target their eventual dominance status in the agonistic network, implying that they possessed a cognitive map of the social structure of their group. Seyfarth (1987) reviewed studies suggesting that monkeys can vocally label different classes of predators as well as studies suggesting that the quality of monkey vocalizations varies according to whether individuals are interacting with kin or non-kin (in the sense of genealogy) or with higher- or lower-ranking individuals. F. de Waal's (1982) work documenting social manipulation, and therefore social concepts, among chimpanzees has been referred to earlier. It is likely that these intriguing findings will stimulate new and innovative research. For the present topic, we could hope that findings will emerge on the causal relation between social interaction and social concept. But still nothing is found in alloprimate behavior equivalent to a proscription that a particular male must establish a formal bond with a particular female because she is his father's sister's daughter (see Chagnon, 1988).

How are social relations built out of social interactions? Robert Hinde (a recent summary is in Hinde, 1983) has given special attention to this question in studies on both alloprimates and humans. Yet a satisfying theory of the pro- cess seems not in sight; most work is descriptive or functional (in the sense of seeking adaptive explanations). Perhaps the approach taken by some human ethologists could be adapted to studies of alloprimates. Particularly exciting are the studies that, through either interview or projective techniques, provide information on an individual's concept of his own position in an interactional social network. The individual's concept of his own status is then compared to his status as revealed through observations of his interactions (see Strayer and Strayer, 1980; Savin-Williams, 1980). Given the ingenuity that is now being directed towards understanding cognition among alloprimates, it is at least pos- sible that comparable findings on monkeys or apes will someday appear.

However, I think that we will still not be very close to an understanding of how status became symbolized, and how the symbols, or their engrams in the brain, came to dominate or at least to influence interaction. This understand- ing waits on the neurosciences to reveal the unique workings of the human brain. Whatever the hints at some specialization of function in the left cerebral hemisphere of monkeys (Hefner and Hefner, 1984), lateralization remains the hallmark of the human brain (Geschwind and Galaburda, 1987). Neural sys- tems unique to humans, implied by studies of split-brain patients (Gazzaniga, 1989), play a role in belief formation and the integration of mood and experi- ence and, according to Gazzaniga (1989), are independent of language. Perhaps a wedding of the tools of the human ethologists with those of the cognitive neuroscientists will someday indicate the connection of human interaction with the symbol- making process. Then the process of status generalization and the phylogeny of the human kind of kinship might be profitably examined. Nevertheless, the systems of kinship as described by the ethnographer will remain products of culture history. Whatever new insights come from studies of the alloprimates, they are likely to illuminate only processes of interaction and nothing about human kinship.

ACKNOWLEDGMENTS

The author thanks James Brown, William Irons, Robert Launay, Frans de Waal, Kerry Knox, and an anonymous reviewer for their constructive comments on a draft of this chapter. In addition, this chapter profited from discussions among graduate students and faculty during a year-long seminar, The Logic of Anthropological Enquiry, at North- western University.

NOTES

1. The term *alloprimates,* the other primates, was coined by Earl W. Count (1958, 1973) as a convenient way of distinguishing humans from the rest of our primate relatives without use of the pejorative "nonhuman" and without committing an unintended tax- onomic judgment (such as all primates except *Homo sapiens sapiens, Homo* sp., and so forth).

2. "Symbolopoiesis," symbol-making, is the process by which the peculiarly human brain associates an abstract meaning with an event or action that in other animals may be purely biological. A full discussion and exploration of this process is given in Count (1973).

3. It is impossible to touch on or even mention here all topics included in the study of human kinship systems and social structure, the study of which has occupied the entire professional careers of not a few anthropologists. Nor would such an exercise advance the more limited purpose of this chapter. One must refer to the extensive literature for an introduction. A detailed, informative, yet not overwhelming entrance into the subject is by Murdock (1960).

4. As is noted below, the exchange of women, a kind of property, between patrilineages has been alleged to have been an important aspect of the social structure of many human groups.

5. The marmosets and tamarins, South American monkeys of the superfamily Ceboidea. Although the gibbons and siamangs (Anthropoidea: Hylobatidae) are said to form monogamous family groups, they seem to lack several characteristics of the Callitrichidae that remind us of human nuclear families, such as caretaking of young by older siblings and continued residence of some offspring within the family group past the age of sexual maturity. Earl Count (1973:101) has argued that "The most significant single process that has produced the human family out of the alloprimate has been the familialization of the male." The Callitrichidae have seemingly achieved an analogous state, and the details of life within their family groups deserve careful scrutiny by anthropologists interested in the evolution of the family.

6. Much of the new information is summarized in Smuts et al. (1987).

7. By picking out elements of social organization scattered throughout a higher-level taxon and treating them as if they were potential parts of a whole, Fox is implicitly using the notion of "generic behavior" first advanced by Haas (1962). Simply stated, the idea is that the behavioral repertoire of a species may include latent elements, some of which are expressed overtly in the behavior of related species. This concept is elaborated by Sade (1987).

8. Do supraindividual structures actually exist in alloprimate societies? This question has been critically addressed by S. A. Altmann (1981), who argues that even dominance hierarchies, thought by many to be a prominent feature of many primate societies, are intellectual artifacts of the observer.

9. This statement ignores the issue of ideas-as-behavior. Pursuing this matter further here would lead us astray from the purpose of this chapter into discussions of the neurological basis of mind, obviously a vast topic. Nevertheless, the resolution of the issues raised here must be sought in that area. Interested readers should see Count's paper, "An essay on phasia. On the phylogenesis of man's speech function" (in Count, 1973).

10. Some statuses may be very easy to establish with a minimum of interaction, the best known being the dominant/subordinate statuses of a pair following agonistic episodes. This implies that an instinctive predisposition to achieve such status is an evolved trait.

11. This has perhaps received most attention from social psychologists interested in the social dynamics of small groups. Omark, Strayer, and Freedman (1980) contains many contributions of importance to the present topic.

12. I mean here biological evolution, not the process of historical change sometimes called "cultural evolution."

REFERENCES

Altmann, S. A. (ed.) (1965). *Japanese Monkeys: A Collection of Translations.* Atlanta: S. A. Altman.

——— (1981). Dominance relationships: the Cheshire cat's grin? *Behavioral and Brain Sciences 4:*430–431.

Breuggeman, J. A. (1973). Parental care in a group of free-ranging rhesus monkeys *(Macaca mulatta). Folia Primatologica 20:*178–210.

Chagnon, N. A. (1988). Male Yanomamo manipulations of kinship classifications of female kin for reproductive advantage. Pp. 23–48 in L. Betzig, M. B. Mulder, and P. Turke, eds., *Human Reproductive Behaviour: A Darwinian Perspective.* Cambridge, England: Cambridge University Press.

Count, E. W. (1958). The biological basis of human sociality. *American Anthropologist 6:*1049–1085.

——— (1973). *Being and Becoming Human: Essays on the Biogram.* New York: Van Nostrand Reinhold.

Daly, M., and M. Wilson (1983). *Sex Evolution and Behavior,* 2nd ed. Boston: Willard Grant Press.

Dasser, V. (1988). A social concept in Java monkeys. *Animal Behaviour 36:*225–230.

Fox, R. (1972). Alliance and constraint: sexual selection in the evolution of human kinship systems. Pp. 282–331 in B. Campbell, ed., *Sexual Selection and the Descent of Man.* New York: Aldine.

——— (1975). Primate kin and human kinship. Pp. 9–35 in R. Fox, ed., *Biosocial Anthropology.* New York: Halsted Press.

Gazzaniga, M. S. (1989). Organization of the human brain. *Science 245:*947–952.

Geschwind, N. and A. M. Galaburda 1987 *Cerebral Lateralization.* Cambridge, MA: MIT Press.

Goodall, J. (1968). The behaviour of free-living chimpanzees in the Gombe Stream Reserve. *Animal Behaviour Monographs 1:*163–311.

——— (1986). *The Chimpanzees of Gombe: Patterns of Behavior.* Cambridge, MA: Harvard University Press.

Gouzoules, S., and H. Gouzoules (1987). Kinship. Pp. 299–305 in B. B. Smuts, D. L. Cheney, R. M. Seyfarth, R.W. Wrangham, and T. T. Struhsaker, eds., *Primate Societies.* Chicago: University of Chicago Press.

Haas, A. (1962). Phylogenetisch bedeutungsvolle Verhaltensanderungen bei Hummeln. Bericht uber Verhaltensstudien en einem Nest mit Arbeiter Konigen (Bombus hypnorum). *Zeitschrift fur Tierpsychologie 19:*356–370.

Hefner, H. E., and R. S. Hefner (1984). Temporal lobe lesions and perception of species specific vocalizations by macaques. *Science 226:*75–76.

Hinde, R. A., (ed.) (1983). *Primate Social Relationships: An Integrated Approach.* Sunderland, MA: Sinauer.

Knox, K. L. (1989). Observations on dominance in a group of emperor tamarins *(Saginus imperator).* Unpublished dissertation. Evanston, IL: Northwestern University.

Koford, C. B. (1963) Rank of mothers and sons in bands of rhesus monkeys. *Science 141:*356–357.

Koyama, N. (1970). Changes in dominance rank and division of a wild Japanese monkey troop in Arashiyama. *Primates 11:*335–390.

Kummer, K. (1968) *Social Organization of Hamadryas Baboons.* Chicago: University of Chicago Press.

Missakian, E. A. (1973). Genealogical mating activity in free-ranging groups of rhesus monkeys (Macaca mulatta) on Cayo Santiago. *Behaviour 45:*225–241.

Morgan, L. H. (1870). Systems of consanguinity and affinity of the human family. *Smithsonian Contributions to Knowledge 17:*1–590.

Murdock, G. P. (1949). *Social Structure.* New York: Macmillan.

—— (ed.) (1960). *Social Structure in Southeast Asia.* Chicago: Quadrangle Books.

Omark, D. R., F. F. Strayer, and D. G. Freedman (eds.) (1980). *Dominance Relations. An Ethological View of Human Conflict and Social Interaction.* New York: Garland Publishing Co.

Sade, D. S. (1965). Some aspects of parent-offspring and sibling relations in a group of rhesus monkeys, with a discussion of grooming. *American Journal of Physical Anthropology 23:*1–18.

—— (1967). Determinants of dominance in a group of free-ranging rhesus monkeys. Pp. 99–114 in S. A. Altmann, ed., *Social Communication Among Primates.* Chicago: University of Chicago Press.

—— (1968). Inhibition of son-mother mating among free-ranging rhesus monkeys. *Science and Psychoanalysis 12:*18–38.

—— (1972). A longitudinal study of social behavior of rhesus monkeys. Pp. 378–398 in R. Tuttle, ed., *The Functional and Evolutionary Biology of Primates.* Chicago: Aldine-Atherton.

—— (1987). Latent processes and the biological bases of social organization. *North Country Naturalist 1:*39–52.

Sade, D. S., B. D. Chapko-Sade, J. M. Schneider, S. S. Roberts, and J. T. Richtsmeir (1985). *Basic Demographic Observations on Free-Ranging Rhesus Monkeys.* New Haven, CT: Human Relations Area Files.

Savin-Williams, R. (1980). Dominance and submission among adolescent boys. Pp. 217–229 in D. R. Omark, F. F. Strayer, and D. G. Freedman, eds., *Dominance Relations: An Ethological View of Human Conflict and Social Interaction.* New York: Garland Publishing Co.

Seyfarth, R. M. (1987). Vocal communication and its relation to language. Pp. 440–451 in B. B. Smuts, D. L. Cheney, R. M. Seyfarth, R. W. Wrangham, and T. T. Struhsaker, eds., *Primate Societies.* Chicago: University of Chicago Press.

Smuts, B. B., D. L. Cheney, R. M. Seyfarth, R. W. Wrangham, and T. T. Strusaker (eds.) (1987). *Primate Societies.* Chicago: University of Chicago Press.

Strayer, J., and F. F. Strayer (1980). The representation of social dominance in children's drawings. Pp. 287–297 in D. R. Omark, F. F. Strayer, and D. G. Freedman, eds., *Dominance Relations: An Ethological View of Human Conflict and Social Interaction.* New York: Garland Publishing Co.

Waal, F. B. M. de (1982). *Chimpanzee Politics: Power and Sex Among Apes.* New York: Harper.

Walters, J. (1980). Interventions and the development of dominance relationships in female monkeys. *Folia Primatologica 34:*61–89.

9

Apes, Humans, and Culture: What Primatological Discourse Tells Us About Ourselves

CALVIN B. PETERS

There is an old story about a gorilla entering an upscale restaurant, seating himself at the bar, and ordering a scotch, neat. Somewhat taken aback, the bartender pours the gorilla's drink and takes the $20 bill that has been placed on the bar in payment. At the cash register, the bartender's darker side gets the best of him and he determines to turn a quick profit, reasoning that despite the unusual nature of the encounter, his customer (now mark) is *only* a gorilla. He provides a single dollar to the gorilla as change and returns to his place near the beer taps. Finally, the simian's silence is too much to bear and the bartender blurts out, "Say, you know, we don't get too many gorillas in here." Without missing a beat the gorilla retorts, "Well, at these prices I can see why!"

Our reactions to the story say much about our attitudes toward apes (the story will, of course, work with the substitution of a chimpanzee or orang-utan for the gorilla), toward ourselves, and toward those things that separate them from us. The humor in the story derives from the deliberate creation of an impossible scene and an equally impossible cast of characters: a restaurant bar that remains unaffected despite the presence of gorilla, a conniving bartender who treats the gorilla's appearance not as the astonishing event it is but merely as an opportunity to make a few dollars from the naivete of a stranger, a gorilla who talks and who maintains his innocence and simplicity in spite of the fact he has been shortchanged.

There is, however, something more to our reaction than amusement at the implausible. We can, without much effort, conjure a mental picture of the situation the narrative creates. What is more, that mental image, despite its impossibility, is not unfamiliar. We have seen too many apes in all-too-human situations—in films, in television programs, in advertisements—to be truly startled by the idea of a gorilla, even a talking one, in a bar. And, at the risk of overinterpretation, there is something familiar, even unconsciously comforting, in the fact that it is the human bartender who is the schemer and the gorilla who is guileless.

THE TROUBLING ATTRACTION OF THE APE

The story has humor and attraction for us quite simply because the great apes (and to a lesser extent monkeys) hold humor and attractiveness for us. Perhaps that humor and attraction are, as Frans de Waal (1982:18) suggests, a "camouflage for quite different feelings—a nervous reaction, caused by the marked resemblance between human beings and chimpanzees. . . . If they are animals,

Fig. 9–1 David Greybeard, Jane Goodall's ape friend, verifies Frans de Waal's claim that chimpanzee eyes reveal self-assured, intelligent personalities. (Photo: Hugo van Lawick. Copyright National Geographic Society.)

what must we be?" Whatever its source, the capacity of apes to fascinate us, to entertain us, to make us "uneasy with great ease"[1] cannot be denied.

Without delving too deeply into the human psyche, the "nervous reaction" of humanity to "our close cousins" (King and Wilson, 1975) is probably the result of the odd amalgam of proximity to and distance from humanness that the ape embodies. On one hand, when we "look straight and deep into a chimpanzee's eyes, an intelligent, self-assured personality looks back at us" (de Waal, 1982:18); on the other, the eyes of that same chimpanzee cannot comprehend the simplest words in a children's story nor appreciate the grace and elegance of the Taj Mahal. At one and the same time, the chimpanzee, the gorilla, the orang-utan seem to be us and not-us.

The boundary between the uncannily human and the clearly nonhuman that appears to run within the ape is fascinating and unnerving because it promises to reflect the very essence of humanity itself. If those things that humans do, think, and feel that apes cannot do, think, and feel can be sorted from those things that we and the apes have in common, we can, it seems, identify with sureness what makes humanity possible. But this is risky business. If, after careful sorting, we are left with nothing that we do that an ape cannot do, or cannot be taught to do, then the question "if they are animals, what must we be," will become much more than a puzzle to the thoughtful zoo-goer.

APES AND HUMANS, NATURE AND CULTURE

Neither the probing of the boundary between human and ape nor concern about the possibility that such a boundary may not, in fact, exist is new. Indeed, the discovery of the chimpanzee by Europeans and its importation to the continent set the process in motion. As early as the mideighteenth century, London was abuzz about the "most surprising creature brought over ... which the Angolans call chimpanzee, or the mockman" (*London Magazine,* September, 1738, cited in Goodall, 1986:6; see also the entry "chimpanzee" in the *Oxford English Dictionary,* 1971), and the French openly speculated about the permeability of the ape–human boundary:

> ... such is the likeness of the structure and function of the ape to ours that I have little doubt that if this animal were properly trained he might at least be taught to pronounce, and consequently to know a language. Then he would no longer be a wild man, but he would be a perfect man, a little gentleman, with as much matter or muscle as we have, for thinking and profiting by his education (Julien La Mettrie [1709–1751] quoted in Limber, 1980:198).

Darwin's publication of *The Origin of Species* imparted a sense of scientific, if not social, respectability to such speculations, and the questions of human origins engendered by *The Descent of Man* and Thomas Huxley's *Man's Place in Nature* ensured that apes, humans, and what if anything separated them would begin to receive systematic attention.

From the perspective of the late twentieth century, many of the early efforts to sort human from ape seem quaint, and slightly ridiculous. In 1909, U. S. psychologist Lightner Witmer determined that Peter, a roller-skating chimpanzee who smoked cigarettes, was a "middle-grade imbecile" (Desmond, 1979:64–65), and a decade later a British news service trumpeted a "very interesting plan of French government scientists to educate generations of primates in the expectation that some day they may talk and act like human beings" (Haraway, 1989:21). Furthermore, there were efforts (failed) to teach both chimpanzees and orang-utans to speak English and abortive investigations into the capacity of the ape to reason abstractly (see Desmond, 1979, for a complete discussion).

Despite the mixture of halting progress and grandiose expectations that accompanied the initial explorations of the ape–human boundary, the questions that animated the investigators continue to be a focus of primatological attention. Can apes use language? Can apes reason in a formal way to solve novel problems? What are the limits of ape intelligence? Over the years, of course, the questions have been refined and embellished—Do apes invent tools? Do apes have a sense of self? Do apes have a theory of mind?—but their thrust has not changed (see Beck, 1974, 1980; Gallup, 1970; Gallup et al., 1977; Premack and Woodruff, 1978; Savage-Rumbaugh et al., 1978).

What has changed is the context in which such questions are asked and answered. The early inquiries took place against a backdrop of unreflective certainty about the distinctive superiority of humanity, not only to apes but to all other creatures as well. Although the medieval sureness that humanity was the "apple of God's eye" was gone, there was broad-based agreement that humanity, especially in its white, male, European incarnation, was the only vessel of "the gift of articulate language,—the power of numbers,—the power of generalisation,—the powers of conceiving the relation of man to his Creator,—the power of foreseeing an immortal destiny,—the power of knowing good from evil, on eternal principles of justice and truth" (*Edinburgh Review* [1863] quoted in Desmond, 1979:20).

In this setting, the plumbing of the ape's mental powers and linguistic abilities resembled nothing so much as early ethnographic accounts of the "rudimentary intelligences" of aboriginal peoples and the "genuinely wild and savage scenes" that characterized their social life (Durkheim, 1965:245–255).[2] Despite their challenges to the Eurocentrism that probed them, both Australian and ape served more to confirm than to contradict the correctness of the *Weltanschauung*. Neither could be more civilized or European than European gentlemen themselves.

It should not surprise us that apes, even those who wore clothes and who struggled to form English words, did not threaten seriously the common sense anthropology of western societies. The world was hierarchically ordered. The line from civilized humanity to the barbarian and savage—from whom we might extract some information about the "elementary forms" of human behavior (see Morgan, 1963; Durkheim, 1965; for an interesting if opinionated discussion, see Harris, 1968)—ran straight to the ape. With "primitive peoples" interposed between the chimpanzee and civilization, humanity was safe. If apes could talk and perform simple tasks it did not make them human; like the "middle-grade imbeciles" they were thought to be, the apes lacked the manners and morals of cultured humanity.

The anthropological context against which investigations of the mind of the ape were conducted changed dramatically in the middle one-third of the twentieth century. Largely as the patrimony of Franz Boas and Alfred Kroeber, professional anthropologists began to abandon the concept of "primitive" and with it the Eurocentric worldview that had insulated the truly human from the animal ape. Kroeber's (1915:286) "profession" that "the so-called savage is no transition between the animal and the scientifically educated man. . . . All men are totally civilized. . . . There is no higher and no lower in civilization," although ahead of its time, signaled a sea change in how the western scientific community understood humanity and its relation to the world.

A major part of the transformation is prefigured in Kroeber's use of the terms "civilization" and "civilized." To the modern anthropologist, although Kroeber's ideas are forward-looking, his language is anachronistic, still redolent of the separation of gentlemen from barbarians and barbarians from savages. What Kroeber accomplished conceptually—that "civilization" separated all humanity from animal—anthropology as a discipline accomplished linguistically through its appropriation of "culture."

"Culture" made it clear that it was humanity itself, humanity in all its expressions, that was distinctly separate from the animal world. Although the squabbles that remain are legion, anthropology seems to have coalesced around culture as the definition and demarcation of humanity:

> Man is unique: he is the only living species that has a culture. By *culture* we mean an extrasomatic, temporal continuum of things and events dependent upon symboling. Specifically and concretely, culture consists of tools, implements, utensils, clothing, ornaments, customs, institutions, beliefs, rituals, games, works of art, language, etc. All peoples in all times and places have possessed culture; no other species has or has had culture (White, 1959:3).

This makes things very different, indeed, for the apes and their examiners. Before, the "best" that could be hoped for was that the ape, through the acquisition of simple skills, could be thought a "savage." Now, the "best" (or perhaps "worst") is something of quite a different sort. If the ape can use language, if the ape can invent and employ tools, if the ape can reflect on its intentions and impute motives to others, if in other words the ape has the capacity do those things that make culture possible, then can the ape be thought to have crossed not into savagery but into humanity?

Gallup and his colleagues (1977), who are convinced that the chimpanzee has a human-like self-awareness, have openly wondered whether, to be "logically consistent," human "political, ethical, and moral philosophy" ought to be applied to the chimpanzee. More starkly, Gallup has asked the disturbing question, "[if] it is becoming increasingly clear that chimpanzees and people share the same conceptual equipment . . : how then do we justify keeping them behind bars?" (quoted in Desmond, 1979:271).

PRIMATOLOGY AS ETHNOGRAPHIC DISCOURSE

The issue of the permeability of the culture boundary between humanity and ape remains, despite Gallup's rhetorical sureness, an open question not only for professional scholars but for the broader society as well. What is at stake, however, runs much deeper than the "imprisonment" of chimpanzees, gorillas, and orang-utans in zoos and laboratories. Indeed, in large measure, the ape has already been "released" to the natural world in both academic inquiry and the public imagination. The now famous field studies of Goodall (1971, 1986), Fossey (1983), and Galdikas (1980) have moved each of the great apes, in turn, from the austerity of the modern laboratory to lush, natural settings rich in enchanting detail. Furthermore, each of them has managed to produce descriptions of the lives of the apes that, like the ethnographies they are, persuade because the author's accounts "convince us that what they say is a result of their having actually penetrated (or, if you prefer, been penetrated by) another form of life, of having, one way or another, truly 'been there'" (Geertz, 1988:4–5).

The ethnographies of Goodall, Fossey, and Galdikas have perhaps had their greatest impact beyond the limits of academic primatology. The documentary films and photographic essays of the National Geographic Society devoted to their investigations have pushed the ape and primatology itself to the center of societal consciousness.[3] The power of the photographic images and the literary characterizations of the narratives that accompany them places the apes "just at or over the line into 'culture,'" and places the readers and viewers "at or over the line into 'nature'" (Haraway, 1989:148; Haraway's account of "apes in Eden, apes in space" is both fascinating and challenging).

The traffic between nature and culture that the National Geogrpahic Society fostered (Haraway, 1989:146) has become commonplace in the minds of the public and in the work of primatologists. For example, in *Peacemaking among Primates,* de Waal (1989) ranges, with great ease, back and forth between ape and human. Forgiveness, reconciliation, and many "conciliatory gestures and contact patterns" are claimed to be the common possession of apes and humans (pp. 270–271); bonobos are said to have a "social life that gave the impression of being ruled by *compassion* [italics in the original] (p. 220); and de Waal himself muses about the frustrations of not being able to administer to chimpanzees questionnaires with items that ask, "Did you recall your morning fight with X when you kissed her this afternoon?" and "Did it make you feel better afterward?" (p. 44).

Although de Waal's unabashed use of anthropomorphic narrative is extreme, his text reveals the power of the image of the ape approaching culture and of humanity approaching nature. Even a critical reviewer (Bramblett, 1989:409), who labeled de Waal's method as "simple homologies . . . [that] fall short of

Fig. 9–2 Goodall's prose and the accompanying photographs combine to convince readers that she has managed to penetrate the lives and the world of the chimpanzees. (Photo: Hugo van Lawick. Copyright National Geographic Society.)

insightful analysis," nonetheless pronouced the "stories absorbing and informative." No doubt, much the same judgment would be made by the bookstore browser who came upon this on the dust jacket: ". . . sex occurs in all possible combinations and positions whenever social tensions need to be resolved. 'Make love, not war' could be the bonobo slogan."

The capacity of de Waal's prose to absorb us is clear, but how are accounts such as these informative, and about what or whom do they inform? There seem to be at least two ways to address the question of the informative value of "primate stories." First, as de Waal's reviewer surely had in mind, accounts of the apes can inform us about the apes themselves—their mental capacities, their social structure, the extent to which their expressions and gestures communicate symbolically, and so on. Furthermore, and surely de Waal's reviewer intended this as well, these accounts can perhaps inform us about ourselves in a relatively direct way through the potential illumination of the evolution of those patterns of behavior that we share with nonhuman primates.

There is no dispute about the capacity of the accounts of de Waal or Goodall or others to inform us about apes. And there is little dispute, although a good deal of caution (hence the criticism of de Waal's outlandishness), about their capacity to inform us about the evolution of humanity. Indeed, as long as "primate stories" inform us in a straightforward way about either the apes or ourselves they are not "stories" at all but "findings," with a sense of facticity not found in stories.

There is, however, a second, less obvious way in which such accounts can inform us about ourselves. The stories—as forms of inquiry, as forms of discourse, as the voices of particular authors—embody considerable informative power. This sort of information is very different from the "findings" that the accounts are thought to convey. When the accounts are conceived as findings, the information is objective, produced not by authors and texts speaking in one of many possible voices but by the facts speaking for themselves. When the accounts are conceived as information because they are "stories" of a particular kind, it is otherwise. An author-saturated narrative of chimpanzee life in which rich description, attention to detail, and finely drawn characterization produce verisimilitude informs us about author and reader in a way that is separate from the content of the story itself. An author-evacuated study of chimpanzee problem-solving abilities in which quantitative data, control of the setting, and replicability of the process create facts informs us about ourselves in a way that the study does not discuss. Goodall's choice (1986:655) to thank "the chimpanzees themselves, all those unique and vivid personalities: Flo and her talented family, Melissa and Goblin, Passion and Pom, Mike and Goliath, and he who will always have first place in my affections, David Greybeard" informs us about ourselves in a way that is quite independent of any findings she has produced. Likewise, Premack's and Woodruff's decision (1979:526) to "thank Sarah's trainer, Keith Kennel, for his collection of the data" rather than Sarah herself adds profoundly, if somewhat disturbingly, to their finding that the chimpanzee does, indeed, possess a theory of mind.

THE INFORMATIVE VALUE OF DISCOURSE

The information that derives from the fact that primatological discourse is the result of particular authorial choices is, of course, not limited to those accounts that concern nature and culture and the potential of humans and apes to cross from one to the other. For the most part, however, the informative value of the discourse as discourse in other areas is overshadowed by what are regarded as findings. Where culture is concerned, things are different. Here, the discourse as discourse often overpowers whatever findings it might contain. Goodall's introduction to *The Chimpanzees of Gombe* (1986) and de Waal's account of chimpanzee personalities in *Chimpanzee Politics* (1982) are simple, but powerful, cases in point. Goodall begins her exhaustive volume with an excerpt from her diary:

> All morning I had searched the Kasakela Valley for chimpanzees. At midday there had been a tropical deluge; now, an hour later, the rain was coming down in a steady, gray drizzle. Perhaps because I was wet and cold I did not see the chimpanzee until I was within ten meters of him. Nor did he see me. . . . Then a sound came from above. A large adult male stared down at me, lips tensed. . . . At once I made out another chimpanzee ahead: two eyes staring in my direction and a large black hand gripping a branch. . . . Above me the male uttered the eerie alarm call of the chimpanzee—a long, wailing *wraaa.* . . . Suddenly the end of a branch hit my head as the large male above me became more excited. Another chimpanzee charged through the undergrowth, straight toward me. At the very last minute he veered away and ran off at a tangent into the forest. A short while later all the black shapes had vanished. The forest was quiet again save for the pattering of rain on leaves—and the thudding of my heart (p. 1).

The elegance of her prose and her ability to make us hear the scene—her thudding heart, the eerie wailing of the chimpanzee, the pattering rain—obscure completely the "findings" that the passage reports. Yet somehow we are informed of the motives of individual chimpanzees and of the capacities for communication and coordination of the chimpanzees as a group. In addition, we are also persuaded (as we have been through many passages such as this) that Goodall can speak authoritatively, on the chimpanzee and on humanity as well.

Chimpanzee Politics achieves the same persuasiveness through a combination of photographic portraiture and, compared to Goodall, sparse, austere narrative. After an array of 20 wallet size photographs of the chimpanzees whose lives he intends to portray, de Waal undertakes his text proper:

> Chimpanzees have outspoken personalities. Their faces are full of character and you can distinguish them one from another just as easily as you distinguish people. Also their voices all sound different, so that years later I can tell them apart by ear alone. Each ape has his or her very own way of walking, lying down and sitting. Even by the way they turn their heads or scratch their backs, I can recognize them. But when we speak of personality of course we think especially of the differences in the way in

which they treat their groupmates. These differences can only be portrayed accurately by using the same adjectives we use to characterize our fellow human beings. Therefore terms such as "self-assured," "happy," "proud," and "calculating" will be used in this chapter of first acquaintance with the individuals. These terms reflect my subjective impression of the apes (p. 54).

FOUR MALES

YEROEN   LUIT   NIKKIE   DANDY

FEMALE SUB-GROUP 'MAMA'

AMBER   MAMA & MONIEK   GORILLA & ROOSJE

FRANJE   FONS

Fig. 9–3 Portraits of de Waal's chimpanzees create powerful impressions that match his own anthropomorphic narrative. (Photos: Frans de Waal.)

As with Goodall, the "findings" de Waal reports are obscured not so much by his prose but by the photographs that precede it. We are informed by the implied narrative of the photographic portrait that chimpanzees are self-assured and calculating, happy and proud.[4] And, because de Waal's subjective impression of the apes accords with our own, we are persuaded that he too can speak with authority about the chimpanzee and us.

It is not just de Waal's authorial presence or Goodall's discursive style that persuades and informs. Their accounts are replete with findings, both those that can be claimed as their own and those of others that they report. But the findings themselves are not independent of the discourse in which they are embedded; it is the discourse that allows us to use the findings as a way of finding out about ourselves.

Premack and Woodruff (1979), for example, explored in a set of experimental circumstances the possibility that chimpanzees can, among other things, "comprehend" lies and "produce" lies. Their conclusion, that "the oldest animal is a successful liar in production and comprehension" was hedged by the disclaimer that "too little [is known] about the inferences the animals may be making to be assured that they are indeed lying" (p. 524). These phrases are the gist of a finding. When embedded in Premack's and Woodruff's account of their study, this finding is interesting and, despite their caveat, reasonably persuasive, though arguable (see the "Commentary" that follows Premack and Woodruff, 1978).

The technical language of "Does the chimpanzee have a theory of mind?" stands in sharp contrast to de Waal's (1982:133) account of a "series of signal disguises" by Luit, an adult male chimpanzee at Arnhem. de Waal relates a sequence of hooting threats directed at Luit by Nikkie, a longtime antagonist. With each threat, Luit grimaces in fear, but uses his "fingers to press his lips together." After three attempts, Luit finally succeeds in "wiping the grin off his face" and turns, finally, to look at Nikkie. de Waal, who reports that he "could not believe his eyes" concludes that "this was a case of *genuine bluffing* [emphasis in the original], in the sense that [the chimpanzees] pretended to be braver and less frightened than they really were."

Although the finding reported by Premack and Woodruff and the finding related by de Waal are similar in what they tell us about the chimpanzee, they are quite different in what they tell us about ourselves. For the reader, whether the "animals" of Premack and Woodruff are "knowing" liars or not makes little difference. They are not characters (and neither are the "benevolent" or "villainous" trainers who interact with them), and the scene, though carefully crafted as an experiment, lacks the sense of risk that is inherent in lying among humans. With Luit, it is very different. He is a character, and as readers we care very much about what happens to him. And, because the scene de Waal paints is fraught with the danger of "having a bluff called," we are drawn subtly to reflect on our own bluster.

Goodall and de Waal speak authoritatively because their accounts, as accounts, convince us not only that they have "been there" but that we have been there as well. Through their images and words, Goodall and de Waal manage to bring to the exotic and alien world of the chimpanzee a cultural logic that familiarizes the apes to such an extent that the meanings of their behaviors seem as accessible as those of our neighbors. This is no small accomplishment. As Geertz (1988:107) points out, the strategy of portraying the alien as familiar and of "conjoining the beliefs and practices of one's most immediate readers" to the wildly exotic is "most often referred to as satire." The humorous effects of humanizing the ape (recall Peter the roller-skating chimpanzee or the talking gorilla who opened this chapter) are very great, yet both Goodall and de Waal succeed in suppressing them.

The fact that we regard these accounts as portraits not parody, science not sarcasm, has informative value that transcends the findings reported. That we

THE FAR SIDE   By GARY LARSON

"Shh. Listen! There's more: 'I've named the male with the big ears Bozo, and he is surely the nerd of the social group—a primate bimbo, if you will.'"

Fig. 9–4 The ease with which Gary Larson creates humor from the circumstances of primatological ethnography is testimony to the danger Goodall and de Waal avoid. (*The Far Side.* Copyright 1985 Universal Press Syndicate. Reprinted with permission. All rights reserved.)

are able, without laughter, to participate in the anthropomorphism of de Waal and to feel deeply the familiar, human affection Goodall has for David Greybeard teaches us about ourselves in the same way that good ethnography does, by reminding us to look into ourselves as we would look into others (see Geertz, 1988:102–128). Goodall (1986:655) thanks the chimpanzees of Gombe for "their *being*" [emphasis in the original]. That we recognize and believe the double entendre is perhaps, her most important finding.

NOTES

1. This expression is derived from Robin Fox's (1980) observations about the ease with which incest arouses our unease. To some extent the antics, particularly those connected to sex, of apes and monkeys have for us the same sort of fascination Fox attributes to incest: We might well be horrified and repulsed, but we want to keep watching.

2. An early "observation" on chimpanzees should bear this out: "The cleared . . . spaces . . . are used by the chimpanzees to build immense bonfires of dried wood. . . . When the pile is completed one of the chimpanzees begins to blow at the pile as if blowing the fire. He is immediately joined by others, and, eventually, by the whole company, and the blowing is kept up until their tongues hang from their mouths, when they sit around on their haunches with their elbows on their knees and holding up their hands to the imaginary blaze. In wet weather they frequently sit this way for hours together (Buttekofer, 1893:337; quoted in Harris, 1968:254).

3. The National Geographic Society has produced five "immensely popular" documentaries about nonhuman primates: *Miss Goodall and the Wild Chimpanzees* (1965); *Monkeys, Apes, and Man* (1971); *Search for the Great Apes* (1975); *Gorilla* (1981); and *Among the Wild Chimpanzees* (1984) (Haraway, 1989:133).

4. The use of photographic portraiture is, itself, a form of primatological story telling. Of course, the use of visual images is, at times, an adjunct to the verbal narrative. Frequently, however, photographs, particularly portraits, stand as a form of discourse themselves. For example, in *The Chimpanzees of Gombe,* 22 color plates, nearly all portraits of chimpanzees, inform the reader in ways more powerful than even Goodall's graceful prose.

REFERENCES

Beck, B. (1974). Baboons, chimpanzees, and tools. *Journal of Human Evolution 3:*509–516.
———(1980). *Animal Tool Behavior.* New York: Garland Press.
Bramblett, C. (1989). Review of *Peacemaking Among Primates. American Journal of Physical Anthropology 80:*409.
Desmond, A. (1979). *The Ape's Reflexion.* New York: Dial Press.
Durkhiem, E. (1965). *Elementary Forms of the Religious Life.* New York: Free Press [1915].
Fossey, D. (1983). *Gorillas in the Mist.* Boston: Houghton Mifflin.
Fox, R. (1980). *The Red Lamp of Incest.* New York: E. P. Dutton.
Galdikas, B. (1980). Living with orangutans. *National Geographic 157:*830–853.

Gallup, G. (1970). Chimpanzee: self recognition. *Science 167*:86–87.

Gallup, G., J. Boren, G. Gagliardi, and L. Wallnau (1977). A mirror for the mind of man, or will the chimpanzee create an identity crisis for *Homo sapiens. Journal of Human Evolution 6*:303–313.

Geertz, C. (1988). *Works and Lives.* Stanford, CA: Stanford University Press.

Goodall, J. (1971). *In the Shadow of Man.* Boston: Houghton Mifflin.

———— (1986). *The Chimpanzees of Gombe: Patterns of Behavior.* Cambridge, MA: Harvard University Press.

Haraway, D. (1989). *Primate Visions.* New York: Routledge, Chapman and Hall.

Harris, M. (1968). *The Rise of Anthropological Theory.* New York: Thomas Y. Crowell.

King, M-C., and A. C. Wilson (1975). Our close cousin, the chimpanzee. *New Scientist 67*:16–18.

Kroeber, A. (1915). Eighteen professions. *American Anthropologist 17*:283–288.

Limber, J. (1980). Language in child and chimp. Pp. 197–220 in T. A. Sebeok and J. Umiker-Sebeok, eds., *Speaking of Apes.* New York: Plenum Press.

Morgan, L. (1963). *Ancient Society.* New York: Meridian Books [1877].

Premack, D., and G. Woodruff (1978). Does the chimpanzee have a theory of mind. *The Behavioral and Brain Sciences 1*:515–526.

Savage-Rumbaugh, E. S., D. M. Rumbaugh, and S. Boysen (1978). Linguistically mediated tool use and exchange by chimpanzees *(Pan troglodytes). Behavioral and Brain Sciences 1*:539–554.

Waal, F. B. M. de (1982). *Chimpanzee Politics.* New York: Harper and Row.

———— (1989). *Peacemaking among Primates.* Cambridge, MA: Harvard University Press.

White, L. (1959). *The Evolution of Culture.* New York: McGraw-Hill.

Suggestions for Further Reading

Chapter 1 Mortifying Reflections

There are several excellent introductory texts available for readers new to primatology, including Linda Fedigan's book, *Primate Paradigms* (1982, Montreal, Eden Press) and the recent volume by Robin Dunbar, *Primate Social Systems* (1988, Ithaca, Cornell University Press). For additional information on the history of primate studies, see Hugh Gilmore's 1981 essay (*American Journal of Physical Anthropology* 56:387–392) and Donna Haraway's *Primate Visions* (1989, New York, Routledge, Chapman and Hall). The difficulties of applying data from animals to human behavioral problems are discussed in Mario Von Cranach's edited work, *Methods of Inference from Animal to Human Behaviour (1976,* Chicago, Aldine).

Chapter 2 Maternal Behavior Among Primates

Among the many excellent studies in the primate literature, Jeanne Altmann's monograph on the ecology of maternal behavior, *Baboon Mothers and Infants* (1980, Cambridge, Harvard University Press), and James McKenna's 1979 review article (*Yearbook of Physical Anthropology* 22:250–286) are two of the best. Altmann, McKenna, and several others also have interesting essays in the recent book, *Parenting Across the Life Span* (Lancaster et al., eds., 1987, New York, Aldine De Gruyter).

Chapter 3 Primate Paternalistic Investment

Pat Whitten's recent review article on infants and their relationships with adult males is highly recommended (in Smuts et al., eds., 1987, *Primate Societies,* Chicago, University of Chicago Press). The interested reader is also directed to David Stein's field study, *The Sociobiology of Infant and Adult Male Baboons* (1984, Norwood, N.J., Ablex), and to Barbara Smuts's book, *Sex and Friendship in Baboons* (1985, New York, Aldine). The latter work describes the connections between paternalistic behaviors and male-female relationships and sexual behavior.

Chapter 4 Ontogeny of Behavior

Although several years old, the collection of essays edited by Suzanne Chevalier-Skolnikoff and Frank Poirier is still useful and instructive (1977, *Primate Bio-social Devel-*

opment: Biological, Social, and Ecological Determinants, New York, Garland). A more recent overview of primate social development patterns can be found in the chapter by M. Pereira and Jeanne Altmann (in E. Watts, ed., 1985, *Non-human Primate Models for Human Growth and Development,* New York, A. R. Liss). Thelma Rowell's book, *Social Behaviour of Monkeys* (1972, Baltimore, Penguin), is a very readable introduction to primatology and includes an excellent chapter on infant development. The volume edited by N. F. White, *Ethology and Psychiatry* (1974, Toronto, University of Toronto Press), contains several interesting essays on the ontogeny of behavior. Also recommended are Nicholas Blurton Jones's book *Ethological Studies of Child Behavior* (1972, New York, Cambridge University Press), and *The Effect of the Infant on Its Caregiver* (1974, New York, John Wiley and Sons) edited by Michael Lewis and Leonard Rosenblum.

Chapter 5 Evolution and Female Sexual Behavior

Solly Zuckerman's book, *The Social Life of Monkeys and Apes* (1981 reissue of 1932 edition, London, Routledge and Kegan Paul), is still worth reading as a historical piece because it played an important role in the development of comparative sexology. Similarly, Ford and Beach's *Patterns of Sexual Behavior* (1951, New York, Harper and Brothers) remains a useful source of data on human sexuality. A more recent cross-cultural overview of human sexual behavior can be found in S. G. Frayser's *Varieties of Sexual Experience* (1985, New Haven, HRAF Press). Don Symons's sociobiological analysis, *The Evolution of Human Sexuality* (1979, Oxford, Oxford University Press), a work that treats many of the same topics discussed in the present chapter, is highly recommended. Also recommended are Sarah Hrdy's study of langur reproduction, *The Langurs of Abu* (1977, Cambridge, Harvard University Press), and her volume on comparative sexuality, *The Woman That Never Evolved* (1981, Cambridge, Harvard University Press). A recent summary of the sexual behavior of wild chimpanzees can be found in Jane Goodall's latest book, *The Chimpanzees of Gombe* (1986, Cambridge, Harvard University Press).

Chapter 6 Male Sexual Behavior

The Evolution of Human Sexuality (1979, Oxford, Oxford University Press) by Don Symons provides a sociobiological analysis of several aspects of men's sexual behavior. For current cross-cultural data, S. G. Frayser's book, *Varieties of Sexual Experience* (1985, New Haven, HRAF Press) is a useful source. Glenn Hausfater's field study, *Dominance and Reproduction in Baboons* (1975, Basel, S. Karger), provides an excellent description of the sexual behavior of wild male savannah baboons. The sexual behavior of captive male chimpanzees is vividly described by Frans de Waal in *Chimpanzee Politics* (1982, New York, Harper and Row). A. F. Dixson's article on genitalia and copulatory behavior in males (1987, *Journal of Zoology 213:* 423–443) is a good review of current research.

Chapter 7 Aggression, Power, and Politics

Konrad Lorenz's *On Aggression* (1969, New York, Bantam) is recommended for readers interested in the history of comparative studies of aggression. Although several years old, the collection of essays edited by Ralph Holloway, *Primate Aggression, Territoriality, and Xenophobia* (1974, New York, Academic Press), is still a useful compilation of information. Aggression and reconciliation in five species of primates (including humans) are described and analyzed by Frans de Waal in *Peacemaking Among Primates* (1989, Cambridge, Harvard University Press). A recent sociobiological analysis of human aggression can be found in Martin Daly and Margo Wilson's book, *Homicide* (1988, New York, Aldine De Gruyter).

Chapter 8 Kinship

Jeffrey Kurland describes the behavioral influences of genetic relatedness in *Kin Selection in the Japanese Monkey* (1977, Basel, S. Karger). Also recommended are Sarah Gouzoules's review article detailing the evidence for kin recognition among nonhuman primates (1984, *Yearbook of Physical Anthropology 27:*99–134) and Sarah and Harold Gouzoules's essay on kinship in the volume edited by Smuts et al. (1987, *Primate Societies,* Chicago, University of Chicago Press). A useful set of essays on social expertise and the evolution of intelligence in primates and humans is contained in the volume edited by R. W. Byrne and A. Whiten, *Machiavellian Intelligence* (1988, Oxford, Clarendon Press). Several interesting essays on the effects of kin selection among humans can be found in Napoleon Chagnon and William Irons's volume, *Evolutionary Biology and Human Social Behavior* (1979, North Scituate, MA, Duxbury Press).

Chapter 9 Primatological Discourse

Clifford Geertz's *Works and Lives* (1988, Stanford, CA, Stanford University Press) is an excellent introduction to the analysis of ethnography as discourse. *Primate Visions* by Donna Haraway (1989, New York, Routledge, Chapman and Hall) provides a sweeping, feminist view of primatology as a form of inquiry. For a general discussion of the problems involved in comparisons between human behavior and animal behavior, Mary Midgley's *Beast and Man* (1980, New York, New American Library) is an excellent beginning point. *The Ape's Reflection* by Adrian Desmond (1979, New York, Dial Press) is a readable discussion of the history of comparative primatology that argues that apes ought to be admitted to the "exclusive club heretofore reserved for humanity alone."

Index